Farms and Territory

SCIENCES

Agronomy and Food Science, Field Directors – Jack Legrand and Gilles Trystram

Agronomy, Subject Heads – Yves Coquet, Marianne Le Bail and Jean Roger-Estrade

Farms and Territory

Coordinated by
Philippe Martin
Solène Pissonnier

iSTE

WILEY

First published 2025 in Great Britain and the United States by ISTE Ltd and John Wiley & Sons, Inc.

ISTE Ltd
27-37 St George's Road
London SW19 4EU
UK

www.iste.co.uk

John Wiley & Sons, Inc.
111 River Street
Hoboken, NJ 07030
USA

www.wiley.com

Library of Congress Control Number: 2025939455

British Library Cataloguing-in-Publication Data
A CIP record for this book is available from the British Library
ISBN 978-1-78945-226-6

ERC code:
LS9 Applied Life Sciences, Biotechnology, and Molecular and Biosystems Engineering
LS9_3 Applied animal sciences (including animal breeding, veterinary sciences, animal husbandry, animal welfare, aquaculture, fisheries, insect gene drive)
LS9_4 Applied plant sciences (including crop production, plant breeding, agroecology, forestry, soil biology)
LS9_5 Food sciences (including food technology, food safety, nutrition)
SH2 Institutions, Values, Environment and Space
SH2_9 Urban, regional and rural studies
SH2_10 Land use and regional planning
SH2_11 Human, economic and social geography

The manufacturer's authorized representative according to the EU General Product Safety Regulation is Wiley-VCH GmbH, Boschstr. 12, 69469 Weinheim, Germany, e-mail: Product_Safety@wiley.com.

Contents

Introduction

Philippe MARTIN and Solène PISSONNIER

Paris Saclay University, AgroParisTech, INRAE, UMR SADAPT, Palaiseau, France

The book you are holding in your hands does not claim to be an exhaustive account of "farms and territories". The purpose is rather to start one on a journey, just as one may embark on journeys to visit sites of significant interest. Each of these journeys is based on a particular point of view and cannot by itself cover the entire subject. The same is true of this book, edited by two agronomists and influenced by their own experiences of the subject, as well as the interdisciplinary collaborations they have had with colleagues at AgroParisTech and via the laboratories with whom they interact with on their research. This book complements others that have proposed their own journeys. This book is not entirely neutral and is part of a broader perspective on the sustainability and the agroecological transition of farmlands in their local territory.

The book begins by taking a look at land use, which supports both farms and the agroecological transition, through the dual lens of a lawyer and a sociologist. The land can be used to grow a variety of crops, as well as livestock. It is this link between livestock farming and the land that is examined in the three following chapters. The first, from an agrarian systems perspective, looks at the dynamics of territories located in foothill and mid-mountain areas, which are not very suitable for arable farming, and where livestock farming is combined with forestry. The second chapter looks at possible synergies between crop and livestock farming at a territorial scale, at a time when regional specialization is still at work, accompanied by the gradual disappearance of mixed crop–livestock systems within the same farm. The third chapter in the series deals with a resource just as important as land, namely genetic resources, and looks at the territorial management in terms of animal

Farms and Territory,
coordinated by Philippe MARTIN and Solène PISSONNIER.

husbandry, bearing in mind that the question also applies to plants, although this is not addressed in this work.

After these three chapters on livestock farming, the journey takes a different route and looks at some relatively atypical farms. Starting off with microfarms, it distinguishes between field and city microfarms. It then moves on to a chapter about gatherers who, like itinerant shepherds, collect their crop production where they have not sown it.

We then shift our gaze to analyze farms and their territorial integration. Here we examine how biodiversity is taken into account within farming territories.

We then return to the question of biomass produced in agriculture, addressed through four chapters. The first sets out the foundations of the territorial metabolism approach, in which territories are considered as organisms that organize the flow of matter from one compartment to another. From this point of view, we shall take a closer look at the food dimension, and thus at transfers from fields to urban areas. We next look at the role of agriculture as a producer and user of organic waste products (including urban waste). In the last point of this series, we make a closer examination of methanization, which produces gas from waste before returning it to the soil. Methanization also raises the question of dedicated agricultural production, which is thus removed from territorial metabolism for human or animal consumption.

Dealing with the links between territory and farm raises methodological issues of data access. The final chapter presents a wide range of databases that can be used to inform these relationships, as well as tools that can be used to enhance the value of these databases.

The diversity of the themes addressed in these 12 chapters is intended to stimulate reflection on the links between farms and territories. We hope that the various examples presented here will enrich readers' knowledge across an expansive range of issues.

1

Which Land Resources to Adopt for an Ecological Transition of Agri-Food Systems? A Cross-View of Law and Sociology

Pierre-Etienne BOUILLOT[1] **and Romain MELOT**[2]

[1] *Paris Saclay University, AgroParisTech, Palaiseau, France*

[2] *Paris Saclay University, AgroParisTech, INRAE, UMR SADAPT, Palaiseau, France*

In this chapter, we present the principal issues in current legal provisions relating to agricultural land resources, taking as our point of observation the tools and dynamics that are contributing to the transition of agri-food systems. We offer a dual legal and sociological perspective on these developments, addressing both the analysis of legal developments and their implementation and mobilization by social actors through empirical studies. We focus mainly on the French institutional framework, supplemented by references to empirical work in developed countries. Without aiming to be exhaustive, our aim is to provide an overview of the main legal instruments that contribute to the greening of agricultural land policies, as greening can be defined as "a process by which the environment is taken into account in public policies, organizations and even professional practices" (Mormont 2013, authors' translation).

Farms and Territory,
coordinated by Philippe MARTIN and Solène PISSONNIER.

1.1. Legal framework and contractual practices in farming: towards a greening of land tools?

While agricultural policies are largely defined at the EU level, one of the particularities of the legal framework for farmland is that it remains largely dependent on the national legal framework. In the context of French law, we can observe a movement towards the greening of land tenure tools, in parallel with the integration of environmental objectives into community law, which can be seen by examining the legal provisions that define the administrative framework for access to land and contractual obligations.

1.1.1. *Forms of authorization over access to land*

1.1.1.1. *The latest on rural land development*

Of the three rural land development procedures provided for under Article L. 121-1 of the CRPM[1] – agricultural and forestry land development, exchanges and amicable transfers of rural real estate and the reclamation of uncultivated land – we focus on developments in the first to understand the greening of rural land development. To this end, we need to return briefly to its origins: land consolidation. This is a relatively old procedure, whose first legal foundations were laid out by a law, dated November 27, 1918. The reforms that this procedure has undergone have each brought new elements to the process, making land consolidation one of the major tools for regional planning in the service of economic and social development. On the one hand, rural land consolidation has been given the objective of increasing farm productivity, and, on the other, it accompanies regional development when combined with major works or urban planning operations. While economic and social considerations originally governed the redistribution of land, the environmental consequences of land consolidation (flooding, erosion, damage to biodiversity) have led the legislator to gradually introduce corrective measures to the procedure.

After the law of July 12, 1983 (no. 83-630), the land survey was supplemented with an environmental impact study. As the reorganization of agricultural land also involves related work (e.g. hedge removal, road modification), it became necessary to analyze the environmental impact of these modifications. The purpose of the inquiry is to inform the public and gather their opinions, suggestions and counterproposals (Lorvellec 1988). The law of February 23, 2005 (no. 2005-157) introduced far-reaching changes to rural land development and, in particular, to the

1. Code rural et de la pêche maritime, France's Rural and Maritime Fishing Code.

various land consolidation procedures. Article L. 121-1 of the CRPM sets a new common goal for the various land development methods, which is "to ensure the development of rural natural areas [...] in compliance with the objectives mentioned in Articles L. 111-1 and L. 111-2" (authors' translation), which include environmental objectives. Article L. 123-1 of the CRPM still states that the main aim of this procedure is to improve the agricultural use of rural properties, so as to create "rural holdings in a single block or with large, well-grouped plots" (authors' translation). While the main aim is still to improve farming operations by rationalizing farm labor productivity, the agricultural land development procedure has been enriched by environmental considerations.

The law of August 8, 2016, for the reconquest of biodiversity (no. 2016-1087) added the possibility that the rural development of the perimeter in which the procedure is implemented could enable "the use of parcels for natural, agricultural or forestry purposes with a view to preserving the environment" (authors' translation). By way of example, the prefect can introduce prescriptions to encourage the conservation of cultivated wetlands as natural grasslands by prohibiting drainage and filling (Guyet 2016). Previously, the only way to legitimize the land consolidation process from an environmental standpoint was through an impact study. This was reinforced by the decree of March 30, 2006 (no. 2006-394) and is divided into two parts. The first is similar to what was previously required, i.e. a public inquiry including an analysis of the land and an environmental study (CRPM, R. 123-9). The second concerns environmental recommendations relating, among other things, to the preservation of natural areas and the prevention of natural hazards. These are among the provisions common to all land development methods (C. rur., R. 121-20). In addition to this development, which above all requires the administration to take into account the impact on the environment before authorizing land reorganization (Struillou 2002), other rules demonstrate a more substantial evolution, such as the extension of the role of the communes in the development of land.

1.1.1.2. *Mechanisms for regulating access to land: farming authorizations*

The administrative regulation of access to land through the granting of farming authorizations, which characterizes the policy known as *le contrôle des structures* (translated here as farming authorizations), is unique to France among developed countries. It is a legacy of the reforms introduced in the early 1960s to regulate the rural land market in the context of agricultural modernization (Boinon 2011). Under Article L. 331-2 of the CRPM, farmers planning to develop an area of land on the same farm that exceeds the threshold set by the regional master plan for agricultural holdings are required to apply for prior authorization. The validity of the lease is subject to obtaining this authorization. The land actually farmed must be taken into account, even if it is provided free of charge (Cass. Civ. 3e, November 26, 2008, no. 07-16.679).

According to Article L. 331-1 of the CRPM, farming authorization is expressly aimed at the installation of new farmers, rather than the expansion of existing farms. The aim is to limit the expansion of farms, which would prevent new installations. In particular, this objective helps to maintain activity in rural areas and the corresponding population in order to occupy the whole territory. However, legislative action has not stemmed the decline in the number of farms. Small- and medium-sized farms are the hardest hit. The disappearance of these farms has led to a concentration of land on the largest farms. More recently, environmental considerations have been added to the initial objective of promoting access to land and limiting land concentration. The reform of the 2014 Law on the Future of Agriculture[2] (no. 2014-1170 of the loi d'avenir pour l'agriculture de 2014) makes it possible to combine economic, social and environmental objectives (CRPM, Art. L. 331-1 et seq.).

A comparative study of farm master plans that define local priorities for the allocation of farming authorizations (first on a departmental scale, then on a regional scale after the 2014 law) shows that this consideration of environmental objectives nevertheless remains variable depending on the territory (Guéringer 2023). Research into the implementation of farming authorizations also highlights that, in the event of competition between farmers for the allocation of authorization to farm, environmental considerations are still rarely mobilized (Piet et al. 2021). In addition, conditions of access to information on the land market are sometimes unfavorable to project developers from non-agricultural backgrounds, who are often committed to the agroecological transition (Hobeika 2013; Baysse-Lainé 2020).

1.1.1.3. *Land development companies (SAFERs): interventions on the land market*

Sociétés d'Aménagement Foncier et d'Établissement Rural (SAFERs, Land Development and Rural Development Companies) were created by the law of August 5, 1960 (no. 60-808). They are private-sector bodies that take the form of limited companies, although they are similar to associations in that they are prohibited from pursuing profit-making objectives. Their mission is to channel the agricultural land market and promote rural development, and to this end, they have been given the means to purchase land freely offered for sale by landowners and resell it to farmers in accordance with legal objectives. As such, they have the right of pre-emption that enables them to acquire land and farms on a priority basis, with

2. Loi d'avenir pour l'agriculture de 2014.

the aim of selling them on. The exercise of this right concerns all agricultural property sales, since the SAFERs are informed by the notary before each sale. The SAFERs therefore play an important role in controlling and guiding the land market, particularly in areas where the majority of farmers are direct tenants. This importance is not only measured when the right of pre-emption is exercised. As this right can be exercised for every sale, sellers often prefer to make an amicable sale to SAFERs rather than try to sell their property to someone else, who would undoubtedly be put off by the SAFERs' intervention. However, this system can be circumvented using corporate techniques (see infra).

In fact, in the event of the sale of an agricultural property for valuable consideration, the seller's right of ownership and contractual freedom are significantly attenuated, but the exercise of the right of pre-emption is strictly circumscribed, since it can only be exercised in accordance with precise objectives. As an instrument of land policy, SAFERs' intervention must comply with the objectives set by this policy. Their delimitation makes it possible to justify the infringement of property rights and contractual freedom. Until the law of July 9, 1999, Article L. 143-2 of the CRPM limited the exercise of the right of pre-emption to the pursuit of economic and social objectives. Whether or not the effects of exercising the right of pre-emption are in line with one of these objectives is strictly authorized by the judicial judge, who verifies the act's conformity with the general interest, as expressed here by the objectives defined by the law. Indeed, as case law has repeatedly affirmed, it is impossible to rely on a reason other than those expressed by the legislator (e.g. Cass. civ. 3e, September 25, 2002, no. 01-11.224).

Up until 1999, SAFERs were required to exercise their right of pre-emption in accordance with the aforementioned objectives, which excluded any reference to environmental protection. The Agricultural Orientation Act of July 9, 1999, amended Article L. 141-1 of the CRPM, concerning the missions assigned to the SAFERs, so as to implement "the land component of the policy for the planning and sustainable development of rural areas" (authors' translation). This law added an eighth purpose to Article L. 143-2 of the CRPM relating to the exercise of the right of pre-emption by SAFERs: "the implementation of landscape enhancement and environmental protection projects approved by the State or local authorities and their public establishments" (authors' translation). Since the law of February 23, 2005, on the development of rural areas (Law no. 2005-157), SAFERs have also been required to contribute to "the diversity of landscapes, the protection of natural resources and the maintenance of biological diversity" (authors' translation). This new mission is reflected in the addition of a new objective to the exercise of the right of pre-emption, namely, to develop peri-urban agricultural and natural areas. Therefore, in partnership with the department, SAFERs can help maintain an agricultural or natural green belt around towns and cities.

As environmental concerns are taken into account in pre-emption tools, the number of holders of this right is expanding beyond strictly agricultural bodies, with local authorities becoming increasingly involved. The departments, which also have pre-emptive rights over sensitive natural areas, have been involved since 2005 in the perimeters for the "protection of peri-urban agricultural and natural areas"[3] (Struillou and Joye 2015). More recently, communes (or their groupings and mixed syndicates) located in a water catchment area are also empowered to apply to the prefect for the institution of an agricultural right of pre-emption for reasons of water quality preservation (Law no. 2019-1461 of December 27, 2019 relating to involvement in local life and the proximity of public action, codified in Art. L. 218-1 et seq. of the Code de l'urbanisme; decree no. 2022-1223 of September 10, 2022).

It should be added that the CRPM does not impose any conciliation or hierarchy between the various economic, social and environmental objectives assigned to the SAFER. For example, Article L. 143-3 of the CRPM merely states that the pre-emption decision must always be justified by reference to one or more of these objectives. Furthermore, case law specifies that not all objectives need to be pursued simultaneously (Cass. Civ. 3e, October 13, 2010).

The regulation of the agricultural market through the right of pre-emption has yet to be empirically documented in the social sciences. Sociological studies have shown that the missions of SAFERs have evolved from their initial objectives of restructuring and consolidating farms in a context of agricultural modernization to a role as local experts in the land market, a development that is not without its tensions (Sencébé et al. 2013). Research in economics shows that their impact has been positive, over the long term, in containing inequalities in farm size (Piet et al. 2012). At the local level, a study carried out in the Île-de-France region on SAFER pre-emptions requested by local authorities concludes that the latter's concerns are often linked to motives other than environmental preservation or local food supply (fight against informal housing) (Bignebat et al. 2022). The broadening of the spectrum of actors with pre-emptive rights does not always mean that greening objectives are gaining ground.

3. Commonly referred to by the acronym PAEN (*Protection des espaces agricoles et naturels périurbains*) in French.

1.1.2. *A renewed framework for contractual land management practices*

The greening of the legal framework governing agricultural land in France is also reflected in the rules governing contractual relations.

This phenomenon primarily concerns rental relationships. While land can be farmed by the owner, they can also grant a third party the right to farm the land, which is the case for the vast majority of land in France. The absolute nature of ownership and contractual freedom has been tempered by rural law. These limitations are mainly to be found in the specific provisions governing rural leases. The general framework of the "status of tenant farming"[4], inherited from the post-war period, establishes a series of protections for the farmer (notably the right to renewal and the right of pre-emption), while regulating the rental market through an administrative mechanism of rent control. The evolution of these rules governing access to land bears witness to the fact that they have been put at the service of economic, social and, more recently, environmental objectives.

For example, the ban on penalizing lessees for adopting environmentally friendly practices was recognized by the July 9, 1999, law on agricultural policy (Law no. 99-574). This prohibits lessors from seeking termination of a lease on the grounds that the lessee has implemented practices that respect the environment, water or product quality, or preserve biodiversity on the leased land.

Since the 2006 law on agricultural policy (Law no. 2006-11), it has also been possible to insert environmental clauses into rural leases, in order to promote the agroecological transition by obliging the lessee to implement environmentally friendly farming practices (L. 411-27 CRPM), such as not turning over grassland, limiting or banning the use of plant protection products or fertilizers, and growing or raising livestock in accordance with organic farming specifications (R. 411-9-11-1 CRPM). The conditions for the inclusion of these clauses, which were initially very limited, have been broadened with each successive reform.

Outside the conditions set out in Article L. 411-27 of the CRPM, the inclusion of environmental clauses seems possible (Bodiguel 2011). A broader integration of environmental clauses into the general framework of the farming statute has been confirmed by the jurisprudence of the Cour de cassation. In two rulings handed down in 2020, it recognized that non-compliance with environmental clauses could constitute grounds for terminating a rural lease (e.g. Cass. 3e civ., February 6, 2020, no. 18-25460, Cass. 3e civ., November 19, 2020, no. 19-21.348 and Cass. 3e civ., November 17, 2021, no. 20-10.934). This development in case law has been

4. *Statut du fermage.*

analyzed as an invitation to lessors to opt for a lease containing environmental clauses outside the framework provided by Article L. 411-27 of the CRPM (Crevel 2021).

Although this is not the qualification provided for by the CRPM, it is referred to as an environmental rural lease. The use of these environmental leases is notably supported by the associative movement, with a movement such as Terre de liens (Land Ties), for example, taking the line of only supporting installation projects within this contractual framework (Lombard and Baysse Lainé 2019).

More recently, the greening of the legal contractual framework governing agricultural land has extended beyond the rental relationship between the lessor and farmer to include the management of rural property as such, through the notion of a real environmental obligation. This obligation takes the form of a voluntary commitment by the owner, formalized in a contract concluded with a public entity (or a private entity acting to protect the environment). The obligation is real in the sense that it is attached not to the person of the owner, but to the property, and thus endures over the long term in the event of sale. The provisions stemming from the 2016 law specify that the purpose of these obligations is "the maintenance, conservation, management or restoration of elements of biodiversity or ecological functions" (authors' translation) and that they can be used "for offsetting purposes" (Law no. 2016-1087 of August 8, 2016, for the reconquest of biodiversity, nature and landscapes, codified at Art. L. 132-3 and seq. in the Environment Code). These obligations can be granted by a landlord and imposed on the lessee under a rural lease, if these obligations are not incompatible with the status of tenant farming. Otherwise, it will be necessary to revise the lease to include environmental clauses as provided for in Article L. 411-27 CRPM (Bosse-Platière and Lucas 2020).

International social science research on the subject frequently aims to assess the effectiveness of these contractualization practices, focusing in particular on the place of owners in the systems of actors. To what extent do stakeholders take advantage of the proposed tools? How effective are contractual solutions or negotiation approaches? Research focusing on contractualization practices with landowners has focused in particular on conservation easements granted by public bodies or associations (land trusts), which are present in common law countries and have inspired the French legal framework for environmental obligations. In the United States, for example, the Uniform Conservation Easement Act defines conservation easements as obligations whose purposes are "the protection of natural resources, the maintenance of air or water quality, or the preservation of historic or cultural features of real property" (Richardson and Bernard 2011, authors' translation).

The limits of contractualization practices lie in the difficulty of reconciling individualized land negotiation with strategic planning. Local agreements, often negotiated on an ad hoc basis, do not always make it possible to agglomerate continuous areas (Stoms et al. 2009). In the areas concerned, agricultural land remains hemmed in by the advance of dispersed urbanization. What is more, protected areas are not always the most strategic, since negotiations depend on the goodwill of landowners, and the effectiveness of the system is sometimes limited (Brinkley 2012). Furthermore, conservation easements can give rise to complex financial arrangements in the context of ecological compensation (Barral 2020). Finally, as an example of the financial arrangements for conservation easements in the United States in the context of ecological offsetting, among the contractual practices associated with agricultural land, we might mention land parcel exchange practices, which can be either permanent or temporary in nature. The practice of exchanging plots of land between farmers has been the subject of empirical research in western (Pauchard et al. 2016) and northern (Dongmo 2023) France, where its contribution to reducing agricultural pressures on the environment is debated.

1.2. Changes in agricultural landholding structures: a challenge for the transition of agri-food systems and generational renewal

1.2.1. *Access to agricultural land for start-ups and the challenge of generational renewal*

The demographic crisis in agriculture – a marked aging of farm managers coupled with a shortage of young project leaders – is a phenomenon that affects all EU member states, and for this reason remains an issue regularly identified by the Common Agricultural Policy in its various evolutions. Compared to the situation in France, this crisis is even more marked in southern and eastern European countries. This is due in particular to the land tenure structure of these countries, where small farms owned by older farmers predominate. This is particularly the case in the Mediterranean basin, where agricultural land is traditionally fragmented, or in certain former Soviet states, where the transition to a market economy in the 1990s led to the creation of agricultural micro-owners (Zagata and Sutherland 2015).

Market characteristics can contribute to this demographic crisis, sometimes for opposite reasons. In the UK, high land values, which are poorly regulated in the absence of regulatory mechanisms such as the right of pre-emption, have been identified as an obstacle to settlement (Maye et al. 2009). Conversely, in areas where land values are low, a different type of strategy can be observed. In Spain, surveys carried out in the Valencia region show that farmers nearing retirement who run small farms with low sales potential prefer to keep them as part of their estate for self-subsistence or leisure purposes. However, the aging of farm managers heralds short-term changes in land ownership structures (Moragues-Faus 2014). In Portugal, small-scale extensive silvopastoral farms are often taken over by multi-activity project owners with no farming background and looking for a rural lifestyle (Pinto-Correia et al. 2015). More generally, the lack of farm takeover prospects is a factor in disinvestment in the farm or simplification of farming systems (Inwood and Sharp 2012). Conversely, research carried out in Austria (Glauben et al. 2004) and Spain (Aldanondo Ochoa et al. 2007) has shown that larger farms are more likely to be taken over within a family framework.

Generally speaking, comparative research carried out in Europe on the profile of new project holders tends to show that they are more involved in agricultural diversification (of both production and activities), often through the search for new markets, particularly in the case of holders from non-farming backgrounds (Sutherland et al. 2015; Suess-Reyes and Fuetsch 2016). They are more concerned with the environmental and social performance of their farms, which echoes other survey results highlighting that organic farmers are proportionally more numerous among project bearers than in the population of incumbent farmers (Lobley et al. 2009).

1.2.1.1. *"Societalization" and outsourcing of agricultural activity and land capital: an obstacle to the transition of agri-food systems?*

Farming and land management can take place in either individual or corporate form. The latter is becoming more common than the former. Apart from family farms with personal liability, rural law provides for a number of different corporate forms: (1) farming companies, such as the SCEA[5], subject to the general provisions of the Civil Code (C. civ., Art. 1832 et seq.), the GAEC[6], subject to Articles L. 323-1 et seq. of the CRPM, and the EARL[7], subject to Articles L. 324-1 et seq. of the CRPM; (2) land management companies, such as the GFA[8], subject to Articles L. 322-1 et seq. of the CRPM.

These corporate forms respond to specific economic, fiscal and social motivations. Depending on the case, they are geared more towards the intra-family transfer of land capital (GAEC) or opening up to shareholdings outside the family sphere (GFA). Based on an analysis of the legal framework, environmental

5. Société civile d'exploitation agricole, Civil Society for Agricultural Exploitation.
6. Groupement agricole d'exploitation en commun, cooperative farming group.
7. Exploitation agricole à responsabilité limitée, private limited liability farm corporation.
8. Groupement foncier agricole, agricultural land association.

motivations cannot be directly linked to these structures. However, the GIEE[9] may be mentioned, as its purpose is to promote environmentally friendly farming practices (Dissaux 2022). Although not formally a company, this structure resulting from the 2014 Law on the Future ensures the recognition of a legal entity "whose members collectively carry out a multi-year project to modify or consolidate their agricultural production systems or methods and their agronomic practices, aiming for performance that is at once economic, social and environmental" (CRPM, Art. L. 315-1, authors' translation). Recognition by the Ministry of Agriculture brings certain benefits, such as increased subsidies. By 2021, 753 GIEEs had been recognized since the scheme came into force in 2015. They bring together more than 12,000 farms, with a wide range of situations in terms of collective size, location, production, partners and project themes.

However, the corporate phenomenon has taken on unprecedented proportions since the increase in the demographic crisis in agriculture and the acceleration in the number of farm managers retiring. In the absence of new farmers, takeovers by neighboring farms already in place have led to an increase in tenanted areas, a significant proportion of which are leased by legal entities (Levesque et al. 2011).

Land regulation tools have failed to prevent a concentration of farmland and may have been hijacked through corporate techniques (Assemblée Nationale 2018). An initial breakthrough was achieved in the Future Law for Agriculture, Food and Forestry of October 13, 2014, which gave SAFERs the option of exercising their right of pre-emption for the acquisition of all the shares in a company whose main purpose is agricultural property. This arrangement did not prevent an agricultural landowner wishing to sell their land from setting up an agricultural company and selling part of their shares to another financial company in order to obtain a more advantageous price. In such cases, SAFERs cannot intervene. The acquiring company can therefore buy back the majority of shares without the risk of the sale being pre-empted.

In 2016, the purchase of 1,700 hectares of cereal land in the Indre region by a Chinese company rekindled the need for reform[10]. The proposed law on the fight against the encroachment of agricultural land and the development of biocontrol was intended to give SAFERs a preponderant role by enabling them to exercise the right of pre-emption in the event of partial transfers of shares in companies whose main

9. Groupement d'Intérêt Économique et Environnemental, Economic and Environmental Interest Group.

10. Written question no. 21719 by Mr. Jean-Claude Leroy on the conditions for the acquisition of agricultural land by foreign companies published in the JO Sénat of May 12, 2016, p. 1964.

purpose is land ownership. This possibility was moreover censured by the Constitutional Council, as disproportionately infringing the right of ownership and freedom of enterprise (French Constitutional Council no. 2017-748 DC of March 16, 2017).

As some legal experts have pointed out, this extension of the right of pre-emption is technically not an appropriate legal means of addressing the issue of land grabbing. They suggest strengthening the judge's power to control fraudulent operations and a mechanism to monitor the activity carried out on the land (Bosse-Platière et al. 2017). This possibility had been proposed by the legislator in 2014 by subjecting direct or indirect shareholdings in agricultural companies to administrative authorization. However, the Constitutional Council censured these dispositions, deeming them equally prejudicial to property rights and entrepreneurial freedom (French Constitutional Council no. 2014-701 DC du October 9, 2014).

Law no. 2021-1756 of December 23, 2021, on emergency measures to regulate access to agricultural land through corporate structures, known as the Sempastous Law, introduces a new mechanism to improve the regulation of land transfers through company shares. The conditions for control by the SAFERs are set out in Articles L. 333-1 et seq. of the CRPM and maritime fishing regulations, which establish control thresholds. The decision to authorize or refuse is entrusted to the prefect, who may suspend the examination of a request in order to encourage applications from new installations as opposed to a concentration. The implementation of this system, whose complexity has been pointed out (Robbe 2023), will enable us to verify its effectiveness in terms of concentration and generational renewal.

On an empirical level, sociological studies have documented another dimension of company structures, namely the development of outsourcing practices through the provision of services. Indeed, the services provided by agricultural contractors, long confined to targeted services, are tending to change in nature to sometimes represent a complete outsourcing of the farmer's activity, in contradiction with the provisions of the statute on tenant farming, which requires the leaseholder to actually operate the farm (Anzalone and Purseigle 2014; Purseigle et al. 2019). These practices are motivated by a variety of strategies (keeping the farm in the family patrimony in the absence of a buyer, circumventing the status of tenant farming and the control of structures). They represent a challenge for farm advisory services and for supporting changes in practices towards transition, in the event that the titled farmer turns out not to be the real farmer.

1.3. The multifunctionality of agricultural land in food systems

The transformation of agricultural activities in areas under urban influence (short circuits, diversification) and initiatives aimed at reconnecting town and country are resulting in greater consideration of agricultural land by fields such as urban planning law, which initially treated it as a lateral object. At the same time, agriculture finds itself at the heart of debates on more systemic approaches to territorial planning, combining land and food policies.

1.3.1. *Agricultural diversification and land planning: from land preservation to urban food planning?*

For a long time reduced to a policy on programming the urbanization of building plots (the main aim of which was to produce affordable housing by capturing agricultural land rent), urban planning was reoriented in the early 2000s, within the framework of the Solidarity and Urban Development law (known as the SRU[11] law), towards a territorial project approach that offers scope for forward-looking thinking. This project-based approach is enshrined in the introduction of a programmatic component, the “sustainable development plan”, within urban planning documents. This includes an agricultural land strategy linked to environmental objectives, since the plan must define “the general orientations of development, equipment, urban planning and landscape policies, the protection of natural, agricultural and forest areas, and the preservation or restoration of ecological continuity” (see Articles L. 151-5 et seq. of the French Urban Planning Code, authors’ translation, resulting from Law no. 2000-1208 of December 13, 2000, on “solidarity and urban renewal”).

Urban planning and rural development policies pay particular attention to the protection of agricultural land. Urbanization and the development of infrastructures of general interest (road networks, installation of water, gas and electricity pipes, photovoltaic panels, etc.) result in the disappearance of agricultural land.

Since the law of July 9, 1999 (no. 99-574), rural development law has been enriched with tools for the specific protection of agricultural land, such as protected agricultural zones. The aim of these zones is to protect a given agricultural area from any change of use or land use that would permanently alter its agronomic, biological or economic potential (C. rur., Art. L. 112-2). However, their use has not curbed the problem of agricultural land consumption. In 2009, the programming law for the implementation of the Grenelle Environment Round Table (Law no. 2009-967),

11. Solidarité et développement urbain.

which sets qualitative and quantitative objectives for State action in the environmental field, incorporated the need to “preserve agricultural land, in particular by limiting its consumption and artificialization” (authors’ translation). With this in mind, the Grenelle 2 law of July 12, 2010 (Law no. 2010-788), and the agricultural modernization law of July 27, 2010 (Law no. 2010-788), have created new tools for developing rural land-use planning policies to prevent the reduction of agricultural land. In other words, the aim is to control changes in the use of rural areas, particularly agricultural land. As far as agricultural land use is concerned, town planning law is not intended to direct land use towards a particular agricultural activity. However, an empirical study carried out on a panel of recent inter-municipal local urban plans shows that, in practice, some inter-municipalities indirectly direct these uses through their regulations on agricultural buildings, by favoring greenhouses for market gardening, for example (Liu et al. 2023).

With a view to limiting the sprawl of agricultural land, the SRU law introduced stricter constraints for building projects in agricultural zones, shifting the requirement for justification from compatibility to necessity (Art. R. 151-23 of the French Urban Planning Code). In the 2000s, the “Council of State” case law interpreted this principle of necessity strictly, ruling that building permits for agritourism projects were illegal. In a reverse move aimed at restoring greater flexibility to the allocation of building rights, the legislator introduced derogatory sectors in agricultural zones (“sectors of limited size and capacity” (authors’ translation), in which construction is authorized on an exceptional basis. Art. L. 151-13 of the French Urban Planning Code, resulting from Law no. 2014-366 of March 24, 2014, for access to housing and renovated urbanism, known as the ALUR[12] law).

In line with an approach aimed at conferring greater flexibility on the allocation of building rights in agricultural zones, the legislator is extending the scope of construction in agricultural zones by explicitly including diversification activities: “constructions and facilities required for the processing, packaging and marketing of agricultural products, where these activities constitute an extension of the act of production” (Art. L. 151-11 of the French Urban Planning Code, authors’ translation, resulting from Law no. 2018-1021 of November 23, 2018, on the evolution of housing, development and the digital economy, known as the ELAN[13] law).

These contradictory developments reflect a tension between objectives to combat urban sprawl (which aim to limit the creation of housing through construction or changes of use in agricultural areas close to farm headquarters, including for

12. Accès au logement et un urbanisme rénové, dite loi.

13. Evolution du logement, de l’aménagement et du numérique.

farmers), on the one hand, and, on the other, the development of diversification projects linked to the re-territorialization of agriculture.

1.3.2. *The new challenges of re-territorializing agriculture: the land dimensions of food policies*

Today, agricultural land policies are being renewed as part of the development of local policies in favor of local food and the re-territorialization of agriculture. Indeed, these policies are eminently interlinked and should be considered together. For the time being, the proposed instruments are not very restrictive, even if there are legal tools that can be mobilized in the direction of an ambitious land policy favoring the relocalization of the French food system (Bosse-Platière 2021). These include initiatives by local authorities to take over land for agricultural projects, such as test areas for market gardening and short-cycle facilities.

In this respect, territorial food projects offer a framework conducive to "the consolidation of territorialized supply chains and the development of consumption of products from short circuits, in particular from organic production [by meeting] the objective of structuring the agricultural economy and implementing a territorial food system" (L. 111-2-2 CRPM, authors' translation). Therefore, even if they take the form of incentive action programs (Bodiguel 2018), territorial food projects can be a programming space for agricultural land policies involving a wide spectrum of actors beyond the farming profession's organizations alone: local authorities, the associative sector, downstream actors. They can support initiatives to design diversified agri-food projects that require innovations in urban planning regulations (on-farm processing, mobile farm buildings). They can also be a forum for discussion on the allocation of space for processing or distribution activities geared towards the re-territorialization of agriculture (supply platforms, vegetable shops, producers' boutiques, etc.) (Liu et al. 2023). This incentive framework comes up against a more restrictive legal framework that limits actions to re-territorialize agriculture. For example, public procurement law can be seen as an obstacle to setting an express clause requiring public catering to be supplied by local circuits. Nevertheless, a number of communes have succeeded through legal arrangements (Darrot et al. 2019).

1.4. Conclusion and perspectives

Old land management tools are now being used again in agricultural policies. In this synthesis, we wanted to focus on the legal tools that can accompany transitions in agri-food systems. The dual approach we have proposed, based on law and sociology, is designed to foster both a conceptual and empirical understanding of legal instruments. As we have shown, their appropriation for the ecological and food transition is both the object of social innovation and the subject of strong dynamics. These dynamics relate to issues of coexistence between agricultural systems and spatial justice in terms of access to the resource of land, as illustrated by the phenomena of concentration and societalization, which call into question the scope of these transitions. On this last point, the reform of agricultural land law, which seemed "heretical" not so long ago, is now the subject of consensus in principle (Grimonprez 2018, 2023). More broadly, such a reform must take into account a wider range of issues than when the foundations of rural law were first laid. Sustainable management of water resources, adaptation to climate change, generational renewal and farm income are all parameters that need to be taken into account when developing rural land tenure rules.

1.5. References

Aldanondo Ochoa, A.M., Casanovas Oliva, V., Almansa Saez, C. (2007). Explaining farm succession: The impact of farm location and off-farm employment opportunities. *Spanish Journal of Agricultural Research*, 5(2), 214–225.

Anzalone, G. and Purseigle, F. (2014). Délégation d'activités et sous-traitance : au service de la transmission de l'exploitation ou d'un patrimoine ? In *L'agriculture en famille : travailler, réinventer, transmettre*, Pierre, G., Jean-Philippe, C., Sandrine, P., François, P., Sylvie, Z. (eds). EDP Science, Les Ulis.

Assemblée Nationale (2018). Rapport d'information déposé par la mission d'information commune sur le foncier agricole. Report, 5 December.

Barataud, F. and Hellec, F. (2015). L'outil foncier, une solution délicate pour protéger les captages d'eau potable. *Économie rurale*, 347(3), 3–20.

Barral, S. (2020). Conservation, finance, bureaucrats : managing time and space in the production of environmental intangibles. *Journal of Cultural Economy*, 1, 15. doi: 10.1080/17530350.2020.1846593.

Barral, S. and Pinaud, S. (2017). Accès à la terre et reproduction de la profession agricole. Influence des circuits d'échange sur la transformation des modes de production. *Revue Française de Socio-Economie*, 18(1), 77–99.

Barthe, L. and Eychenne, C. (2018). Les associations foncières pastorales : un dispositif de gouvernance concrète du foncier dans les territoires de montagne. *EchoGéo*, 43. doi: 10.4000/echogeo.15228.

Barthès, C., Bertrand, N., Borg, D. (2013). Les chartes dans la gouvernance foncière des espaces agricoles. In *Terres agricoles périurbaines. Une gouvernance foncière en construction*, Bertrand, N. (ed.). Éditions Quæ, Versailles.

Baysse-Lainé, A. (2020). Une géographie relationnelle de l'accès au foncier agricole en France. *L'Espace géographique*, 49(3), 193–212.

Baysse-Lainé, A. (2022). Vers un accès au foncier plus inclusif pour les néo-agriculteurs ? Informations, réseaux et autochtonie. *Espaces et sociétés*, 186–187(3–4), 191–210.

Bertrand, N. and Duvillard, S. (2016). Intégration des enjeux environnementaux dans la gestion du foncier agricole : introduction. *Sciences Eaux Territoires*, 19(2), 2–5.

Bertrand, N. and Melot, R. (2016). Terres agricoles au Nord et au Sud. Quelles transformations pour l'action publique ? Introduction. *Économie rurale*, 353–354, 3–8.

Bignebat, C., Melot, R., Moustier, P., Raynaud, E., Soullier, G. (2022). Purchasing land for agricultural projects in France's Île-de-France region. In *Sustainable Food Systems for Food Security: Need for Combination of Local and Global Approaches*, Thomas, A., Alpha, A., Barczak, A., Zakhia-Rozis, N. (eds). Éditions Quæ, Versailles.

Bodiguel, L. (2011). Les clauses environnementales dans le statut du fermage. *Environnement et développement durable*, 8–9(10), 13–19.

Bodiguel, L. (2018). Le développement des projets alimentaires territoriaux en France : quel droit pour quelle relocalisation de l'agriculture et de l'alimentation ? In *XVe congrès mondial de l'Union Mondiale des Agraristes Universitaires, Les évolutions actuelles du droit rural et agroalimentaire : entre globalisation, régionalisation et relocalisation*. UAM, 409–415.

Boinon, J.-P. (2011). Les politiques foncières agricoles en France depuis 1945. *Économie et Statistique*, 444(1), 19–37. doi: 10.3406/estat.2011.9641.

Bonnefond, M. and Fournier, M. (2013). Maîtrise foncière dans les espaces ruraux. Un défi pour les projets de renaturation des cours d'eau. *Économie rurale*, 334(2), 55–68.

Bosse-Platière, H. (2021). Imaginer une politique foncière de relocalisation du système alimentaire ou l'éloge du flexitarisme juridique. In *La relocalisation des systèmes alimentaires : un défi pour le droit*, Aubin-Brouté, R.-J. and Grimonprez, B. (eds). Presses universitaires juridiques de Poitiers, Poitiers, 45–62.

Bosse-Platière, H. and Lucas, A. (2020). L'obligation réelle environnementale et le statut du fermage : le mariage de la carpe et du lapin. *Cahier du CRIDON de Lyon*, 31.

Bosse-Platière, H., Rochard, D., Besson, S., Travely, B. (2017). Loi relative à la lutte contre l'accaparement des terres – L'espoir constitutionnel. *JCP N*, 1122.

Brinkley, C. (2012). Evaluating the benefits of peri-urban agriculture. *Journal of Planning Literature*, 27, 259–269. doi: 10.1177/0885412211435172.

Cavailhès, J., Mesrine, A., Rouquette, C. (2011b). Le foncier agricole : une ressource sous tensions. *Économie et Statistique*, 444(1), 3–18.

Chouquer, G. and Maurel, M.-C. (2019). *Les mutations récentes du foncier et des agricultures en Europe*. Presses universitaires de Franche-Comté, Besançon.

Courleux, F. (2011). Augmentation de la part des terres agricoles en location : échec ou réussite de la politique foncière ? *Économie et Statistique*, 444(1), 39–53.

Crevel, S. (2021). Résiliation du bail – Un bail quasi environnemental. *Droit rural*, 490.

Darrot, C., Maréchal, G., Bréger, T. (2019). Rapport sur les Projets Alimentaires Territoriaux (P.A.T.) en France : Etat des lieux et analyse. Report, Agrocampus Ouest and Cabinet Territoires et Alimentation Terralim, Rennes.

Dissaux, N. (2022). *Droit rural*. LGDJ.

Dongmo Zangue, Y. (2023). Les échanges temporaires de parcelles entre exploitations : motivations agricoles et intérêt environnemental. PhD Thesis, Paris-Saclay University [Online]. Available at: https://hal.inrae.fr/tel-04103206v1.

Duvernoy, I., Jarrige, F., Serrano, J. (2005). Une agriculture multifonctionnelle dans le projet urbain quelle reconnaissance quelle gouvernance. *Cahiers de la multifonctionnalité*, 8, 87–104.

Etrillard, C. and Pech, M. (2015). Mesures de compensation écologique : risques ou opportunités pour le foncier agricole en France ? *VertigO – la revue électronique en sciences de l'environnement*, 15(2). doi: 10.4000/vertigo.16450.

Glauben, T., Tietje, H., Weiss, C.R. (2004). Intergenerational succession in farm households: Evidence from upper Austria. *Review of Economics of the Household*, 2(4), 443–462.

Grimonprez, B. (ed.) (2018). *La réforme du droit foncier rural : demander l'impossible*. LexisNexis, Paris.

Grimonprez, B. (2023). Le droit rural entre déclin et regain. *Revue de droit rural*, August/September.

Guéringer, A. (2008). "Systèmes fonciers locaux" : une approche de la question foncière à partir d'études de cas en moyenne montagne française. *Géocarrefour*, 83(4), 321–329.

Guéringer, A. (2023). Applications différenciées d'une politique foncière agricole : une lecture comparée des SDREA. *Économie rurale*, 383(1), 19–33.

Guéringer, A., Perrin, C., Barthès, C. (2017). Tensions sur l'espace agricole : quand les enjeux fonciers réinterrogent le rapport entre propriété et usage. *VertigO – la revue électronique en sciences de l'environnement*, 17(1).

Guyet, G. (2016). L'environnement dans l'Aménagement foncier agricole et forestier : quelles conditions pour en faire un outil multifonctionnel ? Thesis, DUMAS [Online]. Available at: https://dumas.ccsd.cnrs.fr/dumas-01703024.

Hobeika, A. (2013). La collégialité à l'épreuve : la production de l'unité au sein de la FNSEA. *Politix*, 103, 53–76. doi: 10.3917/pox.103.0053.

Inwood, S. and Sharp, J. (2012). Farm persistence and adaptation at the rural–urban interface: Succession and farm adjustment. *Journal of Rural Studies*, 28(1), 107–117. doi: 10.1016/j.jrurstud.2011.07.005.

Jarrige, F., Thinon, P., Nougaredes, B. (2006). La prise en compte de l'agriculture dans les nouveaux projets de territoires urbains. Exemple d'une recherche en partenariat avec la Communauté d'Agglomération de Montpellier. *Revue d'Économie Régionale & Urbaine*, August(3), 393–414.

Lamoureux, L. (2016). Politique d'acquisitions foncières et protection de la ressource en eau. *Sciences Eaux Territoires*, 19(2), 38–43.

Léger-Bosch, C., Houdart, M., Loudiyi, S., Le Bel, P.-M. (2020). Changes in property-use relationships on French farmland: A social innovation perspective. *Land Use Policy*, 94, 104545.

Levesque, R., Liorit, D., Pathier, G. (2011). Les marchés fonciers ruraux régionaux entre dynamiques des exploitations agricoles et logiques urbaines. *Economie et statistique*, 444–445, 75–98. doi: 10.3406/estat.2011.9644.

Liu, T., Altes, W.K.K., Romain, M., Wallet, F. (2023). Reterritorialisation of agricultural activities in land-use and food planning: Comparing the Netherlands and France. *European Planning Studies*, 21. doi: 10.1080/09654313.2023.2244567.

Lobley, M., Butler, A., Reed, M. (2009). The contribution of organic farming to rural development: An exploration of the socio-economic linkages or organic and non-organic farms in England. *Land Use Policy*, 26, 723–735.

Lombard, P. and Baysse-Lainé, A. (2019). Terre de Liens, un levier foncier militant au service d'un projet politique pour l'agriculture. *Économie rurale*, 369(3), 83–101.

Lorvellec, L. (1988). *Droit rural*. Masson, Paris.

Margetic, C. (2014). Les PEAN en France : des envies et de premières concrétisations. *Pour*, 224, 341–349.

Margetic, C., Rouget, N., Schmitt, G. (2016). Le foncier agricole à l'épreuve de la multifonctionnalité : desseins environnementaux et alimentaires dans les métropoles lilloise et nantaise. *Norois*, 241(4), 87–104.

Maye, D., Ilbery, B., Watts, D. (2009). Farm diversification, tenancy and CAP reform: Results from a survey of tenant farmers in England. *Journal of Rural Studies*, 25(3), 333–342. doi: 10.1016/j.jrurstud.2009.03.003.

Moragues-Faus, A. (2014). How is agriculture reproduced? Unfolding farmers' interdependencies in small-scale Mediterranean olive oil production. *Journal of Rural Studies*, 34, 139–151.

Mormont, M. (2013). Écologisation : entre sciences, conventions et pratiques. *Natures Sciences Sociétés*, 21, 159–160. doi: 10.1051/nss/2013102.

Napoléone, C. and Melot, R. (2021). Farmland management and sustainable development in the Mediterranean: Land use changes, public policies, and collective resources. *Regional Environmental Change*, 21(2), 31.

Nougarèdes, B. (2011). Quelles solutions spatiales pour intégrer l'agriculture dans la ville durable ? Le cas des "hameaux agricoles" dans l'Hérault. *Norois. Environnement, aménagement, société*, 221, 53–66.

Pauchard, L., Madeline, P., Marie, M. (2016). L'échange parcellaire : une nouvelle étape dans l'aménagement foncier de l'Ouest français. *Norois. Environnement, aménagement, société*, 240, 7–24.

Pech, M. and Étrillard, C. (2016). Le foncier agricole, variable d'ajustement ou déterminant de la compensation écologique ? *Sciences Eaux Territoires*, 19(2), 60–63.

Périnet-Marquet, H. (1997). Le droit de préemption dans les espaces naturels sensibles. *Revue Juridique de l'Environnement*, 22(2), 173–182.

Perrin, C. (2013). L'agriculture multifonctionnelle comme écrin pour la ville durable : une vision urbaine des liens entre la ville et l'agriculture en Méditerranée ? *Bulletin de l'association de géographes français*, 90(3), 337–355.

Perrin, C. and Nougarèdes, B. (2020). *Le foncier agricole dans une société urbaine. Innovations et enjeux de justice*. Cardère, Avignon.

Piet, L., Latruffe, L., Le Mouël, C., Desjeux, Y. (2012). How do agricultural policies influence farm size inequality? The example of France. *European Review of Agricultural Economics*, 39(1), 5–28. doi: 10.1093/erae/jbr035.

Piet, L., Melot, R., Diop, S. (2021). What drives competition on the farmland market? A case study in Brittany (France). *European Review of Agricultural Economics*, 48(1), 60–96. doi: 10.22004/ag.econ.329588.

Pinto-Correia, T., Gonzalez, C., Sutherland, L.-A., Peneva, M. (2015). Countryside consumption: Transition towards lifestyle land management. In *Transition Pathways towards Sustainability in European Agriculture*, Sutherland, L.-A., Darnhofer, I., Zagata, L., Wilson, G.A. (eds). CABI, Wallingford.

Purseigle, F., Anzalone, G., Nguyen, G. and Hervieu, B. (2019). Des entreprises agricoles “aux allures de firme”. Mutations des exploitations agricoles françaises et nouveaux modes d’accès au foncier. In *Les mutations récentes du foncier et des agricultures en Europe*, Chouquer, G. and Maurel, M.-C. (eds). Presses universitaires de Franche-Comté, Besançon.

Requier-Desjardins, M., Melot, R., Paoli, J.-C., Rodrigues, O., Riaux, J. (2019). Le foncier en Méditerranée : une dichotomie entre Nord et Sud ? *Développement durable et territoires*, 10(3). doi: 10.4000/developpementdurable.16341.

Richardson, J. and Bernard, A. (2011). Zoning for conservation easements. *Law and Contemporary Problems*, 74, 83–108 [Online]. Available at: https://scholarship.law.duke.edu/lcp/vol74/iss4/5.

Robbe, F. (2023). Loi Sempastous et contrôle de l’accès au foncier agricole à travers les prises de participation sociétaires : modalités d’application. *Revue de droit rural*, 2.

Rouhaud, F. (2012). La protection des espaces ruraux par les documents d’urbanisme : quelle efficacité pour l’exercice des activités agricoles ? *Revue de droit rural*, 399, dossier 3.

Roussel, S., Tardieu, L., Vaissière, A.-C. (2019). Compensation écologique et agriculture : est-ce compatible ? *Revue économique*, 70(1), 123–137.

Sencébé, Y. (2012). La SAFER. De l’outil de modernisation agricole à l’agent polyvalent du foncier : hybridation et fragmentation d’une institution. *Terrains & travaux*, 20(1), 105–120.

Sencébé, Y, Pinton, F., Alphandéry, P. (2013). Le contrôle des terres agricoles en France. Du gouvernement par les pairs à l’action des experts. *Sociologie*, 4(3), 251–268.

Stoms, D.M., Jantz, P.A., Davis, F.W., DeAngelo, G. (2009). Strategic targeting of agricultural conservation easements as a growth management tool. *Land Use Policy*, 26, 1149–1161. doi: 10.1016/j.landusepol.2009.02.004.

Struillou, J.-F. (2002). Vers un remembrement-environnement ? Aménagement de l’espace rural, exploitation agricole et sauvegarde des milieux naturels. *Revue de droit rural*, 324, 342.

Struillou, J.-F. and Joye, J.-F., (2015). Les communautés et le droit de preemption. Étude à caractère juridique. *ADCF – GRIDAUH*.

Suess-Reyes, J. and Fuetsch, E. (2016). The future of family farming: A literature review on innovative, sustainable and succession-oriented strategies. *Journal of Rural Studies*, 47, 117–140.

Sutherland, L.-A., Darnhofer, I., Zagata, L., Wilson, G.A. (eds) (2015). *Transition Pathways towards Sustainability in European Agriculture*. CABI, Wallingford [Online]. Available at: https://www.cabidigitallibrary.org/doi/book/10.1079/9781780642192.0000.

Zagata, L. and Sutherland, L.-A. (2015). Deconstructing the "young farmer problem in Europe": Towards a research agenda. *Journal of Rural Studies*, 38, 39–51. doi: 10.1016/j.jrurstud.2015.01.003.

2

Synergies between Animal Husbandry and Forests for the Development of Agroecological Systems in Middle Mountain and Mediterranean Foothill Regions: Comparative Study of the Limousine Mountain and the Lodevois

Nathan MORSEL[1] and Nadège GARAMBOIS[2]

[1] *Association pour le Pastoralisme de la Montagne Limousine, Millevaches, France*
[2] *AgroParisTech, UMR PRODIG, Palaiseau, France*

2.1. Introduction

Silvopastoralism can be defined as the coexistence, on the same surface, of silvicultural and pastoral farming activities that work in synergy. Grazing, for example, maintains the undergrowth and facilitates silvicultural operations (layering, thinning, etc.), which in turn helps to develop forage resources (Guérin 2008). However, the coexistence of silvicultural and pastoral uses in the same area can cover different situations. For example, when silviculture is dominant and herds

Farms and Territory,
coordinated by Philippe MARTIN and Solène PISSONNIER.

only make occasional incursion, without interacting with silvicultural management, we speak of a "grazed forest". Conversely, when silvicultural intervention is virtually absent and pastoral use predominates, we speak of "wooded rangelands", with the risk of forest cover closure and the disappearance of pastoral resources. It is only when pastoral and silvicultural objectives are linked that we can truly speak of silvopastoralism (Guérin et al. 2010). In France today, silvopastoralism is mostly confined to certain small regions in the south of the country, where grazing herds play an important role in fire control (DFCI[1] areas). It is therefore absent or marginal in many French regions, despite the importance of forested areas.

However, until the 1950s, feeding ruminant herds in France's middle mountain and Mediterranean foothill regions depended to a large extent on grazing rangelands, including wooded areas, which were also used at least by farmers to supply wood for domestic use. Since the 1950s, the major increase in the physical productivity of human work in ruminant farming has combined higher output per animal (milk production, growth rate or average daily gain, prolificacy, etc.) with an increase in the number of animals raised per worker (Mazoyer and Roudart 1997). In middle mountain and mediterranean foothill regions, this trend has led to a sharp increase in the proportion of concentrated feed and forage distributed to herds, to the detriment of grazing, particularly on rangelands, especially wooded ones (Hubert et al. 2010; Aubron et al. 2019; Garambois et al. 2020).

The aim is therefore to understand the causes and symptoms of the low use of silvopastoralism in foothill and middle mountain regions in France today, to study how these practices can be (re)deployed in agroecological systems, and to analyze their benefits for both livestock farmers and foresters. To this end, this research[2] (Morsel 2024) is part of a comparative agriculture approach (Cochet et al. 2007; Cochet 2011). The latter aims to analyze agricultural transformations and compare productive processes, thanks to an agroeconomic approach and the mobilization of specific systemic concepts (agrarian system on the scale of the small agricultural region, production system on the scale of farms), making it possible to apprehend agrarian dynamics in a multi-scalar way and in their complexity and diversity. This research is based on in-depth fieldwork and surveys[3] carried out in two small French

1. Défense des Forêts Contre les Incendies: Defense of Forests Against Wildfire.

2. Developed from a doctoral thesis funded by the Fondation de France.

3. Fieldwork conducted in 2018, then between 2020 and 2022, included (1) 80 interviews, mainly with retired farmers, as well as with a number of local professionals (forest owners, UPRA technicians, manager of the Conservatoire des Espaces Naturels (CEN), DDA research officer, etc.); (2) 126 technical-economic interviews with active farmers, including 23 with farmers in agropastoral systems; (3) 15 technical interviews with breeders–shepherds or salaried shepherds conducted specifically during the herds' guarding period to analyze their conduct in greater detail; (4) 23 interviews with various actors playing a key role in the

farming regions with contrasting forestry dynamics, respectively, located in the middle mountain and Mediterranean foothills: the Limousine Mountain and the Lodevois (Hérault) (see Figure 2.1).

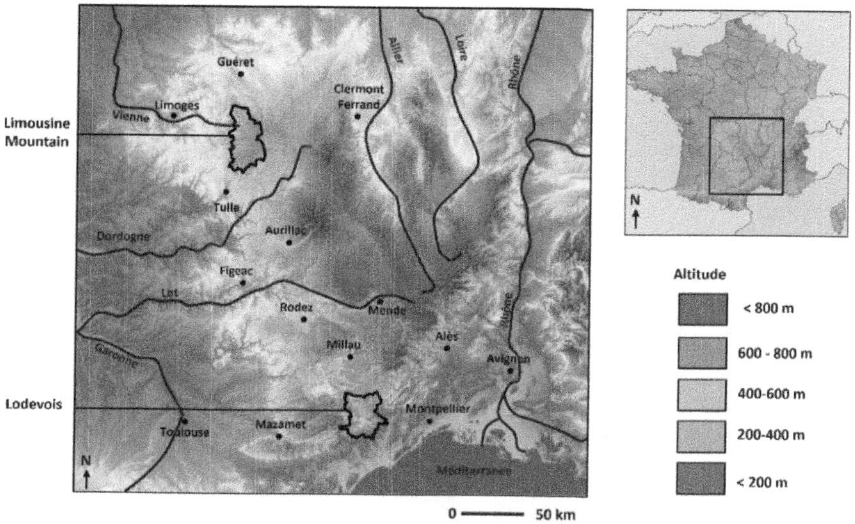

Figure 2.1. *Location of the Limousine Mountain and the Lodevois (source: IGN maps; production: authors). For a color version of this figure, see www.iste.co.uk/martin/farmsandterritory.zip*

Despite the current extent of forested areas in these two regions (see Figure 2.2), silvopastoralism is virtually non-existent, and the links between livestock farming and forestry activities fall into two contrasting situations. On the Limousine Mountain, most of the forests are coniferous and exploited for wood production, where herds only make occasional incursions with no link to the predominant forest management (the "grazed forest"). On the other hand, in the Lodevois, most of the forests are deciduous and are not exploited in any way and play a slightly more important role in feeding the herds ("wooded rangelands"[4]). In these two regions,

development of agropastoral systems (landowners (CEN, foresters, winegrowers), managers of natural areas (PNR, CEN), forest managers (ONF), departmental pastoral services, veterinarians, etc.). Numerous interviews with older farmers have enabled us to reconstruct the past transformations of these two farming systems, going back to the situation in the 1950s. The characterization and results of all the current production systems mentioned or presented in this chapter, i.e. those based on the predominant dynamics of agricultural development as well as those based on an agropastoral approach, are all based on this fieldwork alone (Morsel, 2024).

4. Known in French as "*parcours boisé*".

however, we have been witnessing for the past two decades a movement of pastoral reconquest driven by small groups of farmers who are implementing agroecological systems and mobilizing silvopastoral practices to this end. Wooded ecosystems thus play an essential role in these livestock farms at certain key times of the year.

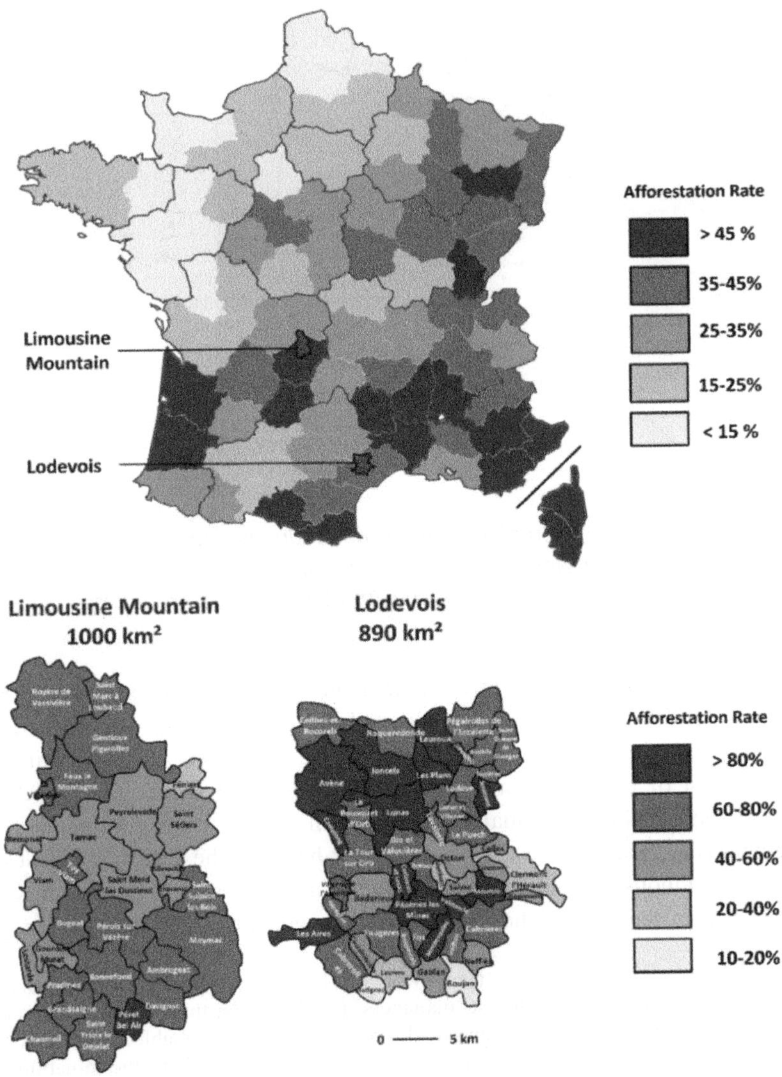

Figure 2.2. *Forest cover of the Limousine Mountain and the Lodevois (source: IGN 2020 data; production: authors). For a color version of this figure, see www.iste.co.uk/martin/farmsandterritory.zip*

This chapter aims to (1) retrace the agricultural and silvicultural evolution of these two regions in order to understand why silvopastoralism has regressed (Lodevois) or never developed (Limousine Mountain), despite the size of the forested areas found in both regions today; (2) clarify the role that these wooded ecosystems can play in the operation of agropastoral systems based on agroecology, which have recently been set up in these two regions; and (3) analyze the current obstacles and levers to the development of these silvopastoral practices in the Limousine Mountain and Lodevois.

2.2. The Lodevois and the Limousine Mountain: a disconnection between animal husbandry and forestry

2.2.1. *Once central pastoral practices, with contrasting uses of wooded ecosystems*

2.2.1.1. *Varied wooded ecosystems with multiple functions for the Lodevois mixed crop–livestock farming systems*

A Mediterranean foothill region, the Lodevois is a transition zone between the wine-growing plains of the Hérault and the Causse of Larzac[5], where dairy sheep farming predominates. With altitudes ranging from 100 to 800 meters, this region features a highly layered vegetation.

In the 1950s, the landscape was shaped by an agriculture closely associating mixed cropping, viticulture and chestnut-growing on plots close to the farms, and vast areas of rangeland from open dry grasslands to coppices of pubescent oak and holm oak, maintained by grazing and, where necessary, pastoral burn. Large areas of holm oak coppice were still exploited for charcoal production, with cycles of 25–50 years. Otherwise, there were only a few rare woodlands resulting from the RTM[6], operations carried out on the Larzac foothills at the end of the 19th century, which led to the creation of state-owned forests (Roque et al. 2011).

Dairy ewes were the predominant livestock farming, whose milk was destined for the production of Roquefort cheese. The flocks were taken to pasture and shepherded year-round, with lambings scheduled at the end of the winter and milk production running into the summer, allowing the peak of the flock's needs in spring to coincide with the availability of forage in the pastures. Woodland vegetation played a key role in the ewes' diet (see Figure 2.3). In autumn and winter, ewes

5. Causse, gorge-gouged limestone plateaus of southwestern France. The name is from *cau*, local form of *chaux*, meaning "lime".

6. Restauration des Terrains en Montagne: restoration of mountain land.

were driven to oak coppices and chestnut groves to graze the undergrowth and consume acorns and chestnuts not harvested for human consumption. Dried chestnuts were used to fatten lambs. In summer, grazing in riparian forests was combined with ash pruning, which could also be stored to feed ewe lambs and rams during the winter.

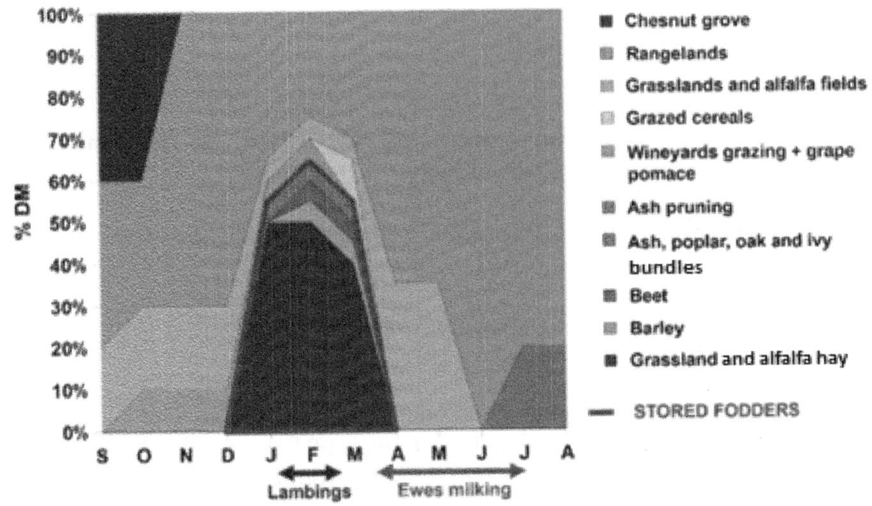

Figure 2.3. *Typical forage calendar for dairy ewes in the Lodevois in the 1950s (source: surveys; production: authors). For a color version of this figure, see www.iste.co.uk/martin/farmsandterritory.zip*

2.2.1.2. *The Limousine Mountain, home to Calluna moors and peat bogs*

A humid, middle mountain region with an oceanic, mountainous climate, the Limousine Mountain is characterized by gently sloping hills, with a low altitudinal range (700–900 meters).

In contrast to the Lodevois, in the 1950s, woodlands and forests were limited to a few coppices and Scots pine plantations planted in the early 20th century. At the time, the landscape was characterized by the predominance of calluna moors and peat bogs, which occupied up to 75% of the area. The remainder was divided between permanent grassland and arable land planted with cereals and fodder vegetables in rotation with grazed, multi-annual grassy wasteland. The farms combined sheep breeding for the production of lean lambs, cattle breeding for the production of veal calves reared under their mother and pig farming. The sheep flocks were grazed all year round: on calluna moors, peat bogs and herbaceous wasteland. Lambing at the end of winter was intended to match the flock's peak needs with the peak of vegetative growth in spring, in particular to fatten lambs on pasture. Cattle feeding, whose calving periods were staggered to allow the cows to act as nursemaids once their calves had been sold, already relied much more heavily on hay and, if not, grazing of grassy wasteland and peat bogs in spring and summer. In contrast to the Lodevois, the rangelands, which were used primarily to feed ruminants, did not include wooded ecosystems (see Figure 2.4).

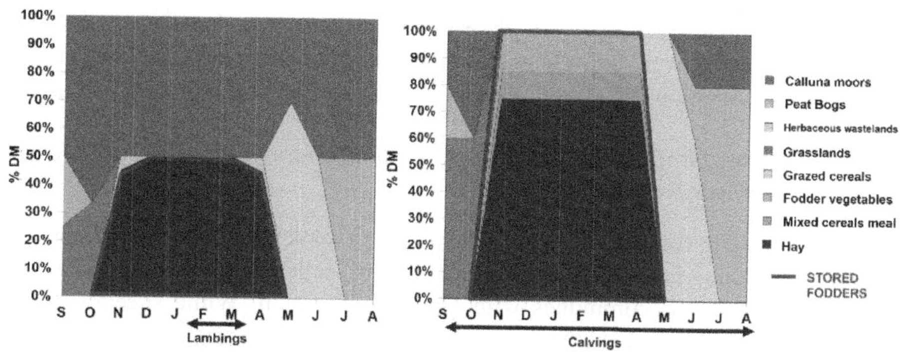

Figure 2.4. *Typical forage calendar for suckler ewes (left) and cows (right) of the Limousine Mountain in the 1950s (source: surveys; production: authors). For a color version of this figure, see www.iste.co.uk/martin/farmsandterritory.zip*

2.2.2. *Unprecedented decline in the use of rangelands and the disconnection between agriculture and forestry*

2.2.2.1. *In the Lodevois, differentiated specialization between the low valleys and the plateaus, with the abandonment of wooded ecosystems*

In the Lodevois, the gradual development (as elsewhere in France) of production systems based on the use of motorized equipment and inputs (fertilizers, phytosanitary products, etc.) (Mazoyer and Roudart 1997) since the 1960s has led to the specialization of farms in the lower valleys in viticulture, accompanied by the cessation of livestock farming and with it the abandonment of grazing land. A few afforestation projects were launched on dried grassland or shrublands, generally with communal status, financed in large part by the National Forestry Fund created in 1946. The high plateaus have seen a specialization in dairy sheep farming, with the abandonment of viticulture and chestnut-growing. The increase in milk production per ewe and the shift of lambing to late autumn, driven by the downstream industry (Roquefort), led to far-reaching changes in the way flocks were fed, based on a growing reliance on distributed fodder and concentrated feed to the detriment of grazing, particularly on rangelands. Farmers concentrated on exploiting motorized areas, even seeking to extend them by rangeland clearing in the 1980s and 1990s, which was largely financed following the creation of authorized syndicated associations (ASAs), supported by the Hérault Departmental Directorate of Agriculture (DDA) in the late 1970s, and which enabled farmers to benefit from numerous subsidies.

Vast areas still used as grazing land in the 1950s have thus been abandoned, both in the lower valleys and on the high plateaus, in particular wooded ecosystems (oak coppices, chestnut groves with the cessation of chestnut harvesting for both human and animal consumption, riparian forests with the cessation of ash pruning, etc.).

2.2.2.2. *On the Limousine Mountain: specialization in suckler cattle and large-scale softwood plantations*

From the 1970s onwards, major land clearing took place on the Limousine Mountain, enabling farmers to convert parcels of calluna moors into grassland, particularly in the few communes with land consolidation. While specializing in ruminant breeding, farms on the Limousine Mountain kept combining sheep and cattle farming until the 1980s. They completed their specialization in the 1990s, for the most part in grazer calves breeding, which developed following the creation of the common beef market in 1968, to the detriment of veal calves reared under their mother, which was more labor-intensive and barely more remunerative in the 1960s–1970s (Pfimlin et al. 2009). Since the 1970s, a small number of farms have specialized in sheep breeding for the production of heavy lambs, fattened in sheepfold with concentrated feed mainly purchased. The widespread predominance of cattle farming in the region has been accompanied by a general refocusing of grazing on permanent grasslands, to the detriment of moors and bogs.

At the same time, and thanks to the work of the Fond Forestier National (National Forestry Fund), created in 1946 to rebuild France's forestry potential and ensure the country's self-sufficiency in lumber and paper pulp (Dodanne 2009), a large-scale afforestation movement took place in this region. While this mainly concerned the non-motorized plots of land of farms that stop their activities – plots that were of no interest to farms still in operation at this time – softwood plantations were sometimes created on arable land. This development of softwood plantations was carried out by active farmers on the former non-motorized rangelands they no longer farmed, as well as by retired farmers and their descendants, or by notables as an investment (Larrère 1978).

2.2.3. *The new forests of the Lodevois and the Limousine Mountain: contrasting structure and composition, but both disconnected from agriculture*

Although these two regions have a similar afforestation rate (just over 60%), the forests of the Lodevois and the Limousine Mountain present substantial differences.

In the Lodevois, the decline of farming activities based on the grazing of dried grasslands, shrublands and undergrowth led to the closure of the landscape and an extension of the pre-existing pubescent and holm oak coppices, increasingly dense and closed. In most of these forests, there is no silvicultural land use.

The few softwood forests that have been planted as regular high forest (even-aged timber management), often on former communal land, have proved to be of mediocre quality, and most have still not been exploited. These forests, which are subject to a Mediterranean climate and located in areas with steep slopes and poor soil fertility, are, with rare exceptions, not used for anything other than hunting.

On the Limousine Mountain, the livestock farmers also focused on motor-mechanized areas which led to the abandonment of many rangelands, in this case calluna moors. The latter have been mostly planted and are today covered by softwood plantations, although part has simply been abandoned and is now covered by forests of sessile and pedunculate oak and beech.

Climatic conditions (in particular, rainfall is more abundant and better distributed than in the Lodevois), soil type and low relief are favorable to silviculture (Beynel 1998).

The establishment of this coniferous forest, essentially monospecific plantations managed as regular high forest, initially enabled the maintenance of small multi-activity farms. Indeed, the farmers found in forestry work complementary to the agricultural activity calendar another source of income enabling them to invest in agricultural equipment.

The transformation of forestry operations, marked by the strong development of motorized mechanization since the 1980s, put an end to this pluriactivity in favor of a specialization of forestry operators (Palluet 2019).

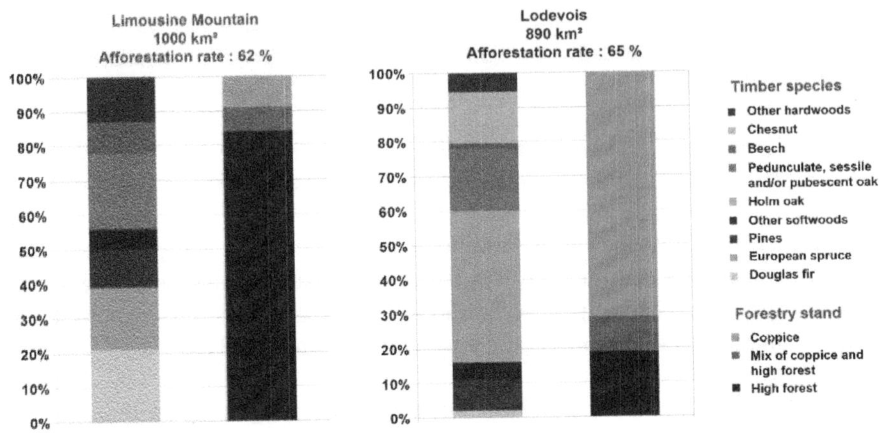

Figure 2.5. *Composition and structure of forests in the Lodevois and Limousine Mountain (source: IFN 2020; production: authors). For a color version of this figure, see www.iste.co.uk/martin/farmsandterritory.zip*

The refocusing of specialized production systems in these two regions on motor-mechanizable land and the decline in grazing on herd feed have thus led to a gradual decline in the use of rangelands, abandoned in the Lodevois and afforested on the Limousine Mountain. The resulting wooded ecosystems, whether spontaneous or artificial, are no longer (or were never) seen as potential forage resources for livestock.

However, over the past 20 years, small groups of livestock farmers now structured into two local CIVAM[7] groups have gradually settled on the Limousine Mountain and the Lodevois and go against this dynamic. Indeed, these farmers have set up systems based on maximizing grazing, with intensive use of rangelands, including wooded areas.

These systems are based on a significant reduction in production costs, thanks to a much more frugal use of inputs and the use of motorized machinery and can therefore be described as *frugal* agropastoral systems.

7. Centres d'Initiatives pour Valoriser l'Agriculture et le Milieu rural: Centers for Initiatives to Promote Agriculture and the Rural Environment.

2.3. Frugal agropastoral systems: putting grazing back at the heart of herd feeding, especially in forest ecosystems

2.3.1. *Substituting mixed grazing of diverse vegetation as the main feed*

In the Lodevois, these agropastoral farms were most often created from nothing, gaining access to land through simple oral agreements with landowners, and more rarely through rent or grazing agreements. On the Limousine Mountain, these agropastoral farmers have generally taken over small farms that were going out of business. Most of them set up outside the family framework with limited or no access to permanent grasslands and arable land – mostly controlled by other livestock farmers. Consequently, these new farmers have developed frugal agropastoral systems in the Lodevois and the Limousine Mountain where the ruminant herds are fed as much as possible through grazing. This strategy relies in particular on rangeland, including during periods of high demand (lactation, fattening, etc.) in order to reduce the need for distributed forage and concentrated feed to the strict minimum. As with the few permanent meadows to which these farmers have access, grazing on rangelands can be managed in fenced parks, but a growing number of them shepherd their flocks all or part of the year, either themselves or with the seasonal support of a salaried shepherd.

In the Lodevois, these agropastoral systems have been developed for suckler cattle, suckler sheep or small dairy ruminants; on the Limousine Mountain, these are currently mainly used for suckler sheep. Consequently, a comparative study of these suckler sheep agropastoral systems in these two regions offers a good insight into their operating logic and the central role played by woodlands in feeding the flocks.

In these two regions, lambing is grouped in late winter and early spring, respectively, to coincide with the peak needs of the ewes (end of gestation, then lactation) and the peak availability of forage in the pasture. In the Lodevois, thanks to agreements with winegrowers, the ewes graze the herbaceous cover at the foot of the vines, rich in grasses and legumes, at the end of winter (at the time of parturition), before grazing herbaceous fallow land and shrublands throughout the spring. The ewes spend the summer in summer pasture in the Cévennes under the care of a salaried shepherd, then spend the autumn and most of the winter in the undergrowth. On the Limousine Mountain, these agropastoral breeders rely on fenced pastures of permanent grasslands in spring and autumn, calluna moors and forest clearings in winter. In summer, farmers send their flocks to the tops of the Limousine Mountain, where summer pastures consist mainly of peat bogs and undergrowth in the summer, and where they are guarded by salaried shepherds. To ensure that the growth of lambs and ewe lambs is also based as much as possible on grazing, including the vegetation of the rangelands, the fattening period is extended (with sales from the following autumn to the newt spring). The age at first lambing is also delayed for renewal ewe lambs, in order to adapt to their slower growth than in other systems in the region, where feeding is based on a more energy-dense diet (temporary grasslands hay and concentrates). However, the climatic conditions and vegetation available on the Limousine Mountain in winter do not, for the time being, make it possible to do without foraging in these frugal agropastoral systems. For this reason, for several years now, some of these farmers have become transhumant, moving their flocks in the winter to more southerly regions of the Massif Central, in the Causses of Quercy (Lot). Livestock farmers and their herds leave for two to three months, where they graze on calcareous grasslands and undergrowth of pubescent and holm oaks, whose acorn production is much more regular and with less risk of intoxication than that of the sessile and pedunculate oaks of the Limousine Mountain.

Taking the example of suckler-sheep farming on the Limousine Mountain, Figure 2.6 shows the central role of grazing and rangeland use in the diet of ewes in the double transhumant agropastoral system, compared with the forage system most commonly practiced on other suckler-sheep farms in the region, with late autumn lambing and winter lactation in order to sell lambs fattened entirely in the sheepfold at Easter.

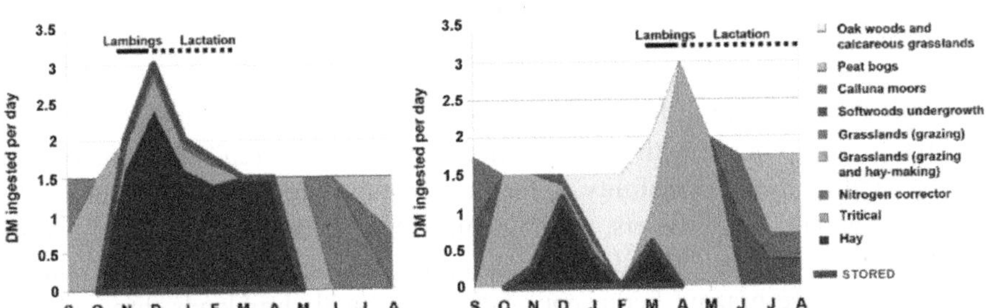

Figure 2.6. *Comparative forage calendars between the predominant system (left) and economical agropastoral systems (right) for suckler-sheep farming on the Limousine Mountain (source: surveys; production: authors). For a color version of this figure, see www.iste.co.uk/martin/farmsandterritory.zip*

2.3.2. *The key role of wooded ecosystems*

2.3.2.1. *Central resources during winter and summer forage low periods in the Mediterranean region*

In the mediterranean part of the Lodevois, during the two periods of low-grazing resource availability (winter and summer), wooded ecosystems are key resources for these frugal agropastoral systems. In winter, the flocks graze into the undergrowth of holm oak, pubescent oak and kermes oak, and into the few remaining chestnut groves in the region, where the animals can feed on acorns and chestnuts. Their consumption alternates with that of more palatable species (Montpellier aphylla in particular) to stimulate the ingestion of these fibrous resources. In summer, animals that are not transhumant graze into undergrowth and riparian forests to graze the protected herbaceous stratum beneath the tree cover. These grazed woods are communal or private, with access essentially by oral agreement, wherein the owners are somewhat known.

On the high plateaus of the Lodevois, wooded rangelands are essential in summer for farmers that do not have access to a summer pasture in the Cévennes and therefore depend on the undergrowth of pubescent oak forests, which can be grazed in alternation with *dried calcareous* grasslands or late grasslands. The undergrowth here is essentially state-owned, and breeders gain access to it through grazing agreements or contracts with the ONF[8], the main forestry stakeholder here, which facilitates the creation of vast islands of undergrowth grazing.

2.3.2.2. *Grazing in coniferous undergrowth to strengthen peat bog grazing on the Limousine Mountain and secure summer resources*

On the Limousine Mountain, the summer pastures are largely made up of peat bogs managed by the CEN[9]. The latter has sometimes been able to partially acquire them thanks to funding from the former Limousine region and the Adour Garonne water agency, which explains the imbalance of rangelands to the detriment of wooded areas. The MAEC[10] scheme, which initially paid salaried shepherds and for which support was higher in wetlands, may also have encouraged the creation of summer pastures mainly composed of peat bogs. Nevertheless, summer grazing of the herbaceous vegetation of the coniferous plantations' undergrowth is highly complementary to that of peat bogs. Peatland vegetation rarely starts to grow before the end of June. It can therefore not be grazed before early July and loses all

8. Office National des Forêts: National Forest Office.

9. Conservatoire des Espaces Naturels: Natural Spaces Conservancy.

10. Mesures Agri-Environnementales et Climatiques: Agro-Environmental and Climate Measures.

palatability with the first frosts in September. The beginning and end of the summer pasture grazing season therefore relies on other types of vegetation, such as undergrowth. In addition, although the purple moor-grass present in the peat bogs is the central forage resource for the summer, it remains coarse and unpalatable. This means that long grazing sequences in peat bogs need to be alternated with short sequences in areas such as the undergrowth of softwood plantations, where the ewes find fine, nitrogen-rich grasses, brambles and scotch broom, in order to boost their appetite (Meuret 2010).

Undergrowth grazing also offers a key adaptation lever in a context of increasingly uncertain summer climate: in very dry years with high temperatures, the animals eat better in the undergrowth than in the peat bogs, where they are exposed to full sun; in very wet years, on the other hand, the peat bogs are less accessible for the herd and the shepherd, due to their waterlogging, and the role of the undergrowth in the herd's diet is all the more central.

2.3.3. *Economical systems focusing on economic labor productivity thanks to agroecological operation*

By basing their herds' feed as much as possible on pasture, particularly during the lambing, lactation and fattening phases, farmers are able to drastically reduce their need for stored fodder and distributed concentrates, which in turn enables them to limit their purchases of inputs (fertilizers, seeds, concentrated feeds, etc.) and their investments in farm equipment. Thanks to the increased use of pastoral resources, frugal agropastoral systems are thus part of a techno-economic strategy that favors wealth creation over the pursuit of increased volumes produced per worker, i.e. seeks to boost economic productivity of work rather than the physical productivity of work. In this sense, these agropastoral systems are indeed part of an agroecological approach that is "*not interested in maximizing the production of a particular commodity, but rather in optimizing the agroecosystem as a whole*" and where the "*emphasis is on ecological sustainability rather than short-term productivity*" (Altieri 1986, authors' translation).

The case of the suckler-sheep agropastoral system of the Limousine is illustrative of the results obtained by most of the frugal agropastoral systems identified in the two regions studied. Indeed, a comparison of two suckler-sheep systems with autumn lambing and fattening in the sheepfold, one conventional with full purchase of concentrates (O1) and the other organic with self-production of cereals (O2), with a frugal agropastoral system (O3), all three identified on the Limousine Mountain, shows that the latter has the lowest production costs per ewe. The aggregate of intermediate consumption (IC) linked to feed purchases and mechanization costs

(IC agricultural contract work) and average annual depreciation of fixed capital linked to equipment (DK farm equipment) do not exceed 20% of gross product in the agropastoral system, compared with 40–50% in the other two production systems. This frugal system thus enables us to maintain a net added value per ewe equivalent to almost 30% of gross income, well above that of the other production systems, which is even negative in some cases.

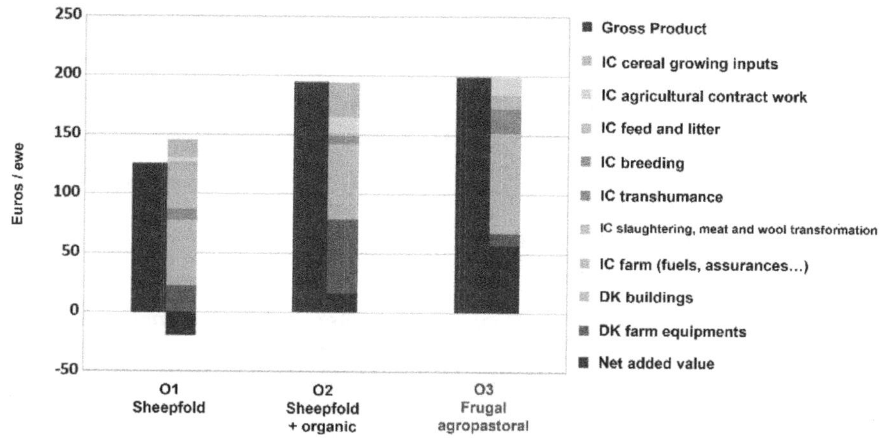

Figure 2.7. *Comparison of the formation of net value added per ewe for three suckler-sheep production systems on the Limousine Mountain (source: surveys; production: authors). For a color version of this figure, see www.iste.co.uk/martin/farmsandterritory.zip*

On the other hand, these frugal systems receive significantly less support from the Common Agricultural Policy, 30–50% lower per worker than other farms on the Limousine Mountain (see Figure 2.8). This is due in particular to the lower number of ewes and hectares per permanent worker, which translates into significantly lower amounts of coupled and decoupled aid. While farm income per permanent worker (including subsidies) is usually quite viable, it is often lower than that of other, more supported production systems.

These frugal systems nevertheless create jobs by using less land per worker. Above all, they are based on land that is much more made up of rangelands and much less of motor-mechanized surfaces, the most limiting in these middle mountain and Mediterranean foothill regions.

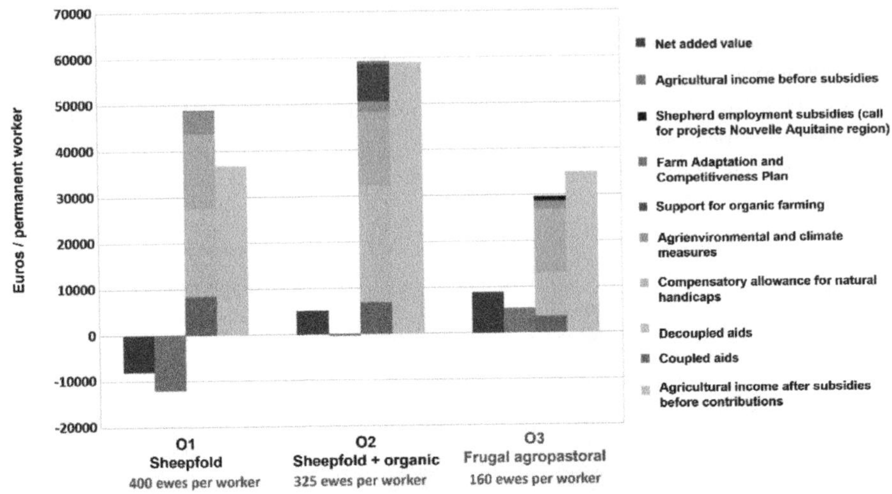

Figure 2.8. *Comparison of incomes per family worker and their formation for three suckler-sheep production systems on the Limousine Mountain (source: surveys; production: authors). For a color version of this figure, see www.iste.co.uk/martin/farmsandterritory.zip*

2.3.4. *Original systems resulting from the combination of several technical frameworks*

In the Lodevois, as on the Limousine Mountain, these frugal agropastoral systems have been based on a techno-economic strategy against the prevailing agricultural development movement and the technical advice associated with it. Despite the central role of the shepherd practices and the very limited use of motorized machinery, these systems are far from a return to those of the 1950s. Analysis of their operating logic reveals a high degree of technical sophistication. Farmers combine different grazing management methods (rotational grazing techniques in mobile paddocks and shepherd practices) and rely on their in-depth knowledge of the wide diversity of plants grazed and their main characteristics (seasonality, capacity to maintain palatability over time). This knowledge and know-how enable them to play on the complementarities between vegetation to set up coherent grazing chains, both on a daily and annual basis. The construction of these chains also requires in-depth work in herd selection, based on criteria that are rarely considered in most current selection schemes (resistance/tolerance to parasitism and certain pathologies such as foot rot, ability to withstand periods of food restriction, etc.), or even against these criteria (lower prolificacy in sheep breeding, for example). Particular attention is also paid to the dietary education of young animals, to accustom them from birth to grazing high-fiber vegetation. These breeders have therefore drawn on other technical references and know-how and have generally had socio-professional trajectories different from those of other breeders.

The introduction of collective structures linked to pastoralism (such as Pasture Groups) and shepherding know-how by breeders who had previously worked as salaried shepherds in regions where a more pastoral approach was maintained, including undergrowth grazing, has thus been decisive in the development of these frugal agropastoral systems. Some of the farmers at the head of these systems are former employees of Parcs Naturel Régional (Regional Nature Parks) or Parcs National (National Parks), institutions for which the link between livestock farming and rangeland maintenance is central, such as the Cévennes National Park. Federated into two local associations belonging to the CIVAM network, and for some members of the Patur'Ajust network, these farmers have from the outset placed the operation of their system within a different paradigm of agricultural development (giving priority to economic rather than physical productivity of agricultural work) and developed a different view of semi-natural vegetation and its use in herd feed. Finally, the presence of naturalist associations in both regions, such as the CEN (Limousine) or the Écologistes de l'Euzière (Lodévois), has also helped to refine some of the farmers' empirical knowledge of the ecological functioning of grazed ecosystems (e.g. the growth dynamics of purple grass-moor in peat bogs), and thus to adjust grazing practices.

Despite the technical reference framework and the promising results already achieved by these farmers in the frugal agropastoral system, they come up against various obstacles in their pastoral practices, particularly concerning the use of wooded areas.

2.4. Obstacles to the use of wooded areas on the Limousine Mountain and the Lodevois: an overview

2.4.1. *Pastoralism and silvicultural management on the Limousine Mountain: rejections by most forestry stakeholder*

Softwood plantations on the Limousine were not initially designed for grazing, and until recently, their silvicultural management meant that they offered no grazing resources in the undergrowth. Indeed, until the 1980s–1990s, the density of plantations and the absence of thinning practices made it impossible for undergrowth to develop. Following joint action by the ADELI[11], the CRPF[12] and technicians from forestry cooperatives (CFBL, Unisylva and Alliance Bois Forêt), this thinning practice became more widespread in the 1980s–1990s, in parallel with the gradual replacement of spruce by Douglas-fir in reforestation and the reduction in plant density at planting (Beynel 1998). A vegetation of wavy hairgrass and brambles appetizing to herds has thus gradually developed in thinned plantations.

Despite the central role that forest plantations can play for frugal agropastoral systems on the Limousine Mountain, the passage of herds through the undergrowth remains at present punctual and informal, over small areas, and on condition that any damage is avoided and does not interfere with silvicultural management. The situation on the Limousine Mountain therefore remains that of "grazed forest".

Livestock farmers with frugal agropastoral systems face almost universal opposition from forestry technicians and owners to grazing animals in undergrowth, for fear of damage, including to mature plantations. There are several possible explanations for this opposition. Softwood plantations on the Limousine Mountain were created against a backdrop of abandoned rangelands, and agriculture and forestry were never seen as being able to work in synergy; the two sectors are sometimes even in competition for land. This is in stark contrast to the situation in south-eastern France, where for several decades, the challenges of defending forests against fire (DFCI) (Etienne 2001) and the marginalization of forests from the timber industry (Guérin 2008) have encouraged the undergrowth grazing. This opposition also reflects the segmentation of professions, with foresters having little knowledge of livestock farming (a fortiori pastoral) and vice versa (Michon 2015), and the time lag between these two activities: while forest management is considered on the scale of years and decades, grazing management is planned on the scale of the day, season or year (Etienne et al. 1994).

In addition to these initial obstacles, which are linked to the history of forestry on the Limousine Mountain and the segmentation of livestock farming and forestry, there is now the damage caused by wild ungulates (roe deer and red deer), which has been on the increase for several decades, notably due to the aging of the hunting population and the proliferation of fallow land that can be used as shelters. Plantations are also subject to wood pests such as large pine weevil, which lives in the stumps of recently felled trees and feeds on young seedlings (Chinery 2012), and to the initial effects of climate change. Spruces are particularly hard hit, as they are more sensitive to drought and bark beetles (weakness parasites) than Douglas-fir

11. Association pour le Développement des Éclaircies en Limousin: Association for the Development of Thinning in Limousin.
12. Centre Régional de la Propriété Forestière: Regional Forest Property Center.

(Dupuy et al. 2015), whose main parasites have not yet arrived in Europe (Nageleisen et al. 2010). Forest owners and managers thus limit the possibility of letting herds through the woods, fearing to add further pressure on their plantations.

2.4.2. *Tensions in the Lodevois: pastoral agreements with the ONF for fire prevention clash with hunting competition*

In the Lodevois region, undergrowth grazing is a far older practice, since forest cover has never been as low as it has been on the Limousine. However, since the early 1950s, forest grazing has completely disappeared in the Mediterranean part of the region, which is now almost entirely devoted to viticulture. As a result, most of the region has experienced a strong dynamic of landscape closure, leading to the disappearance of undergrowth forage resources, particularly in the particularly dense holm oak coppices (Etienne et al. 1994). These holm oak coppices, which cover thousands of hectares in the Mediterranean zone of the Lodevois, are not exploited for silvicultural purposes. As herds are few and grazing pressure often insufficient, these areas are threatened with closure, with an almost complete disappearance of the shrub and herb stratum.

Unlike other departments in south-eastern France (Var, Vaucluse, etc.), there are no DFCI fire prevention contracts in the Lodevois, which would enable farmers to be paid for maintaining wooded areas. Nevertheless, the risk of forest fires is sufficiently high enough to encourage forest managers to sign agreements with farmers. However, most of the woodlands to which livestock farmers have access are considered non-productive, mainly due to the slope, which makes motorized logging impossible, and there are no tracks for transporting wood away from the plots. While forest managers do not provide farmers with technical resources (brush clearing, fencing, etc.), they do allow them to clear the undergrowth themselves in order to install mobile fences and reopen the environment to encourage the development of the herbaceous stratum, as well as prune certain fodder trees such as ash to make their foliage more accessible to the herds. Such labor-intensive investments are only possible thanks to grazing agreements, which secure access to the land and ensure that farmers benefit from their efforts in subsequent years. However, these partnerships are currently under threat. Some forest managers have reported that their superiors are questioning some of the agreements signed between ONF and breeders, because of their low profitability in relation to the time spent by foresters managing them. However, these partnerships make it possible to limit the risk of fires in the woods at lower cost and, if they were developed further, could also facilitate the maintenance of forest roads, an action that is otherwise costly for the ONF.

Lastly, the introduction of silvopastoral practices can find itself in competition with hunting, particularly big-game hunting, which has developed strongly in the Lodevois on gradually abandoned rangelands. Hunters, who are often non-breeders, frequently object to herd grazing, accusing it of scaring away game, either directly through their presence, or indirectly through the reopening of the environment. When these hunters are elected representatives, they may even oppose the implementation of a grazing project in communal forests, preferring to keep the areas closed, even if it means increasing the risk of fire, rather than considering the decline in game populations. This opposition between farmers and hunters is much less marked on the Limousine Mountain, where most hunters are active or retired farmers.

2.4.3. *The 2015 PAC reform: new competition for rangelands and a "pro-rata" system disadvantaging wooded rangelands*

The 2015 PAC reform included rangelands that were not previously eligible for first-pillar decoupled aid payments. This helped to make these areas more attractive to farmers, although they are still not making any significant changes to their herd diets in order to mobilize these semi-natural forage resources. In contrast to the intensive use made of these areas by farmers in frugal agropastoral systems, the vast majority of farmers in both the Limousine Mountain and the Lodevois regions continue to under-utilize these rangelands, where low-need animals graze with low stocking densities.

The inclusion of rangelands in first-pillar aids has been accompanied by a "pro-rata" system, which involves applying a corrective coefficient to the amount paid per hectare for rangelands, depending on the composition of the vegetation occupying the plots. Declared areas are not counted in their entirety, but rather on a so-called eligible share, i.e. estimated as effectively grazed after the application of a pro rata. Massive use of shredding and burning is favored, since the pro-rata system deducts bushes and forests larger than 10 acres from eligible surfaces (except in areas where oak and chestnut groves can be declared), as well as linear elements such as hedges, and species considered inedible by herds such as ferns, brambles or boxwood (Ministère de l'Agriculture 2018). To avoid losing eligible areas, gyro-spinning, burning or even in some more extreme cases land clearing, stone crushing, liming and chemical weed control (on the fern moors of the Lodevois) are therefore mobilized to keep areas as open as possible and above all contain the development of shrubby and overgrown areas.

2.5. Ways to promote the (re)connection between livestock and forests

2.5.1. *Synergies between livestock farmers and foresters already in place on the Limousine Mountain*

On the Limousine Mountain, surveys carried out among shepherds and the few foresters interested in developing silvopastoralism show that the operating logic of agropastoral systems, particularly as they are based on herds being shepherded during the summer, is well suited to the idea of re-articulating agriculture and forestry. Indeed, when grazing is carried out under the watch of a shepherd, the flock never remains static in the undergrowth, which limits the risks of overgrazing in young plantations. The shepherds also identified the periods and weather conditions when the risks were greatest for young trees. The shepherd can take the flock out of the woods at the slightest sign of damage, which often reassures the forest manager or owner compared with the fence-based grazing system. Lastly, by influencing feed selection and the daily composition of the herd's ration, shepherds have also identified which plants to graze to prevent the animals from shifting their consumption to young conifers. This last point is crucial when managing irregular high forest (uneven-aged timber management), where all stages of development coexist within the same plot, and where the risk of damage is greater than in regular stands.

A number of avenues could be explored for more advanced silvopastoral management, based on the initial experiments carried out by a number of foresters, livestock farmers and shepherds. To compensate for the lack of grazing resources at the start and end of the summer pastures, herds could be given access at the start of the summer to young plantations (where the trees are over 1.50 meters high), with the aim of controlling the shrub stratum competing with the young conifers. Access reserved for shepherded herds would limit the risk of trampling, while sufficient diversity of forage resources would help reduce the impact on young conifers (see Figure 2.7). This type of management is already practiced by some shepherds of the Limousine, with no significant damage.

After the first thinning (after 20–25 years), the risk of tree damage by animals is virtually nil, and the reduction in tree cover allows the development of undergrowth. In order to contain its development, in particular that of the shrub stratum, and to anticipate future mechanical interventions, complete consumption of the herbaceous and shrub layer could be set up, in the mid-season or even in winter for herds whose farms are located nearby, otherwise during the summer for herds kept or led in fenced parks. This is currently the most common practice in summer (see Figure 2.9), albeit mostly on an informal basis. However, the surface areas currently available would make it possible to secure and reinforce this practice, while extending it to seasons other than the summer period.

Figure 2.9. *Sheep herding on the Limousine Mountain: on a young softwood plantation (left), in recently thinned plantations (right) (photo: authors)*

In the case of natural regeneration of coniferous forests through seeding cuts to avoid replanting, it is necessary to limit competition between undergrowth and young coniferous trees. The undergrowth must therefore be thinned as much as possible, with good control of wavy hairgrass, bramble and eagle fern. A fixed park in summer could be installed on an area to be regenerated, as a night park or midday resting park, or even on rainy days or the shepherd's day off, in order to put very high grazing pressure on this area. This type of management, already implemented in other countries (Darabant et al. 2007), was the subject of initial experiments on the Limousine in the early 2000s, but in the absence of long-term monitoring by farmers at the time, these experiments were not repeated. The same logic could be used to facilitate the control of spontaneous vegetation in forest clearings between the final felling and replanting (in this case, artificial replanting). Currently, foresters wait two or three years before replanting, in order to limit the damage caused by large pine weevil, following the ban on lindane, the organochlorine insecticide used until the late 1990s to destroy them. Rather than a costly land-clearing operation, the installation of a fixed park in summer would enable the growth of shrubs to be limited by grazing. As the aim is to maintain open wasteland with low vegetation, night parks would be preferable to avoid over-exposing the animals in the heat of summer. Generally speaking, foresters would benefit from the passage of herds, which would also limit the cost of maintaining forest tracks (Guérin and Macron 2005; Guérin et al. 2010).

However, these various avenues for silvopastoral development of the Limousine are conditional on the formation of grazing blocks located close to the current summer pastures, or sufficiently extensive to constitute fully fledged summer pasture districts, in all cases forming blocks that are sufficiently coherent to facilitate shepherding or the installation of a park. In this respect, the fragmentation of forest land on the Limousine Mountain is a major obstacle, with several thousand owners, more than two-thirds of whom have less than 5 ha (PNR Millevaches 2020). Collaboration with the ONF in the few state, communal or sectional forests, in the forests of the Coast Conservancy[13] (around the Vassivière Lake), or with the Forestry Syndicate Groups[14] (found mainly in Creuse) and forestry experts or technicians from forestry cooperatives would therefore be preferable, in order to limit the number of interlocutors. Such collaboration would also offer greater land safety: in the vast majority of cases, access to the undergrowth remains precarious for the time being, with the risk of losing this access from one year to the next and thus unbalancing the organization of grazing.

2.5.2. *In the Lodevois, the interplay between livestock farming and forestry is limited by the timber market and property fragmentation*

In the forest environment of the Lodevois, a silvopastoral management method has already been proposed and tested for several decades (Guérin and Hubert 1987; Bellon and Guérin 1993; Etienne et al. 1994) and is mentioned by several farmers with frugal agropastoral systems: the selective thinning of holm oak coppices. This would produce firewood, regenerate the undergrowth and preserve clumps beyond the reach of the herd, which could eventually be used for timber. However, this type of management encounters several major obstacles: holm oak lumber is very heavy and cracks when dried (Ourcival 2013), but has good calorific value, which is why it is mainly used as firewood (Marchal 1995). However, for holm oak coppice harvesting to be profitable (at the price of wood and according to current silvicultural management standards), clearcuts are most often favored by forest operators over selective cutting. The unsuitability of holm oak products for the timber market (Guérin 2008) therefore still limits this type of management. Finally, even if this type of management were possible, the fragmentation of the plots and the consequent multiplicity of owners severely limit silvicultural interventions.

However, there have been attempts to develop holm oak coppices, on the initiative of the few livestock farmers who own plots of woodland. By selectively cutting for firewood, they have cleared holm oak crowns to stimulate autumn acorn production. In addition, to facilitate flocks shepherding in this type of environment, which is very closed and dominated by very coarse resources, they also sought to

13. Conservatoire du Littoral.
14. Groupements Syndicaux Forestiers.

create *soupades*[15]. In these more open areas, interspersed between the holm oak forests, the animals find very limited but highly palatable forage resources, enabling the shepherds to boost the appetite of their flocks (see Figure 2.10). In these areas, created by land clearance or the addition of topsoil in the middle of the garrigue (e.g. from nearby construction sites), farmers allow spontaneous herbaceous vegetation to take root, or sow forage legumes such as sainfoin. These developments, which also stimulate the consumption of coarser resources of the holm oak forest, could be paid for by local authorities, as part of the herds' contribution to controlling overgrowth and fire risk.

Figure 2.10. *Selective cutting of oak groves to stimulate tasseling (left) and creation of a grassy area nearby (right) in the Lodevois (photo: authors)*

2.6. Conclusion

In the 1950s, the Limousine Mountain was characterized by very low forest cover and the predominance of calluna moors, widely used for grazing ruminants. As a result of extensive specialization in suckler cattle breeding, based above all on the use of motor-mechanized areas, part of the moors was cleared for grassland, if not completely forested, under the impetus of the FFN[16]. The forest thus emerges in the landscape at a time when grazing herds are refocusing on the most productive areas. By contrast, in the 1950s, the Lodevois was characterized by much more extensive forest cover and, at the time, the extensive use of wooded ecosystems to feed the herds. The specialization of farms in small ruminant dairy farming or viticulture was accompanied here by the abandonment of wooded rangelands, and silvopastoralism was gradually abandoned.

15. Occitan word referencing to "supper" and designating small plots of grasslands, where the flocks grazed traditionally begore going back to the sheepfold at the end of the day.
16. Fond Forestier National: National Forestry Fund.

The frugal agropastoral systems that have developed over the last 20 years in the Lodevois and Limousine Mountain against this trend favor an agroecological approach characterized by a particularly intensive use of semi-natural rangeland vegetation, including woodland, to feed the herds. They offer concrete, viable examples of how livestock and forests can be reconnected in these regions. In many respects, the pastoral resources present in the different types of forest ecosystems of the Limousine Mountain and the Lodevois are central to the functioning of these agropastoral systems, as they can be used flexibly and mobilized at different key times of the year.

However, the introduction of silvopastoral practices, which would benefit breeders, forest owners and managers alike, is currently limited by the structuring of the timber industry, which demands standardized products that can be exploited by motorized mechanical means (Guérin 2008), by the ease with which fully mechanical maintenance can be carried out on the most productive forest areas, compared to the animal's teeth, by the pressure exerted by certain stakeholders who prefer to maintain closed environments, and by the initial effects of climate change, which are having an increasingly harsh effect on forests. The increasing risk of forest fires throughout France (Chatry et al. 2010), the rising costs of motorized-mechanized forest maintenance and the development of new outlets for wood that does not conform to market standards (wood chips that can be used for heating) will perhaps encourage a greater collaboration between livestock farmers and foresters in the coming years. This new context may also call for more proactive public policies.

2.7. References

Altieri, M.A. (1986). *L'agroécologie. Bases scientifiques d'une agriculture alternative*. Debard, Paris.

Aubron, C., Latrille, M., Lhoste, V. (2019). Dynamiques agraires dans les *Causses* et Cévennes. Limites d'un développement centré sur l'accroissement de la productivité physique du travail et émergence d'alternatives. In *L'économie a gropastorale revisitée*, Aubron, C., Garambois, N., Nozières Petit, M.-O. (eds). Association Française de Pastoralisme/Cardère, Saint-Martin-d'Hères/Avignon.

Bellon, S. and Guérin, G. (1993). Raisonner l'utilisation sylvo-pastorale du chêne vert. *Forêt méditerranéenne*, XIV(4), 296–305.

Beynel, C. (1998). *Forêt et société de la montagne limousine*. Presses Universitaires de Limoges, Limoges.

Chinery, M. (2012). *Insectes de France et d'Europe occidentale*. Flammarion, Paris.

Cochet, H. (2011). *L'agriculture comparée*. Éditions Quæ, Versailles.

Cochet, H., Devienne, S., Dufumier, M. (2007). L'agriculture comparée, une discipline de synthèse ? *Economie rurale*, 297–298, 99–112.

Darabant, A., Rai, P.B., Tenzin, K., Rober, W., Gratzer, G. (2007). Cattle grazing facilitates tree regeneration in a conifer forest with palatablebamboo understory. *Forest Ecology and Management*, 252(1–3), 73–83.

Dodanne, C. (2009). Les nouvelles forêts du Massif Central : enjeux sociétaux et territoriaux. Ces hommes qui plantaient des résineux pour éviter la friche. Thesis, École normale supérieure Lettres et Sciences Humaines – ENS-LSH Lyon.

Dupuy, J.L., Boivin, T., Duché, Y., Martin-StPaul, N., Pimont, F., Rigolot, E. (2015). Impact du changement climatique sur les risques en forêt : le cas de l'incendie et de ses interactions avec la sécheresse et les pullulations d'insectes. *Innovations Agronomiques*, 47, 29–50.

Étienne, M. (2001). Aménagement de la forêt méditerranéenne contre les incendies et biodiversité. *Revue forestière française*, 53(sp), 149–155.

Étienne, M., Hubert, B., Msika, B. (1994). Sylvopastoralisme en région méditerranéenne. *Revue forestière française*, 46(S), 30–41.

Garambois, N., Jallot, L., Morsel, N. (2020). Les systèmes agropastoraux économes. Vers un développement agricole plus durable en moyenne montagne et piémont. In *L'économie agropastorales revisitée*, Aubron C., Garambois, N., Nozières Petit, M.-O. (eds). Association Française de Pastoralisme/Cardère, Saint-Martin-d'Hères/ Avignon.

Guérin, G. (2008). De la forêt pâturée au sylvopastoralisme. *Forêt Méditerranéenne*, XXIX(4), 491–496.

Guérin, G. and Hubert, B. (1987). Problèmes et perspectives pour une approche multi-usages des espaces forestiers. *Fourrages*, special edition "L'animal, les friches et la forêt II", 271–280.

Guérin, G. and Macron, M.-C. (2005). Sylvopastoralisme. Les clés de la réussite. Techniques pastorales. Institut de l'Elevage et Institut du Développement Forestier, Paris.

Guérin, G., Aubron, C., Boutonnet, J.-P., Gallion, B., Sajdak, G. (2010). Entre sylviculture et pâturage dans les espaces boisés de la forêt méditerranéenne française. *Innovation and Sustainable Development in Agriculture and Food – Actes du Symposium ISDA*, 28 June–1st July.

Hubert, B., Deverre, C., Meuret, M. (2010). Deux siècles de changement radicaux pour les parcours du Sud de la France. In *Un savoir faire des bergers*, Meuret, M. (ed.). Editions Quæ, Versailles.

Larrère, R. (1978). Désertification ou annexion de l'espace rural ? L'exemple du plateau de Millevaches. *Études rurales*, 71–72, 9–48.

Marchal, R. (1995). Propriétés des bois de chêne vert et de chêne pubescent 1ère partie. Caractéristiques physiques. *Forêt Méditerranéenne*, XVI(4), 425–438.

Mazoyer, M. and Roudart, L. (1997). *Histoire des agricultures du monde*. Éditions du Seuil, Paris.

Meuret, M. (2010). Des troupeaux dans la broussaille : un comportement inattendu qui incite à changer de paradigme scientifique. In *Penser le comportement animal*, Burgeat, F. (ed.). Éditions de la Maison des sciences de l'homme, Paris.

Michon, G. (2015). *Agriculteurs à l'ombre des forêts du monde. Agroforesteries vernaculaires*. Actes Sud, IRD Editions, Arles.

Ministère de l'Agriculture et de la Pêche, Agence des Services et de Paiement (2018). Guide national d'aide à la déclaration du taux d'admissibilité des prairies et pâturages permanents.

Morsel (2024). Les systèmes agro-pastoraux économes : élevage et agroécologie en régions de moyenne montagne et de piémont méditerranéen. Thesis, AgroParisTech.

Nageleisen, L.M., Piou, D., Saintonge, F.X. (2010). *La santé des forêts – Maladies, insectes, accidents climatiques. Diagnostic et prévention*. Ministère de l'Agriculture et de la Pêche, Département de la Santé des Forêts, Paris.

Ourcival, J.-M. (2013). L'exploitation forestière. In *Collectif des garrigues. Atlas des garrigues – Regards croisés*. Editions Ecologistes de l'Euzières, Prades-le-Lez.

Palluet, B. (2019). Pourquoi la forêt du Plateau de Millevaches s'attire les critiques et que faire aujourd'hui. *L'AGORA : Points de vue, Forestopic.*

Pflimlin, A., Faverdin, P., Béranger, C. (2009). Un demi-siècle d'évolution de l'élevage bovin. Bilan et perspectives. *Fourrages, Association Française pour la Production Fourragère*, 200, 429–464.

PNR Millevaches (2020). Renouvellement de la charte forestière de territoire. Diagnostic – Forêt et Filière Bois.

Roque, J-L., Izard, R., Lacan, M., Lacroix, P. (2011). *Voyage dans les forêts de l'Hérault*. Biotope, Mèze.

3

Crop–Livestock Interaction at the Territorial Scale: Tools and Methods to Support Its Development

Solène PISSONNIER
Paris Saclay University, AgroParisTech, INRAE, UMR SADAPT, Palaiseau, France

3.1. Introduction

Developing interactions between animal and plant productions not only contributes to farm autonomy, but also develops ecosystem services (Bonaudo et al. 2014). That being said, mixed crop–livestock farms are in decline. The processes of specialization underway for several decades have pushed farmers to expand, specialize and intensify their practices, which today have negative impacts on the health of ecosystems and populations (Mignolet et al. 2012).

It is thus becoming urgent to change practices and re-establish interactions between productions to help mitigate these negative impacts. However, diversifying farms and integrating new activities implies facing a number of obstacles, such as a lack of resources, manpower, knowledge and skills. These obstacles can be overcome by reconnecting production on a territorial scale (Martin et al. 2016). The aim is to create connections between specialized and non-specialized farms, to limit the negative impacts of specialization and rediscover the benefits of mixed farming on a territorial scale. Setting up connections on this territorial scale nevertheless

Farms and Territory,
coordinated by Philippe MARTIN and Solène PISSONNIER. © ISTE Ltd 2025.

raises major technical and organizational issues, and requires a review of the tools and methods available to support their development.

In this chapter, we propose to look at a number of different themes in order to understand how these systems operate on a regional scale. First, we examine the phenomena of specialization and their consequences. We next analyze the characteristics of systems integrating crops and livestock, and the benefits of this reconnection. We also look at how these systems are organized in time and space, and the actors involved. We conclude with an analysis and a reflection on existing and future methods for diagnosing, designing and evaluating these systems.

3.2. Specialization and its ramifications

For many years, territories and farms have been specializing. After World War II, farmers were urged to produce so that France could regain its food self-sufficiency (Mazoyer and Roudart 1997). Following the Treaty of Rome, the CAP[1] came into being in 1962, proposing a system of guaranteed prices to boost production. Europe then specialized in cereal production to the detriment of legume proteins. This specialization was encouraged by subsidies, by supporting the price of wheat and certain field crops.

Gradually, farmland is being reparceled and artificialized, always with the aim of increasing production. Drainage methods were developed in the 1980s, particularly in areas such as Brie and Flanders. This was accompanied by the replacement of meadows and the development of irrigation (Mignolet et al. 2012). As a result, between 1960 and 2010, the area dedicated to field crops increased by 20% (Chatellier and Gaigné 2012). Of course, soil and climate factors also come into play: to maximize yields, we make the most of soil and climate aptitudes. Conversely, livestock production is maintained in less-favored areas unsuitable for arable farming, such as wetlands and mountainous regions. Livestock farming is also concentrated and intensified in certain areas, such as Brittany (see Figure 3.1), which accounts for 21% of French dairy cattle and 56% of pigs. It is France's leading slaughtering region (Agreste Mémento 2022). The development of soilless systems is accompanied by heavy dependence on imports of plant proteins from North and South America. This dependence was reinforced when soybean imports were exempted from customs duties in the 1960s under the GATT Dillon Round (Hache 2015). In 1992, the Blair House agreements enabled the United States to limit the European surface area dedicated to rapeseed, sunflower and soybean crops

1. Common Agricultural Policy.

to 5.1 million hectares[2]. In 2000, the ban on meat-and-bone meal reinforced Europe's dependence on American proteins.

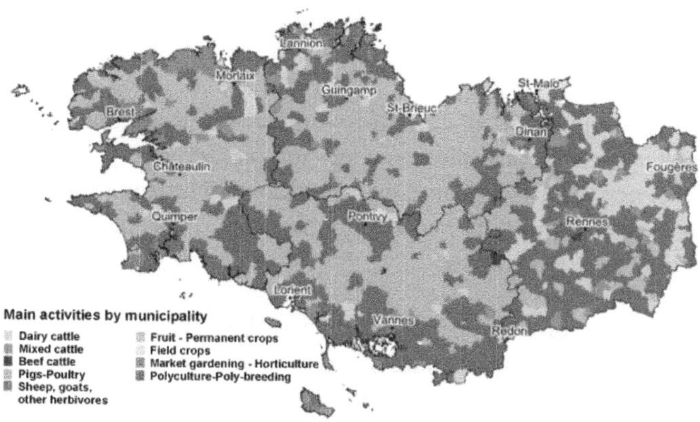

Figure 3.1. *Specialization in livestock production in Brittany: OTEX by municipality in 2020 (Agreste Memento 2022). For a color version of this figure, see www.iste.co.uk/martin/farmsandterritory.zip*

Companies in the animal production sector are also concentrated in certain areas, notably around the ports where soya is imported, resulting in a concentration of production. Brittany, for example, accounts for 40% of France's livestock feed production (Deschamps et al. 2016). The geographic concentration of companies enables agglomeration economies and productivity gains. These gains then become central to the specialization of territories. Indeed, the concentration of production makes it easier to develop informal relationships and encourage the sharing of knowledge and information, thus facilitating the development of technical and organizational innovations, for example, through the sharing of investments or manpower. The local relationships developed facilitate relations between customers and suppliers. Costs are also reduced (economies of scale, low transport costs), enabling prices to be lowered, thereby increasing demand and generating larger-scale production (Chatellier and Gaigné 2012).

At the same time, research, technical institutes, cooperatives, seed companies and input suppliers are specializing in the development of seeds, phytosanitary products, fertilizers, cropping systems, technical itineraries, etc., to increase yields. Falling energy costs encourage farmers to turn to synthetic products, especially fertilizers. The transport of these inputs is facilitated by the development of new

2. See: https://www.terresunivia.fr/decouvrir-terres-univia/l-interprofession/historique.

infrastructures (Gaigné 2024). Efforts to develop knowledge, techniques, inputs, etc., are focused on a reduced number of species, favoring the main field crops and neglecting more anecdotal crops. Within cooperatives, logistics (transport, silo storage, etc.) are organized around the common crops, accelerating the decline of certain diversification crops (Meynard et al. 2013).

In terms of the farming population, the workforce is shrinking as people migrate to urban centers. Farmers are also looking for less drudgery and are equipping themselves with new tools to reduce their workload. The equipment developed by manufacturers is adapted to larger and larger surfaces. This equipment is costly and must be profitable over large areas.

The consequences of this specialization can be seen in today's OTEX[3] (see Figure 3.2). OTEX is calculated using standard gross production (SGP). A farm is specialized in a particular production if its SGP exceeds 2/3 of its total SGP. This figure shows specialized areas: field crops in the Paris Basin, dairy cattle in Normandy, poultry and pigs in Brittany, beef cattle in the center region, etc.

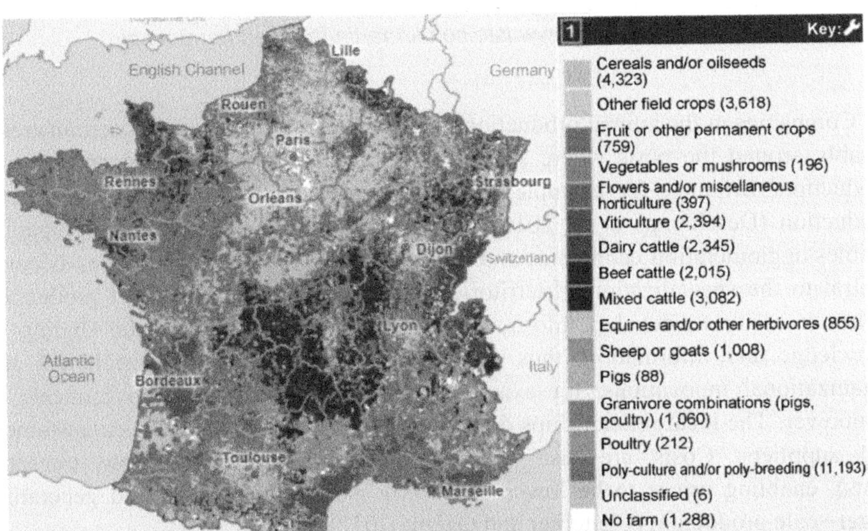

Figure 3.2. *OTEX, "Technical and Economic Orientation of Farms", in France, RGA 2020 data, Agreste (see: https://agreste.agriculture.gouv.fr/agreste-web/disaron/Carte-RA2020-otex/detail/). For a color version of this figure, see www.iste.co.uk/martin/farmsandterritory.zip*

3. Orientation Technico Economique des Exploitations Agricoles: Technical and Economic Orientation of Farms.

This trend towards specialization is also evident throughout Europe. In most of Northern, Western and Central Europe, as well as in the USA, mixed farming accounts for less than 10% of utilized agricultural area (UAA) and farm numbers. Most farms are now "segregated". The trend seems to be less marked in Central Europe: in Poland, Romania and Slovakia, 20% of farms are mixed farms, and more than double that figure in Latvia and Lithuania (Garrett et al. 2020).

On a local level, these phenomena of specialization and intensification of practices have harmful consequences. For example, a study carried out between 1970 and 2000 in the Seine watershed (an area specialized in field crops) by Mignolet et al. (2012) showed that the number of farms specialized in mixed farming fell by 11%, while farms specialized in field crops increased by 16%. As a result, permanent grassland acreage fell by 30%, while wheat acreage increased by 52%. Crop successions have been simplified, with rapeseed and wheat more frequently returning to the same plots. Technical itineraries have also evolved: we are seeing increased use of pesticides, with an increase in the number of treatments, linked to the development of rapeseed and short rotations that make it difficult to manage certain bio-aggressors.

The disappearance of legumes and livestock effluents is accompanied by an increase in the number of applications and doses of mineral nitrogen, both on field crops and in the soil. Pressure from inorganic nitrogen and increased pesticide use pose a risk not only to ground and surface water resources, but also to biodiversity, ecosystem health and human health. Nitrate concentrations measured in catchments in the Seine Basin show a steady increase between 1970 and 2000 (Mignolet et al. 2012).

Brittany is another good example of these dynamics. Between 2010 and 2020, the region lost a quarter of its farms, while the average UAA per farm rose from 48 to 62 ha (RGA 2010 and 2020). In 2020, 63% of farms were specialized in livestock production, and 10% in mixed farming. This specialization is not without consequences. Green tides began to appear in the 1970s. It was not until the 1980s that the link with farming practices became apparent, in particular the over-fertilization and poor management of effluents, which led to nitrates being transferred to the water tables. In the 1990s, associations alerted the European Commission to the existence of water catchments above the 50 mg/L nitrate standard. The various action plans put in place by France seemed ineffective, and eutrophication of environments and impacts on aquatic ecosystems continued (Durand et al. 2014).

The study by Dourmad et al. (2017) lists the negative effects associated with Breton territories. These effects are of various types and are linked to inputs (use of land, energy, fertilizers and pesticides for the production of concentrates), labor and employment (competitiveness of sectors and sometimes poor working conditions in agri-food companies), markets (livestock farmers subject to fluctuations in international prices for raw materials and energy), social and cultural issues (health risks due to high animal densities, odor nuisance, air and water pollution, antibiotic resistance), and finally the environment and climate (risk of nitrate pollution of groundwater, impacts on ecosystems).

One possible solution to limit the negative consequences of specialization and intensification of practices is to recreate interactions between animal and plant production.

3.3. Characteristics of systems integrating crops and livestock

Various authors propose ways of describing and representing crop–livestock systems at the farm or territory level (Sumberg 2003; Bell and Moore 2012). They describe the levels of interaction between livestock and crop production, which they classify according to two "physical" dimensions: space and time, and an "organizational" dimension to describe who manages which activity (same group/actor or not). Space is used to describe the geographical distance between productions, colocation being the most integrated form: presence of both productions in the same place. Time is used to describe whether the activities take place at the same time or over different time periods. The authors postulate that benefits are greater when productions are coordinated in time and space, thus reducing logistics (transport, storage).

Building on this work, Moraine et al. (2017a) propose a new representation based on three interacting spheres: crops, animals and grasslands (see Figure 3.3). The intersection of two spheres is illustrated by practices taking place in the same place at the same time (spatial and temporal coordinations): grazing forage crops corresponds to the intersection of the "animals" and "crops" spheres, grazing grasslands corresponds to the intersection of the "animals" and "grasslands" spheres, grazing temporary grasslands included in rotations or interculture cover corresponds to the intersection of all three spheres. When the spheres do not intersect directly, there may be flows between them: manure from the "animals" sphere to the "crops" sphere, or hay or grass silage from the "grassland" sphere to the "animals" sphere. In this case, there is indeed an interaction, but it is delayed in time or space and may require a storage phase.

Optimizing interactions between the three spheres would limit inputs (pesticides, fertilizers, concentrates, fodder, etc.) and outflows (non-usable waste and pollution). The aim is therefore to optimize these interactions to loop the various flows and benefit from the key ecosystem services listed in Figure 3.3: biological regulation, improved soil fertility, reduced erosion, water filtration, carbon sequestration, etc.

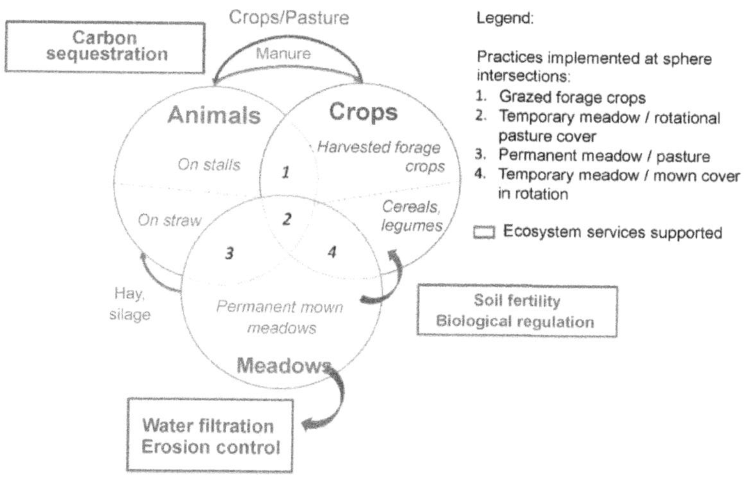

Figure 3.3. *Representation of systems integrating crops and livestock based on biophysical components, ecosystem processes and flows (adapted from Moraine et al. 2017a). For a color version of this figure, see www.iste.co.uk/martin/farmsandterritory.zip*

3.3.1. *Services provided by crop–livestock integration*

Crop–livestock interactions promote the diversification of cropping systems. And diversification helps to "generate and regenerate ecosystem services" (Kremen et al. 2012, translated by the author). Martin et al. (2016) carried out a literature review to identify the benefits associated with the development of different crops and cover crops, in the context of crop–livestock interactions. Two main types of services were analyzed: soil fertility services and regulatory services.

Some resources cultivated to provide feed for animals contribute to the development of soil fertility services, as they help to loop the carbon, nitrogen and phosphorus cycles. This is the case with *cover crops*, which increase soil carbon content over the long term and reduce nitrogen leaching. Pure or mixed legumes also loop the C, N and P cycles, thanks to symbiotic fixation. The root system of legumes improves soil structure and permeability, facilitating water infiltration. *Grass/legume* associations are very interesting mixtures for animal feed, thanks to their protein content, which is sometimes higher than that of pure crops. Thanks to their complementary nature, they can reduce mineral nitrogen inputs. For example, the consumption of mineral nitrogen by wheat in superficial soil horizons "forces" the symbiotic fixation processes of legumes (Justes et al. 2014). Last but not least, *temporary grasslands* can also loop the C, N and P cycles, particularly when they are composed of 30–50% legumes. They sequester carbon and reduce nitrate leaching.

Another practice that promotes ecosystem services linked to soil fertility is the use of effluents in various forms. Manure brought into plots improves the diversity of soil microbial populations and thus soil fertility. Grazing, on the other hand, enables manure to be used directly on the plot. Animals transform the biomass they consume into fertilizing elements rapidly available to plants, modulo a low-intensity animal load (Soussana and Lemaire 2014).

Regulating services are determined by functional biodiversity[4] (within and around the plot of land) and by the level of ecological intensification. Functional biodiversity is influenced by landscape heterogeneity. Once again, it is through diversification of cropping systems that crop–livestock interactions can promote certain regulatory services. For example, the inclusion and conservation of natural and semi-natural *grasslands* can promote landscape heterogeneity (Tichit et al. 2012). What is more, the diversity of crops and semi-natural habitats around a plot helps to encourage beneficials. For example, including alfalfa in rotations provides nectar for aphid parasitoids. Landscape diversity also favors arthropods (which thus benefit from a diversity of resources throughout the year) and helps reduce pest pressure through resource dilution phenomena, or the push–pull effect: moving pests away from the crop ("push") to attract them elsewhere ("pull").

That being said, there are still many gaps in our knowledge of the link between practices and regulating services. We do not know precisely all the processes and practices that can be beneficial or, on the contrary, detrimental to certain ecosystem services. The benefits of this diversification depend on the practices implemented and their level of intensification. In other words, not all forms of interaction are positive from an environmental point of view and can also lead to negative intensification phenomena. In all cases, coordination between actors in the same territory is necessary to manage resources and thus promote ecosystem services.

4. "The set of species that contribute to ecosystem services in an agroecosystem," https://dicoagroecologie.fr/dictionnaire/biodiversite-fonctionnelle/.

3.4. Spatial and temporal coordinations on a territorial scale

There are therefore many advantages to reconnecting crop and livestock farming. Despite this, mixed livestock farms are in decline because there are many obstacles at the farm level: economic obstacles (ability of farms to invest in new equipment, buildings, machinery, etc.), organizational obstacles (ability to absorb the workload associated with developing a new activity), cognitive obstacles (ability to rapidly develop knowledge and skills on the new activity), structural obstacles (ability to provide enough resources in labor, inputs, land, etc.) (Ryschawy et al. 2018). This is why territory-wide studies are developing (Martin et al. 2016). That said, it is impossible to completely ignore the farm scale, within which interactions between productions necessarily take place (particularly within livestock farms). This is why the interactions presented here focus on the territorial scale, but take into account processes at the farm level, formulating the hypothesis of first strengthening autonomy at the farm level before opening up at the local level.

In all cases, these systems are integrated into a socio-economic system, and therefore depend on economic, institutional and social resources. A wide range of actors may be involved, and their coordination is essential to promote exchanges. Actors share objectives, ideas, practices and experiences. To represent the different types of coordination, Moraine et al. (2017a) suggest positioning the previous figure (see Figure 3.3) in time and space (see Figure 3.4). This figure cuts across both farm and territory scales: specialized farms can be represented by a single sphere (e.g. crop sphere for cereal farms) or by a combination of different spheres (crops, meadows, animals for mixed farmer–breeders). Interacting spheres can therefore represent interactions between activities on a farm, as well as interactions between farms.

Type 1 corresponds to a system of coexistence. It involves exchanges of raw materials (manure, forage, straw, etc.) between specialized farms. These exchanges are governed by the international market and can be organized by contracts to determine a price. This is the most common type of exchange today. For example, when a farmer buys grain from the other end of France via a broker or cooperative.

Type 2 corresponds to a logic of complementarity: exchanges take place between spheres on a farm or regional scale. Crop systems are designed to meet the need for fodder and concentrates for animal feed, and livestock effluents can be used for fertilization. For example, cereal farmers grow alfalfa for their herds and get back manure. There is little spatial coordination, but these practices improve the circulation of flows (of nitrogen, for example). These interactions require coordination between actors when they occur on a territorial scale. Their interactions may be opportunistic or determined by long-term contracts.

Type 3 aspires to "local synergy", stronger spatial and temporal interactions are established between the three spheres: intercropped forages, stubble grazing, temporary grasslands integrated into rotations, etc. The system benefits from ecosystem services and synthetic inputs are thus reduced.

Type 4 aims for "territorial synergy" by strengthening temporal and spatial interactions between the three spheres and between farms. This type requires strong coordination between actors to locally adapt outlets and coordinate resource allocation. The interactions created generate other types of exchange (knowledge, labor, etc.). Exchanges within and between farms are organized to reduce inputs and make the most of ecosystem services.

These types correspond to different forms of integration and therefore different practices. Consequently, they do not favor the same types of services. Martin et al. (2016) in their literature review have positioned the expected services according to the different types of spatio-temporal coordination. Type 1 (coexistence) would not enable ecosystem services to be provided due to the disconnection of productions not favoring diversity within cropping systems. Type 4, on the other hand, involves territorial synergy, and could maximize diversification on a large scale, boosting the presence of semi-natural habitats, increasing the area under grassland, cover crops and legumes, spreading manure, etc., maximizing services linked to fertility and biological regulation. This level of coordination implies a high level of organizational complexity, discussed in the next section. For intermediate types, it depends on the practices implemented. Type 2, "complementarity", can initiate diversification and manure spreading, while type 3 confirms this diversification by reinforcing the maintenance of grasslands and semi-natural habitats.

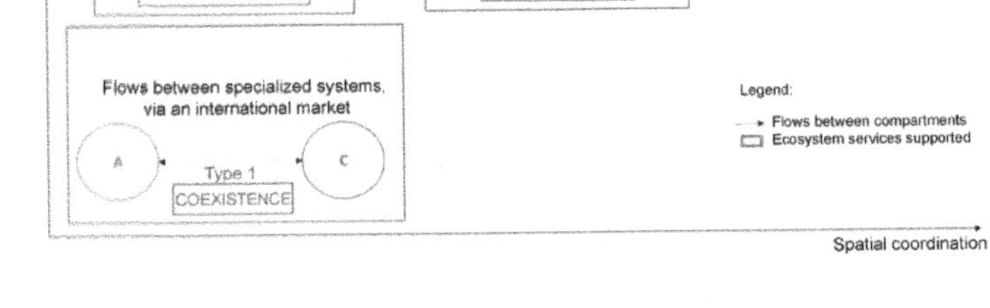

Figure 3.4. *Spatio-temporal coordination of systems to define four types of interactions (adapted from Moraine et al. (2016))*

3.5. Types of actors involved and forms of partnership

Developing crop–livestock interactions on a regional scale requires coordination between actors. The first to be involved are, of course, livestock farmers and cereal growers. A study of existing partnerships between cereal growers and organic livestock farmers in Denmark showed that the determining factors for the smooth functioning of the partners were not automatically linked to the distance between them, but rather to supply and demand, environmental regulations (e.g. livestock farmers in structural surplus having to export their manure to be spread elsewhere), social connections and farm access. Trust, communication, the ability to develop long-term relationships and flexibility are paramount to maintaining these partnerships (Asai et al. 2014a, 2014b).

3.5.1. *Two contrasting examples involving different actor networks*

Livestock farmers and cereal growers are not the only actors involved in crop–livestock exchanges. Box 3.1 presents the example of lowland shepherds in the Île-de-France region, who aim to maintain their business by maximizing the added value of their products and minimizing their costs. To achieve this, they develop partnerships with farmers and local actors who provide them with accessible biomass at different times of the year: managers of natural areas, communities of municipalities, the ONF[5], etc. They depend on a variety of resources and may need to set up complex networks of actors. There are a variety of exchange arrangements: service contracts with natural areas, informal agreements with cereal growers, etc. Relationships of trust can develop over time, leading to concerted strategies for choosing the composition of cover crops, for example. Shepherds seek to be independent of markets, while at the same time being part of complex networks in which the benefits of grazing are optimized, and decisions are taken collectively.

Box 3.2 presents the example of the Grignon farm (the AgroParisTech experimental farm), whose aim is to maximize production while limiting the impact of its activity on the environment through the development of programs such as Grignon Energie Positive[6]. The farm seeks to access a diversity of raw materials to feed its herds. It is developing local partnerships with cereal growers but is also looking to integrate into longer supply chains to diversify its supplies and build optimized rations that will meet production targets. The exchange arrangements include long-term contracts, service provision, digestate sales, etc. These

---

5. Office National des Forêts: management for public forests in France, i.e. forests owned by the French State and by local authorities.
6. See: https://www.cereopa.fr/outils-et-actions/grignon-energie-positive/.

arrangements involve different actors than in the previous example: brokers, cooperatives, agri-food industries, etc. These two examples show us different logics that remain consistent with the objectives of each system: on the one hand, an itinerant shepherd seeking to limit investments, equipment and buildings, and who is part of a network of actors who co-design a grazing itinerary that can help everyone's practices evolve; on the other, a large-scale farm seeking to integrate into a complex socio-economic network in order to diversify its supplies and achieve its production objectives while limiting its impacts. The resulting forms of the crop–livestock interaction are different: grazing cover crops and cereals early in the plant cycle for one, mowing alfalfa for the other.

3.5.2. *Actors with additional motivations*

Systems integrating crops and livestock on a territorial scale are studied by a wide range of actors in the agricultural world. A growing number of projects aim to describe, design, evaluate or support these systems and have developed knowledge about their performance, as well as methods for developing them. Table 3.1 presents some of these projects.

	Dates	Location	Coordinator	Systems Studied
LUZCO	2016–2019	All of France	FR CUMA Grand Ouest	Collective initiatives to develop forage legumes
BREBIS LINK	2018–2020	Grand Sud-Ouest	Chambre agriculture Dordogne	Sheep grazing of additional areas (orchards, vineyards, wooded pastures, canopies, etc.)
Crop–livestock complementarity	2014–2020	Hauts de France	AgroTransfert	All systems
POSCIF	2018–2021	Île-de-France	Agrof'Ile (CIVAM IDF)	Sheep grazing on cereal lands
ARPEEGE	2019–2022	Grand Est	Chambre régionale Grand Est	Protein autonomy for livestock farms
Feder'Acteurs	2020–2024	Aube and Yonne	AgroParisTech and Syndicat AOP Chaource	Interactions between dairy farmers, cereal growers and industry actors
Inter-AGIT	2022–2025	Sud-Ouest	IDELE and chambre d'agriculture de Dordogne	Sheep and cattle grazing of intercrops and additional areas

RECIT	2023–2024	Charente-Maritime	Civam du Pays Ruffecois	Support for reconnection initiatives
SAGITERRES	2022–2024	Minervois (Aude and Hérault)	INRAE and BioCivam de l'Aude	Collective resource exchange strategies between breeders and farmers

Table 3.1. *Examples of recent projects on crop–livestock interactions*

Moraine et al. (2020) have carried out an analysis of eight case studies (projects, agricultural equipment cooperatives, investment and management companes, etc.; different from Table 3.1) aimed at developing crop–livestock interactions. This study and Table 3.1 show that the theme of crop–livestock interactions involves a wide range of actors: technical institutes, chambers of agriculture, research institutes, agricultural equipment cooperatives, associations, protected designation of origin (PDO) unions, etc. The partners in these projects also include economic partners such as cooperatives and animal feed companies. A wide range of regions in France are involved.

These projects have been launched to limit the negative impacts of farm and regional specialization. The *cereal, wine and fruit growers* involved in these projects are seeking to (i) implement alternatives to the chemical or mechanical destruction of their cover crops, (ii) make up for deficits in soil organic matter, (iii) manage weed cover by limiting the use of herbicides or mechanical tools, (iv) diversify their cropping system with crops for animal feed for which they have no local outlets and (v) enhance or maintain their meadows. For their part, *livestock farmers* wish to (i) reinforce their autonomy, particularly when they are subject to thresholds imposed by specifications (PDO zones), and (ii) find resources locally rather than depend on imported raw materials whose prices fluctuate. These strategies are sometimes necessary to maintain their breeding activity. *Industry actors, technical institutes, unions*, etc. seek to (i) maintain livestock farming and grasslands in local areas, to maintain their activity and/or diversify their services, (ii) diversify cropping systems and adopt agroecological approaches, (iii) encourage the development of local outlets and (iv) revitalize local industries. These non-exhaustive lists show that the initial questions raised by the actors behind these crop–livestock initiatives can be varied and linked to other agronomic, economic or social issues, such as maintaining livestock production in a given area.

In the Île-de-France region, a network has gradually formed around sheep grazing on lowlands, supported by the Agrof'Ile association and the Champs des Possibles business incubator. On the one hand, cereal growers were looking to make the most of their winter cover crops and forage legumes, while, on the other, shepherds were looking to set up or diversify their resources for their flocks. This is how the Pâturage Ovin en Systèmes Céréaliers d'Ile de France (POSCIF) project came into being in 2018. The project aimed to develop knowledge about this type of grazing by analyzing the impacts at agronomic, zootechnical and economic levels. Trials, evaluations and workshops were carried out during the project to address the technical and organizational aspects of these systems.

Numerous resources could potentially be grazed in Île-de-France. For cereal growers, these include cover crops, stubbles, legumes, certain crops at the beginning of their cycle, etc. Other resources are also present in natural areas, forests, urban zones, etc., managed by the ONF, communities of municipalities or regional nature parks. But many questions remain unanswered: What is the impact on soils? On yields? On animals? How could the various actors be coordinated? Various practices were implemented and evaluated. In particular, the project showed that there was no risk of soil compaction during grazing, that the impact on yields (after grazing a cover crop) was nil, and that the nitrogen input when grazing a cover crop was low (6 uN), because even if the ewes contribute nitrogen via their dejecta, they also take it up by consuming the biomass.

The project has made it possible to study a special case, that of itinerant or landless shepherds. These shepherds depend entirely on the resources of other actors (cereal growers, land managers, etc.). Figure 3.5 shows an example of an itinerary followed in 2018 by an itinerant shepherd with 250 ewes and Figure 3.6. shows an example of ewes grazing on a cover crop. This shepherd may be required to practice transhumance over radii of more than 10 km over a given period. He has no contracts with the cereal growers. Shepherds and grain farmers work on a win-win basis. The cereal growers offer access to water and electricity, as well as a place under a shed for the shepherd's caravan to follow his flock. The shepherd discusses the composition of the cover crops with the cereal growers, so that they meet both zootechnical and agronomic criteria. In the Fontainebleau forest, on the other hand, the shepherd signs an agreement and is eligible for funding.

Although they only graze on organic plots, these shepherds cannot be certified as organic farmers because they have no surface area to declare, which makes them ineligible.

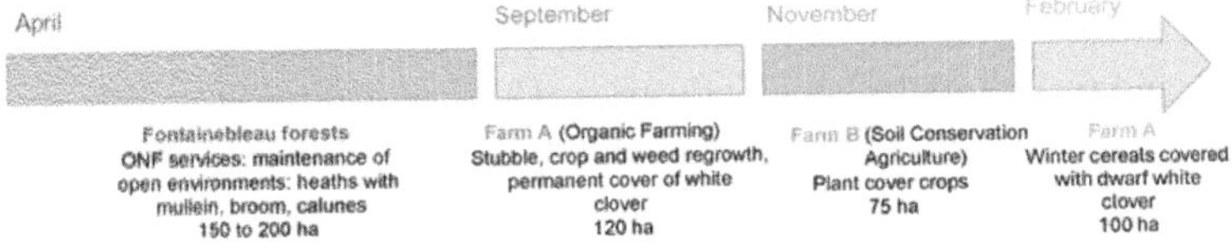

Figure 3.5. *Itinerary of an itinerant shepherd in Île-de-France in 2018*

Figure 3.6. *Ewes grazing on a cover crop, POSCIF © project*

Box 3.1. *Lowland grazing in the Île-de-France region*

The Grignon farm (part of AgroParisTech) is located in the Yvelines region of France. It includes 400 hectares and raises 180 dairy cows and 600 ewes. It processes almost half its milk on site, thanks to a processing unit that produces yoghurts. The farm has been awarded the Haute Valeur Environnementale (High Environmental Value) Level 3 label. It recently built a methanizer. The farm's strategy is not to be 100% self-sufficient; on the contrary, it aims to be anchored in its local area and in the current supply chains, and to diversify its supplies. It is therefore developing more or less local partnerships to ensure the supply of fodder and concentrates for its livestock. Maize and barley silage are self-consumed. Rapeseed is also produced on the farm and crushed by a neighboring cereal grower who sells their oil locally. Oil cake is collected by the Grignon farm. Alfalfa is purchased from various cereal growers in the area (10–15 km), some of whom are organic. All the cereal growers are interested in integrating alfalfa into their rotation. Alfalfa is priced per ton of dry matter, depending on quality. Contracts are generally signed for 2–3 years, and alfalfa is purchased as standing crops. The Grignon farm then harvests the alfalfa with its own equipment. Beet pulp is purchased from agri-businesses that are generally located further away (approximately 100 km). Quantities and prices are contractualized each year, according to needs and supply. The farm also buys potatoes, a very interesting market, especially during the Covid-19 crisis, when the catering industry was closed and the farmers only had to pay for transport. At present, supply is variable and depends on the purchase price. Potatoes are bought by a broker in 30-ton trucks. Where they come from is not always known.

So, there are a wide range of partnerships to manage, with a diversity of actors that goes beyond a simple agreement between a livestock farmer and a cereal grower. Contracts may be signed on a more or less long-term basis, and operations may be required (alfalfa mowing, transport, etc.). These resources can vary from one year to the next, depending on price trends, herd requirements and yields achieved on the farm (maize, for example, can be purchased from neighbors if necessary). It is a strategy that depends on opportunities, as well as on economic, zootechnical and agronomic factors, and on the resources available on the farm (capacity to transport, store, labor available to carry out these operations, etc.).

Figures 3.7 and 3.8 summarize the flows in and out of the farm to feed the dairy herd and operate the anaerobic digestion plant.

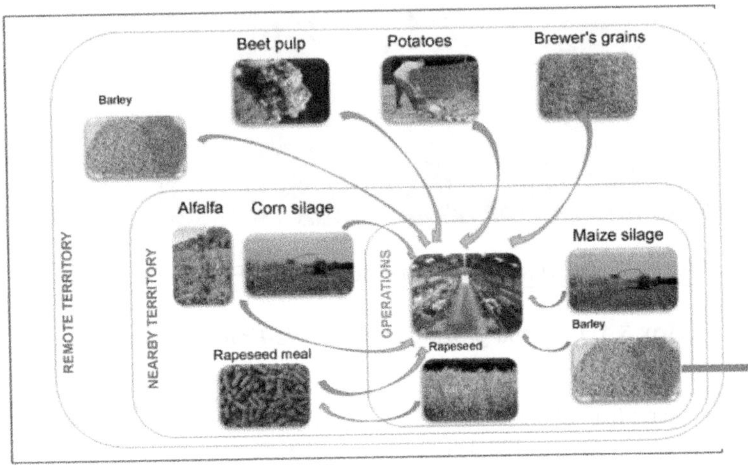

Figure 3.7. *Flows associated with the raw materials used in dairy cow rations at Grignon farm*

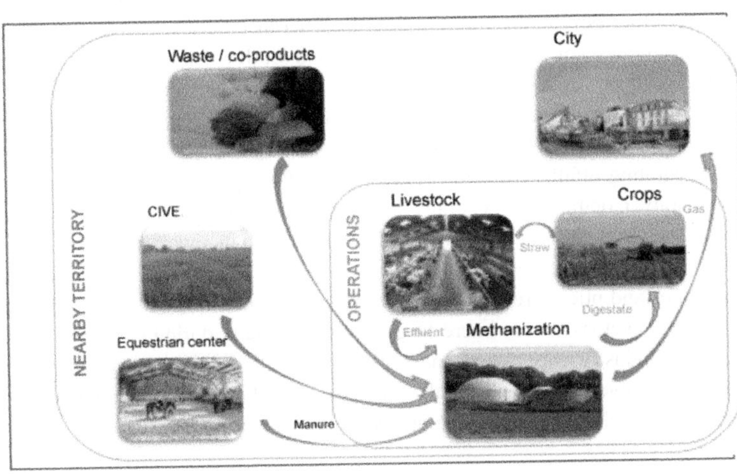

Figure 3.8. *Flows associated with the development of a methanization unit on the Grignon farm*

Box 3.2. *Developing local partnerships to feed dairy cows at Grignon farm*

To strengthen and perpetuate these projects and initiatives, Moraine et al. (2020) point out the importance of collectively defining common objectives, setting out on a long-term trajectory and benefiting from appropriate support. Sharing technical knowledge and taking economic issues into account are crucial. It is also important to rely on "relay" actors, able to help formalize or reformulate ideas and contribute specific skills, to help build a common identity. Public policy actors are generally not very involved in these projects, even though they could play a decisive role in building financial or material support systems.

3.6. Methods for diagnosing, designing, evaluating and supporting the development of these systems

3.6.1. *Methodological challenges*

Agronomists have long been developing tools and methods at plant, animal, plot, herd and farm levels. More systemic approaches on the scale of territories (PDO, catchment area, production basins, etc.) are more recent because they are more complex. According to Husson (2007), a territory is made up of three sub-systems: the system of actors, that of the activities generated by these actors and that of the dynamics and interactions created by these activities. The characteristics of systems integrating crops and livestock show complex particularities at these different levels, which need to be taken into account:

(i) As we have seen, the actors involved are very diverse: natural area management bodies, cereal growers, livestock farmers, economic operators in the sector, development actors, etc. These actors have different objectives, strategies and practices. They also approach the theme of interactions from different angles, which sometimes converge with difficulty within collective projects.

(ii) The activities implemented cross several organizational scales: those of the agricultural plot, the farm, the supply chain, the territory, etc., and must operate in a complementary manner.

(iii) The dynamics and interactions generated are sometimes difficult to maintain over the medium term, due to severe organizational and logistical constraints: absence of contracts, lack of manpower or equipment, supply–demand imbalance, etc.

Due to the diversity and complexity of the actors and the activities and dynamics that characterize them, these systems are associated with environmental, agronomic, economic and social issues. Developing methods capable of taking this complexity and these issues into account represents a major methodological challenge, particularly for the agroecological transition. Indeed, Meynard et al. (2018) and

Boulestreau et al. (2021) explain that to bring about changes in agricultural practices, systemic approaches are needed. Changes are no longer limited to plot- or herd-level practices alone but affect several scales and actors. However, the approaches of advisors, researchers and organizers are sometimes specialized. For example, an advisor or researcher in crop production might propose a cover crop optimized to promote biological activity in the soil, but which would have no interesting feed value for animals. Our frameworks therefore need to be rethought to be more systemic.

Table 3.2 shows some of the methods and tools used to diagnose, design, evaluate or support the development of these systems. It is not intended to be exhaustive, but to show a few concrete examples.

	Work	Work examples and links
Diagnosis	**Territorial metabolism** Analysis of flows (of nitrogen, energy, dry matter, etc.) passing through a territory, enabling the analysis of territorial autonomy. This work can be carried out on limited territories or at national level.	Articles: Garnier et al. (2016) and Le Noë et al. (2018)
Diagnosis	**Semi-structured interviews** Understand farmers' motivations, constraints, obstacles and levers for setting up interactions, operating methods (contractual, informal, price, etc.), etc. Surveys can be set up as part of a more complex methodology, such as the *socio-technical diagnosis* presented in section 3.6.2.	C. Meunier's thesis work on farmers' perceptions of the obstacles and levers regarding the development of interactions (Meunier et al. 2024)
Design	**Method for co-designing scenarios** A specific method has been developed to design exchange scenarios between cereal farmers and livestock breeders. It is based on surveys, workshops, simulations (quantities exchanged) and a map (to geolocate exchanges). The DYNAMIX method is described in Box 3.3.	Articles: Moraine et al. (2017), Moraine et al. 2017b, Ryschawy et al. (2018, 2022) and Moojen et al. (2022)
Design	**Serious games** The main principles of the method	Work carried out in the POSCIF and inter-*AGIT+* projects by S. Pissonnier

	Work	Work examples and links
	DYNAMIX described in Box 3.3 have been adapted to the case of sheep grazing on a cereal grassland, in the form of a serious game representing grazeable plots, the resources present (in the form of tokens positioned on the plots) and the flocks (in the form of counters). Participants must agree on a grazing itinerary.	(AgroParisTech); see Box 3.3 and the serious game video[7] (in teaching context).
Evaluation	**Multi-criteria evaluation** Using a combination of tools specific to crops (SYSTERRE), livestock (e.g. CAP2ER) or the farm (work balance method)	POSCIF project, work carried out by E. Emonet (ACTA): multi-criteria evaluation of the Montacquoy farm, using ewes to destroy cover crops[8]
Evaluation	**Experiments** Field measurement of the impact of these systems. For animals: measurements during grazing of health, fattening, behavior, etc. For crops: level of compaction, impact on bio-aggressors, impact on yields, biomass consumed, etc. Trials of new crops to be developed locally, such as soya. The aim is also to identify the most beneficial practices	Agrof'ile, POSCIF project: grazing intercropping cover crops
Evaluation	**Modeling** Modeling the carbon and nitrogen in grazed cover crops	POSCIF project, work carried out by F. Levavasseur and B. Vasseur (INRAE): internship thesis on the effects of sheep grazing on plant cover and winter crops on carbon and nitrogen cycles
Support tools	**Contract templates, legal guides and pricing** Standard contracts have been drawn up to facilitate agreement between the parties involved, and to clarify the aspects that need to be addressed to make the partnership work in the best possible way for all parties. A calculator to calculate the equivalence between straw and manure has been put on	Legal guide from the CEREL project[9] Straw-manure calculator produced by Arvalis[10]

7. See: https://www.youtube.com/watch?v=7giZ2gFiNzM.
8. See: https://www.agrofile.fr/poscif/.
9. See: https://vienne.chambre-agriculture.fr/fileadmin/user_upload/Nouvelle-Aquitaine/104_Inst-Vienne/Documents/mes_productions/Agroecologie/Cooperation-Innovation/GUIDE_JURIDIQUE_Light.pdf.
10. See: https://paille-fumier.arvalis-infos.fr/.

Work	Work examples and links
line by Arvalis, enabling an informed price to be determined.	
Testimonials (factsheets and videos) Enabling interested farmers to see how certain partnerships work	Testimonial "Exchanging with my neighbor, a plus for my livestock farm", CCE Project, Agrotransfert[11]
Technical data sheets On certain crops (agronomic characteristics), integrating plants into the ration (feed value, for example)	Cereal grazing data sheet – POSCIF[5] project
Raw material exchange sites and resource centers Websites and exchange platforms	Resource center for lowland shepherds, Agrof'Ile[12] Resource center for developing complementarity, CCE project, AgroTransfert[13] Resource center for the development of legumes, LUZCO project[14]

Table 3.2. *Examples of methods and tools to diagnose, design, evaluate or provide support*

This table shows that there are two main types of methods: those created specifically to take into account the complexity of these systems, and those borrowed from other frameworks and adapted to these systems. For example, for multi-criteria evaluation, a number of existing tools specific to crops (SYSTERRE[15]) or livestock (CAP2ER[16]) have been combined to assess agronomic, zootechnical and economic performance. This combination poses major methodological challenges, in terms of correctly interpreting indicators, avoiding duplication, facilitating the filling-in of tools, etc.

Conversely, co-design methods have sometimes been created specifically to develop interactions. This is the case of the DYNAMIX method presented in Box 3.3. DYNAMIX was designed specifically to take into account the complexity of crop–livestock systems on a territorial scale, combining individual and collective phases and mobilizing different tools and methods (mapping, surveys, workshops, etc.). That said, it is highly complex to produce methods capable of taking into

11. See: https://www.youtube.com/watch?v=VkY4J6RvNF8.
12. See: https://bergerdeplaine.agrofile.fr/?Accueil.
13. See: http://www.agro-transfert-rt.org/cce-accueil/.
14. See: https://luzco.fr/.
15. See: https://www.arvalis.fr/outils-et-services/outils-et-fiches/systerre.
16. See: https://idele.fr/detail-article/cap2err.

account all the characteristics of these systems outlined above: diversity of actors, scales, technical and organizational criteria, perceptions, issues, etc. Most methods tend to focus on farmers, and do not integrate those of the value chain, for example. What is more, the methods need to be adapted to the constraints and challenges of the area under study, in order to be as relevant as possible to the actors concerned, which can be time-consuming (updating data on cropping or livestock systems, for example). The challenges involved in developing such methods are therefore considerable.

In most projects and studies, there is very little follow-up to changes in practices. What is more, the methods developed stop at scenarios and discussions, which do not always lead to concrete changes. The actors involved in these methods sometimes find it difficult to seize on the co-constructed solutions to bring about changes in their practices. The various studies carried out show that this phenomenon is notably due to a lack of group facilitation, or more precisely a lack of facilitators and advisors who are often too specialized in crops or livestock, and who are therefore unable to address the cross-cutting links between productions as discussed above (Moojen et al. 2023). Various actors could be called upon to play this role in order to initiate changes in practices and monitor interactions: livestock advisors, field crop advisors, cooperative advisors, economic and environmental interest group coordinators, associations, etc., provided that they have the time to dedicate to this activity, in addition to other advisory activities or the coordination of pre-existing groups. If a facilitator is found, they will have to ensure that individual expectations and objectives are taken into account, as well as each individual's ability to make their own decisions (Ryschawy et al. 2022).

Serious games are particularly relevant tools for supporting changes in practice. The use of game boards, geographical maps or game cards, models, etc., enable participants to project themselves and exchange more easily (Duru et al. 2015). Games also make it possible to build collective strategies while appropriating new knowledge (Klerkx et al. 2012).

Based on this observation, Ryschawy et al. (2018, 2022) have developed a participatory method based on a serious game, with a view to developing interactions between cereal growers and livestock farmers. This method has been used in Ariège with two groups: one made up of four livestock farmers and four cereal farmers on the lowlands, and the other made up of four cereal farmers and five livestock farmers on the hillsides. The method consists of six stages:

Stage 1: collective definition of the problem and objectives.

Stage 2: two-hour technical–economic interviews with each farmer to gather data on practices, work organization, available equipment and motivations.

Stage 3: collective design of purchase/sale scenarios. The farmers' feed requirements and the cereal growers' manure requirements are quantified, in order to find a collective balance. Potential obstacles to collective operation are listed. This stage includes the positioning of actors on a map representing plots of land and farm headquarters, in order to spatialize flows.

Step 4: once the exchange scenarios to be tested have been defined, a simulation model is used to produce supply–demand balances and assess the level of potential or actual autonomy. This stage takes into account individual farms (characterization of the cropping and livestock systems) and then the collective level. This multi-criteria evaluation is inspired by the IDEA method and aims to test the scenario and its consequences on individual farms and the collective. These indicators address the economic, environmental and social dimensions. The collective dimension also includes an assessment of the sustainability of exchanges.

Stage 5: return to the collective to re-discuss the scenarios, with the possibility of iterating between design and evaluation.

Stage 6: monitoring of changes in practices following the workshops.

Facilitators/researchers are on hand to provide knowledge if required, and to help redesign systems. Secondly, participants gather around the territory map to consider the logistics of exchanges, notably via storage and transport counters.

Box 3.3. *The DYNAMIX method*

3.6.2. *Example of a project combining different methodologies*

Methods can also be combined within research projects, as in the Feder'acteurs project coordinated by AgroParisTech and taking place in the Chaource PDO area.

Chaource is a PDO soft cheese with a molded rind produced between Aube and Yonne. It is produced by 53 farmers, subject to specifications that impose autonomous criteria: 75% of the dry matter in the ration must come from the farm, and 85% from the PDO area. These self-sufficiency criteria are becoming increasingly difficult to meet due to the consequences of climate change, which are making forage yields more fragile: dairy farmers sometimes struggle to find local resources. Partnerships between local actors therefore need to be strengthened to secure supplies for farmers and ensure self-sufficiency. It was against this backdrop that the Feder'Acteurs research project was set up by AgroParisTech and the

Syndicat de l'AOP Chaource to analyze and strengthen crop–livestock partnerships in the area. The objectives were to (i) diagnose existing crop–livestock interactions and identify the obstacles and levers to their development and (ii) develop suitable methods for strengthening and/or creating new interactions.

To diagnose crop–livestock interactions in the area, we mobilized the DST[17] method. DST is used to "understand the practices, strategies and networks of actors involved in innovation processes" (Belmin et al. 2018; Casagrande et al. 2023, translated by the author). It helps prepare the design process by identifying the obstacles and levers on the development of certain innovations, in this case crop–livestock interactions. These obstacles and levers are influenced by the functioning of socio-technical systems, defined in the Casagrande et al. (2023) guide as systems "composed of a set of networked actors who share practices, knowledge, technologies, collective representations and formal or informal rules that guide their practices […]. Sociotechnical diagnosis consists in describing the functioning of a sociotechnical system built around a dominant technology, with the aim of understanding the systemic phenomena that frame the innovation process and identifying the obstacles and levers to the development of alternative technologies" (author's translation). According to the methodological guide, two types of socio-technical systems can be described: socio-technical regimes, also referred to as "dominant" regimes, and socio-technical niches based on alternative technologies to those of the dominant regime. The aim of the project was to identify crop–livestock interactions considered as niches in relation to the dominant regime of production specialization.

Between 2021 and 2022, 70 surveys were carried out with a wide range of actors: dairy farmers, cereal growers, advisors, feed manufacturers, cooperatives, watershed associations, etc. This provided an overview of all the actors in the area and their positions on the issue of crop–livestock interactions. The results of the DST were presented in January 2022 to complete the information and build a collective dynamic on the territory.

The DST highlighted three types of crop–livestock interactions: (i) bilateral exchanges: exchanges or purchases of raw materials between cereal growers and dairy farmers; (ii) multi-lateral exchanges: purchase and centralization of a raw material (hay or straw) by a farm, as well as resale; (iii) traced alfalfa and beet pulp channels: organization of a group of actors (sugar refinery, dehydration plant, PDO syndicate, etc.) to trace raw materials from the PDO zone to livestock farms. These three types of interaction are subject to constraints at the farm level (e.g. development of conservation agriculture that returns straw to the soil), at the

17. Diagnostic Socio Technique: socio-technical diagnosis.

industry level (e.g. diversification into human food) and at the regional level (e.g. competition for outlets for beet pulp or grass with the development of methanization). Feedback from the diagnosis to the actors surveyed confirmed the analyses and launched a collective reflection on desirable futures for the region. Several themes came to the fore during these discussions: the desire to make the region more attractive, as well as more self-sufficient.

Following this diagnosis, we entered the design phase. A first workshop was organized in April 2022, bringing together approximately 20 participants from diverse backgrounds. The aim was to encourage participants to build crop–livestock partnerships. The workshop was organized using the Knowledge, Concept, Project (KCP) method, which enables knowledge to be shared and participants to be "challenged", thus encouraging innovative proposals (Elmquist and Segrestin 2009; Hatchuel et al. 2009). The aim was to fill in "project sheets", clearly identifying the actions to be carried out, the project leaders and the potential obstacles. The proposed projects would then have been implemented as part of the Feder'Acteurs project.

The workshop organized in April 2022 built on these reflections on attractiveness and autonomy. Unfortunately, the project sheets could not be filled in accurately enough and were not usable. A number of difficulties were raised: the excessive diversity of participants and the difficulty of agreeing on a common objective. The collective of participants created during this workshop was not mature enough to propose roadmaps to follow and create partnerships. Discussions with the PDO union following the workshop led to the construction of a simulation tool (created in Excel) that would enable us to more directly test scenarios involving changes in the rations of PDO dairy cows, and to discuss the corresponding crop rotation, on the scale of the PDO territory.

A smaller working group was formed and met in May 2023 to create more direct partnerships. It included three livestock farmers and three cereal growers. The aim was to get participants to identify tradable resources and discuss the impacts on their system. Discussions were based on the results of Excel simulations. Three scenarios were prepared in advance and co-constructed with the PDO syndicate coordinator: a scenario "strengthening alfalfa in the ration", a scenario "which resources are impacted by climate change" and a scenario "development of protein peas in the ration". Complementary documents were gathered to fuel the discussions (feed value analyses, standard contracts, testimonials, etc.). The workshop enabled the three dairy farmers and the three cereal growers to share knowledge about their systems, identify interesting resources (first-cut of alfalfa produced for seeds, maize) and discuss potential opportunities (creation of a traced rapeseed meal channel). All participants also emphasized the importance of a strong regional leadership to maintain and concretize these reflections. A WhatsApp group is currently being set up.

This work shows the following:

– The DST method is well suited to describing crop–livestock interactions on a territorial scale and was able to highlight the different types of interactions and the obstacles and levers to their development, taking into account the diversity of actors involved.

– The KCP method was not used effectively, as the participants had not previously formed a collective working towards a common goal.

– The workshops organized between dairy farmers and cereal growers, using a simulation tool, enabled very concrete aspects of crop–livestock interactions to be addressed, and highlighted the importance of leadership in coordinating the actors involved.

– The methods used to diagnose and design interactions must take into account the complexity and diversity of the systems under consideration, while adapting to the expectations of the actors concerned.

3.7. Conclusion

Connecting crops and livestock is a possible solution for initiating the agroecological transition of territories. There is a strong demand from farmers and agro-industrial and territorial actors to develop various forms of interaction to address complex, systemic issues. In recent years, a great deal of work has gone into understanding these interactions and developing methods and tools to support their development. Future advisors and facilitators will need to be trained in the complexity of these systems and methods, to be able to adapt to the situations encountered, while capitalizing on what has already been done to continue filling existing knowledge gaps.

3.8. References

Agreste Bretagne (2016). Draaf – Tableaux de l'Agriculture Bretonne [Online]. Available at: https://draaf.bretagne.agriculture.gouv.fr/IMG/pdf/TAB_COMPLET_cle416585.pdf.

Agreste France (2022). Mémento 2022.

Asai, M., Langer, V., Frederiksen, P. (2014a). Responding to environmental regulations through collaborative arrangements: Social aspects of manure partnerships in Denmark. *Livest. Sci.*, 167, 370–380. doi: 10.1016/j.livsci.2014.07.002.

Asai, M., Langer, V., Frederiksen, P., Jacobsen, B.H. (2014b). Livestock farmer perceptions of successful collaborative arrangements for manure exchange: A study in Denmark. *Agric. Syst.*, 128, 55–65. doi: 10.1016/j.agsy.2014.03.007.

Bell, L.W. and Moore, A.D. (2012). Integrated crop–livestock systems in Australian agriculture: Trends, drivers and implications. *Agric. Syst.*, 111, 1–12. doi: 10.1016/j.agsy.2012.04.003.

Belmin, R., Meynard, J.-M., Julhia, L., Casabianca, F. (2018). Sociotechnical controversies as warning signs for niche governance. *Agron. Sustain. Dev.*, 38, 44. doi: 10.1007/s13593-018-0521-7.

Benoît, M., Deffontaines, J.-P., Lardon, S. (2006). *Acteurs et territoires locaux : vers une géoagronomie de l'aménagement.* Inra-QUAE, Paris. doi: 10.4000/rge.226.

Bonaudo, T., Bendahan, A.B., Sabatier, R., Ryschawy, J., Bellon, S., Leger, F., Magda, D., Tichit, M. (2014). Agroecological principles for the redesign of integrated crop–livestock systems. *Eur. J. Agron.*, 57, 43–51. doi: 10.1016/j.eja.2013.09.010.

Boulestreau, Y., Casagrande, M., Navarrete, M. (2021). Analyzing barriers and levers for practice change: A new framework applied to vegetables' soil pest management. *Agron. Sustain. Dev.*, 41, 44. doi: 10.1007/s13593-021-00700-4.

Casagrande, M., Belmin, R., Boulestreau, Y., Le Bail, M., Navarrete, M., Meynard, J.-M. (2023). *Guide méthodologique pour le diagnostic des freins et leviers sociotechniques aux processus d'innovation dans des systèmes agri-alimentaires.* INRAE, Paris. doi: 10.17180/w78m-dn95.

Chatellier, V. and Gaigné, C. (2012). Les logiques économiques de la spécialisation productive du territoire agricole français. *Innovations in Agronomy*, 22, 185–203 [Online]. Available at: https://hal.science/hal-01208860v1/document.

Deschamps, L., Frétière, K., Maillochon, A., Molina, V. (2016). La Bretagne : première région française pour la production et la transformation de viande. *Insee Analyses Bretagne*, 32.

Dourmad, J.-Y., Delaby, L., Boixadera, J., Ortis, C., Méda, B., Gaigné, C., Dumont, B. (2017). Diversité des services rendus par les territoires à forte densité d'élevages, trois cas d'étude en Europe. *INRA Productions Animales*, 30(4), 303–320.

Dourmad, J.-Y., Delaby, L., Boixadera, J.-L., Ortis, C., Méda, B., Gaigné, C., Dumont, B. (2018). Diversité des services rendus par les territoires à forte densité d'élevages, trois cas d'étude en Europe. *INRA Productions Animales*, 30, 303–320. doi: 10.20870/productions-animales.2017.30.4.2261.

Durand, P., Munier-Jolain, N.G., Colinet, L., Gaunand, A. (2014). Gestion de la pollution d'origine agricole en Bretagne. Technical Report.

Duru, M., Therond, O., Martin, G., Martin-Clouaire, R., Magne, M.-A., Justes, E., Journet, E.-P., Aubertot, J.-N., Savary, S., Bergez, J.-E. et al. (2015). How to implement biodiversity-based agriculture to enhance ecosystem services: A review. *Agron. Sustain. Dev.*, 35, 1259–1281. doi: 10.1007/s13593-015-0306-1.

Elmquist, M. and Segrestin, B. (2009). Sustainable development through innovative design: Lessons from the KCP method experimented with an automotive firm. *Int. J. Automot. Technol. Manage.*, 9(2), 229–244.

Gaigne, C. (2024). Transitions agricoles et dynamiques de spécialisation. *Innov. Agron.*, 93, 1–26. doi: 10.17180/CIAG-2024-VOL93-ART01.

Garnier, J., Anglade, J., Benoit, M., Billen, G., Puech, T., Ramarson, A., Passy, P., Silvestre, M., Lassaletta, L., Trommenschlager, J.-M. et al. (2016). Reconnecting crop and cattle farming to reduce nitrogen losses to river water of an intensive agricultural catchment (Seine basin, France): Past, present and future. *Environ. Sci. Policy*, 63, 76–90. doi: 10.1016/j.envsci.2016.04.019.

Garrett, R.D., Ryschawy, J., Bell, L.W., Cortner, O., Ferreira, J., Garik, A.V.N., Gil, J.D.B., Klerkx, L., Moraine, M., Peterson, C.A. et al. (2020). Drivers of decoupling and recoupling of crop and livestock systems at farm and territorial scales. *Ecol. Soc.*, 25, art24. doi: 10.5751/ES-11412-250124.

Hache, E. (2015). Géopolitique des protéines. *Rev. Int. Stratég.*, 97, 36–46. doi: 10.3917/ris.097.0036.

Hatchuel, A., Le Masson, P., Weil, B. (2009). Design theory and collective creativity: A theoretical framework to evaluate KCP process. In *International Conference on Engineering Design ICED'9*, 24–27 August, Stanford.

Husson, J.-P. (2007). Review of: M. Benoît, J.-P. Deffontaines, S. Lardon, Acteurs et territoires locaux : vers une géoagronomie de l'aménagement: Paris, Inra-QUAE, 2006. *Revue Géographie de l'Est*, 47(2). doi: 10.4000/rge.226.

Justes, E., Bedoussac, L., Corre-Hellou, G., Fustec, J., Hinsinger, P., Jeuffroy, M.-H., Journet, E.-P., Louarn, G., Naudin, C., Pelzer, E. (2014). Les processus de complémentarité de niche et de facilitation déterminent le fonctionnement des associations végétales et leur efficacité pour l'acquisition des ressources abiotiques. *Innov. Agron.*, 40, 1–24. doi: 10.17180/BDVP-HA42.

Klerkx, L., Van Bommel, S., Bos, B., Holster, H., Zwartkruis, J.V., Aarts, N. (2012). Design process outputs as boundary objects in agricultural innovation projects: Functions and limitations. *Agric. Syst.*, 113, 39–49. doi: 10.1016/j.agsy.2012.07.006.

Kremen, C., Iles, A., Bacon, C. (2012). Diversified farming systems: An agroecological, systems-based alternative to modern industrial agriculture. *Ecol. Soc.*, 17, 44. doi: 10.5751/ES-05103-170444.

Le Noë, J., Garnier, J., Billen, G. (2018). Phosphorus management in cropping systems of the Paris Basin: From farm to regional scale. *J. Environ. Manage.*, 205, 18–28. doi: 10.1016/j.jenvman.2017.09.039.

Martin, G., Moraine, M., Ryschawy, J., Magne, M.-A., Asai, M., Sarthou, J.-P., Duru, M., Therond, O. (2016). Crop–livestock integration beyond the farm level: A review. *Agron. Sustain. Dev.*, 36. doi: 10.1007/s13593-016-0390-x.

Mazoyer, M. and Roudart, L. (1997). *Histoire des agricultures du monde. Du néolithique à la crise contemporaine*. Éditions du Seuil, Paris.

Meunier, C., Martin, G., Barnaud, C., Ryschawy, J. (2024). Bucking the trend: Crop farmers' motivations for reintegrating livestock. *Agric. Syst.*, 214, 103820. doi: 10.1016/j.agsy.2023.103820.

Meynard, J.-M., Messéan, A., Charlier, A., Charrier, F., Fares, M., Le Bail, M., Magrini, M.-B., Savini, I. (2013). Freins et leviers à la diversification des cultures : étude au niveau des exploitations agricoles et des filières. *OCL*, 20, D403. doi: 10.1051/ocl/2013007.

Meynard, J.-M., Charrier, F., Fares, M., Le Bail, M., Magrini, M.-B., Charlier, A., Messéan, A. (2018). Socio-technical lock-in hinders crop diversification in France. *Agron. Sustain. Dev.*, 38, 54. doi: 10.1007/s13593-018-0535-1.

Mignolet, C., Schott, C., Benoit, M., Meynard, J.-M. (2012). Transformations des systèmes de production et des systèmes de culture du bassin de la Seine depuis les années 1970 : une spécialisation des territoires aux conséquences environnementales majeures. *Innov. Agron.*, 22, 1–16.

Mignolet, C., Schott, C., Foissy, D., Benoît, M. (2013). Changement d'usage des sols par l'agriculture du bassin de la Seine : une spécialisation des territoires permise par un usage accru des intrants de synthèse. Rapport d'activité PIREN-Seine Phase 6 – 2012. hal-02802052.

Moojen, F.G., De Faccio Carvalho, P.C., dos Santos, D.T., Neto, A.B., Vieira, P.C., Ryschawy, J. (2022). A serious game to design integrated crop-livestock system and facilitate change in mindset toward system thinking. *Agron. Sustain. Dev.*, 42, 35. doi: 10.1007/s13593-022-00777-5.

Moojen, F.G., Grillot, M., De Faccio Carvalho, P.C., Ryschawy, J. (2023). Farm advisors play a key role in integrating crop-livestock at the farm level: perceptions and experiences in Brazil and France. *The Journal of Agricultural Education and Extension*, 1–25. doi: 10.1080/1389224X.2023.2254308.

Moraine, M., Grimaldi, J., Murgue, C., Duru, M., Therond, O. (2016). Co-design and assessment of cropping systems for developing crop-livestock integration at the territory level. *Agricultural Systems*, 147, 87–97. doi: 10.1016/j.agsy.2016.06.002.

Moraine, M., Duru, M., Therond, O. (2017a). A social-ecological framework for analyzing and designing integrated crop–livestock systems from farm to territory levels. *Renew. Agric. Food Syst.*, 32, 43–56. doi: 10.1017/S1742170515000526.

Moraine, M., Melac, P., Ryschawy, J., Duru, M., Therond, O. (2017b). A participatory method for the design and integrated assessment of crop-livestock systems in farmers' groups. *Ecol. Indic.*, 72, 340–351. doi: 10.1016/j.ecolind.2016.08.012.

Moraine, M., Ryschawy, J., Napoleone, M., Ramonteu, S., Choisis, J.-P. (2020). Complémentarités culture-élevage à l'échelle territoire : facteurs de déverrouillage et de pérennisation des projets collectifs. *Innov. Agron.*, 80, 99–112.

Pissonnier, S. (2024). Diagnostiquer et concevoir des interactions cultures élevage à l'échelle du territoire : bilan de trois ans de partenariats entre AgroParisTech et l'AOP Chaource. *RMT SPICEE*, March, Montpellier [Online]. Available at: https://interactions-culture-elevage.colloque.inrae.fr/content/download/721/5961?version=1.

Ryschawy, J., Charmeau, A., Pelletier, A., Moraine, M., Martin, G. (2018). Dynamix, un "jeu sérieux" pour concevoir des scenarios d'intégration culture – élevage à l'échelle du territoire. Une application en Ariège. *Fourrages*, 207–212.

Ryschawy, J., Grillot, M., Charmeau, A., Pelletier, A., Moraine, M., Martin, G. (2022). A participatory approach based on the serious game Dynamix to co-design scenarios of crop-livestock integration among farms. *Agric. Syst.*, 201, 103414. doi: 10.1016/j.agsy.2022.103414.

Soussana, J.-F. and Lemaire, G. (2014). Coupling carbon and nitrogen cycles for environmentally sustainable intensification of grasslands and crop-livestock systems. *Agric. Ecosyst. Environ.*, 190, 9–17. doi: 10.1016/j.agee.2013.10.012.

Sumberg, J. (2003). Toward a dis-aggregated view of crop–livestock integration in Western Africa. *Land Use Policy*, 20, 253–264. doi: 10.1016/S0264-8377(03)00021-8.

Tichit, M., Magda, D., Durant, D., Lauvie, A., Lécrivain, E. et al. (2012). Systèmes d'élevage et biodiversité : des antagonismes aux synergies. *Rencontres Recherches Ruminants*, 19, 1–8.

4

Genetic Choices, Farming Systems and Territories

Etienne VERRIER[1] and Anne LAUVIE[2]

[1] *Paris Saclay University, AgroParisTech, INRAE, UMR GABI, Palaiseau, France*
[2] *Institut Agro Montpellier, INRAE, CIRAD, UMR SELMET, Montpellier, France*

4.1. Introduction

All livestock species display varying degrees of genetic diversity. This diversity can be split into two components: (i) diversity between populations, which have differentiated as a result of domestication, selection by breeders and the movement of human communities, and (ii) diversity within each population. It is these two components of diversity that are used by breeders and their organizations to choose, at a given time, the breeding stock and/or genetic types that best correspond to selection and production objectives. In referring to the choices made by breeders and their organizations, we emphasize that these genetic choices involve both decisions taken at the farm level and reasoned choices at the level of populations as a whole (it is at this second level that decisions have the greatest impact on the evolution of populations). Because these levels of organization are interconnected, it is particularly relevant to consider both breeding practices at the farm level and at the level of animal population and territories. How do the genetic choices made at these two levels of organization and scale fit together? To what extent do the characteristics of the territories where animals are raised play a part in these choices? How much latitude do breeders have in these choices, and what weight do

Farms and Territory,
coordinated by Philippe MARTIN and Solène PISSONNIER. © ISTE Ltd 2025.

they carry in collective decisions taken at the population level? This chapter provides some answers to these questions, without claiming to be exhaustive, mainly based on the situation in France.

4.2. Organization of animal breeding and the role of farmers, farming systems and links to the territory

The mid-20th century saw the establishment of two significantly different forms of animal breeding organization (Fugeray-Scarbel 2020; Verrier 2020). The main factors that led to the establishment of one or the other form were the cost of maintaining animals, the fecundity of females and the speed of generation renewal, all of which conditioned whether or not a private company could invest in breeding herds (Table 4.1).

In this field, we distinguish between a pyramidal organization specific to “small” animals and a mutualist organization specific to “large” animals. In both cases, it should be pointed out that investment in selection (animal identification, pedigree recording, performance recording, genotyping) does not concern all animals, but is concentrated within a nucleus, closed in the case of pyramidal organization, open in the case of mutualist organization.

“Small” animals include pigs (their speed of renewal offsets their rather high maintenance costs), rabbits, poultry and fish. For these species, selection is carried out by private companies (a few collective companies still exist for pork), which distribute breeding animals via a network of breeder–multipliers. In the context of a clear segmentation of the business, farmer-producers do not make any selection decisions, but buy “ready-to-use” animals from multipliers (only pig and rabbit farmers decide whether or not to cull their sows or rabbits).

Except in the case of fish, “commercial” animals, i.e. those not intended for breeding, are derived from double-stage crosses involving four different grandparental lines (poultry), or two grandparental lines and one paternal line (pigs). Farmers are not interested in the identity of the sires, but in whether they belong to one of these lines, and in the “composition” of the proposed cross. In the case of poultry and fish, farmers do not even breed their animals: these farmers, known as “growers” in aquaculture, do not carry out all the tasks usually associated with breeding.

Characteristics	Species or group of species						
	Equids	Cattle	Sheep and goats	Pigs	Rabbits	Poultry	Fish
Animal care costs	+++	+++	++	++	+	+	+
Fecundity of females	≈	≈	+	++	++	+++	++++
Rapid generational turnover	≈	≈	+	++	+++	+++	++
Organization of the breeding program	Mutualist	Mutualist	Mutualist	Pyramidal	Pyramidal	Pyramidal	Pyramidal
Selection nucleus status	Open	Open	Open	Closed (open)	Closed	Closed	Closed
Status of breeding organizations	Collective	Collective	Collective	Private (collective)	Private	Private	Private
Status of "commercial" animals	Purebred	Purebred (crossbred)	Purebred or crossbred	Crossbred	Crossbred	Crossbred	Purebred
Important concept	Sire	Sire	Sire (genetic type)	Genetic type	Genetic type	Genetic type	Genetic type
Farmers make selection decisions (self-renewal of females)	Yes	Yes	Yes	No	No	No	No
Farmers manage animal reproduction	Yes	Yes	Yes	Yes	No	No	No

Table 4.1. *Some characteristics of the more commonly domesticated species and the organization of the breeding programs*

"Large" animals include ruminants and equids, for which it is impossible for a single legal entity to invest in very large female herds, as the prospects of return on investment are too remote. Breeding programs are therefore based on a pooling of resources between farmers, of varying numbers, who are both producers and contributors to the programs. Selection is thus organized on a mutualist basis, with farmers (on a voluntary basis) being more or less closely involved in the main decisions. Farmers also ensure the self-renewal of their female herd, which leads them to make herd-wide selection decisions. In the French context, the majority of milked ruminant females are purebred (i.e. the sire and dam of an animal belong to the same breed), although the proportion of crossbred goats in the herd is not negligible and that of crossbred cows is slowly but steadily increasing. Among dairy cattle, the practice of crossing with beef breeds is increasing for calves not destined to remain on the farm, whether male or female. Suckling ruminant females are also predominantly purebred, but crossbreeding to produce "commercial" animals is common among non-specialized breeds, such as Aubrac and Salers cattle and Mérinos d'Arles sheep. On the whole, horse and cattle farmers remain very sensitive to the identity of the sires, while small ruminant breeders are relatively less so, given the generally large size of their flocks.

In fact, herbivores differ from other animal species in the way their breeding program is organized, and this difference has nothing to do with their diet. On the contrary, differences in diet lead to very marked differences in the nature and intensity of the biophysical link to the territory.

By their very nature, herbivores make the most of grasslands, and their use is set to increase in the future as part of the necessary sustainable agriculture transition. The use of fodder crops (silage maize, alfalfa, etc.) and concentrates bought in from outside is obviously very important, particularly in the major dairy basins (cattle, sheep, goats), but fodder is overwhelmingly produced on the farms themselves (the protein autonomy of herbivore farms in France is estimated to be 75% (Sailley et al. 2022)). The nature of the land and its agronomic characteristics therefore have a strong influence on the breeding program implemented and the aptitudes sought in the animals. In addition, ruminant breeding relies heavily on on-farm performance recording, which ensures that selection decisions are adapted to the conditions in which the offspring of breeding animals will be raised.

Excluding aquaculture, animals of other species are overwhelmingly (>95%) reared in buildings, with ambient temperature control. Most of the feed for these animals is purchased from outside sources, although some farmers make their own feed, at least in part, from crops grown on their own farm. While the territorial dimension in its human components (networks of actors) may play a part in the genetic choices made for animal populations of these species, the biophysical nature of the territory where these farms are located has little influence on genetic choices, all the more so as these choices are made within a closed selection nucleus maintained in highly controlled facilities. However, this does not mean that these "off-farm" activities have no impact on the local environment: for example, the management of livestock effluents on a territorial scale and the highly heterogeneous geographical location of these farms, both in France and across Europe, raise major questions that are beyond the scope of this chapter.

Fish, on the contrary, are raised in tanks, ponds or cages at sea, and it would be interesting to explore the specific features of these aquatic environments in a territorial approach. Water quality and temperature variations over the course of the year are obviously essential conditions for the establishment and operation of fish farms. However, the question of genetic types adapted to the average increase in water temperature, which has already been observed, essentially arises at the level of breeding companies.

As the territory cannot be reduced to its biophysical dimensions, territorial dimensions may of course also be relevant to pig and poultry breeding. Nor are we unaware of the fact that these species are also farmed in certain areas and with certain local breeds, the oldest example being the Bresse Protected Designation of Origin (PDO) poultry breed (see Verrier et al. 2005). However, in the remainder of this chapter, we focus on herbivore farming.

4.3. Collective breeding goals and territories

In France, as throughout Europe and much of the rest of the world, domestic herbivore species are organized into breeds. An animal breed is not a natural subdivision of the species but is the result of a human project and is based on the existence of a collective of breeders who agree on the orientation and selection of the breed (Lush 1994; Société d'Ethnozootechnie 1982; Audiot 1995; article 2.2 of Regulation 2016/1012 of the European Parliament and of the Council, dated June 8, 2016). Vissac (1993) points out that "we can [...] consider the animal population as an object of collective breeding goals and practices that modify, voluntarily or not, its biological content" and that "in this sense, animal populations can be considered as a reflection or biocultural indicator of the relations between a society and its territory" (authors' translation).

In France, the breeder collectives that manage the breeds are very generally set up as associations under the French Law of 1901, known generically as *Livres généalogiques* (stud-, herd-, flock-, etc. books), and they were central actors in animal breeding from the late 19th century to the 1960s. Today, the key actors are Selection Organizations (SOs), which bring together operators with different statuses (insemination cooperatives, breeders' associations, etc.). Taking into account the diversity of actors involved in the management of an animal population within these SOs potentially links breeders of a given breed to other actors in the territories where it is bred. For a given breed, there is generally only one SO, but some breeds with very large actual population sizes have several SOs, working on sub-populations of the breed: two SOs for the Holstein and Charolais cattle breeds, three SOs for the Montbéliarde cattle breed. Given a more liberal regulatory framework than was previously the case, and the evolution of selection tools, we can expect to see an increase in the number of SOs for breeds with large numbers of animals, as is already the case in several European countries.

For a given breed, the SO is responsible for carrying out all the stages of a breeding program but may delegate some of them to a third-party organization under contract. Among these various stages (see Verrier 2020), the first is the definition of breeding goals, which in practice remains the prerogative of the SO. Breeding goals are conceived as the list of traits on which a selection effort is desired and the hierarchy that is made between them (see, for example, Phocas et al. 2013). They are generally expressed in the form of a total merit index, i.e. a linear function that combines the values of animals for the selected traits assigned a relative weight that, more often than not, is defined on the basis of economic considerations (relative impact on an income function).

When we select a young breeding animal for renewal, whatever the criteria used to make this decision, we are not interested in them, but in their future progeny. When we decide to keep a breeding animal on a farm, we also consider their own aptitudes (adaptation to the environment, reproduction, production, etc.). In both cases, the selection criteria and objectives depend intimately on the breeding environment(s) and associated systems. This is an essential link to the biophysical and technical dimensions of the territory. It is easy to understand, for example, that the aptitudes sought in animals will not be exactly the same in an oceanic region with a high potential for forage crops, in a mid-mountain grassland area or on dry Mediterranean pastures. However, this link is expressed differently depending on the diversity of the areas farmed and the number of animals in the breed. Cultural dimensions may also come into play, particularly for all the selection criteria linked to a breed's standard.

Among the breeds of a given species, some are qualified as "local" within the meaning of article D.653-9 of the French Rural et de la Pêche Maritime Code: "a breed that is predominantly linked, by its origins, location and breeding method, to a given territory" (see Table 4.2; for the situation of pigs and poultry, see Verrier et al. 2023). With the exception of horses, the breeds recognized in France are mostly local breeds. However, the majority of livestock (80% of cattle, 95% of goats) are non-local breeds, with the notable exception of sheep, whose farming in France is mainly based on local breeds. Since the majority of local breeds are distributed over a limited territory (a geographical area ranging from a small agricultural region to one or two French departmental administrative districts), their numbers are generally moderate (tens of thousands of breeding females), or even low (thousands or hundreds), to the point where some are considered as "threatened with abandonment for agriculture". However, there are local breeds whose numbers can be counted in the hundreds of thousands (Aubrac and Salers cattle breeds, Blanc du Massif Central and Mérinos d'Arles sheep breeds), or even in the millions (Lacaune sheep breed). In practice, the management, selection and development of these local breeds show strong links to the land, which can take various forms and are the subject of particular development in the next section of this chapter.

	Species				
	Horses	Donkeys	Cattle	Sheep	Goats
Recognized breeds	49	8	54	59	15
Local breeds	18	8	32	48	11

Table 4.2. *Number of recognized breeds and local breeds of domestic herbivores in France in 2022 (source: French Ministry of Agriculture)*

The selection of non-territorialized or more territorialized "mainstream breeds" raises the question of how to take into account the great diversity of environments and farming systems. To illustrate these issues, we analyze the case of France's two main dairy cattle breeds, namely Holstein and Montbéliarde breeds (the third in terms of numbers, not considered here, is the Normande breed, which is also not qualified as local).

Out of a total of 3.6 million dairy cows in France in 2022, 2.3 million are from the Holstein breed (63%). This highly specialized breed is in fact the world's leading dairy breed, and is found in most countries, under extremely diverse farming conditions. Breeding goals are therefore highly country-dependent and can vary within a given country. Genetic management of this breed is carried out on a global scale, thanks in particular to an independent reference body, Interbull, based in

Uppsala (Sweden), which routinely carries out an international genetic evaluation of insemination sires, based on data sent in by each country. All over the world, therefore, collective breeding organizations can draw on a global pool of sires, with the USA, France, Germany, the Netherlands and Canada being the "heavyweights" in this field. For this cosmopolitan breed, the adaptation of breeding animals to different environments and farming systems can only be ensured if the genetic profiles of the sires used for insemination are sufficiently diverse, which can only be guaranteed by the international genetic evaluation mentioned above.

Despite its international success, the Montbéliarde breed is managed on a French scale (640,000 cows in France in 2022, i.e. 18% of the dairy herd). The historical region of origin of the breed is the Franche-Comté region, where it now accounts for some 95% of the dairy stock. In this region, the breed is associated with farming systems that are fundamentally grass-based, and with a predominantly cheese-making approach, notably in the context of several PDOs (Comté, Morbier, Mont d'Or, Bleu de Gex). Today, however, this region represents only about a third of the national herd of Montbéliarde cows. The breed has undergone a major geographical expansion, with a first wave starting in the 1950s in the Grand Est, Auvergne and Savoie (regions where it is on a par with the Holstein breed, or even in the majority), and a second wave starting in the late 1970s in the Grand Ouest (where it remains in the minority compared with the Holstein breed). Schematically, we can consider that the proportion of Montbéliarde cows in a given territory decreases when the environment and the farming system move away from what prevails in Franche-Comté. As a result, and because selection operators are located at the heart of the region of origin (even though their action extends far beyond it), the breeding goals for the Montbéliarde breed are still mainly guided by the Franche-Comté "model", with, in particular, a higher relative weight than in any other breed to the milk protein content in the total merit index, in line with the breed's cheese-making vocation. This does not exclude the breeding companies offering a varied range of insemination bulls to satisfy the diverse demands of their users.

4.4. Breeds, farmers' collectives and other territorial components: diversity of relationships

The term "local" used for many breeds refers directly to the place in which they are raised. As we saw in the previous section, according to the regulatory definition, local breeds are predominantly or very predominantly established in a defined territory and, with a few exceptions (see above), have moderate to low actual population size. In many cases, particularly in the dairy sector, their breeding program uses genomic tools. For breeds with low numbers of breeding animals, programs are essentially aimed at preserving intra-population genetic variability and limiting the rate of inbreeding, and, with a few exceptions (such as the Vosgienne cattle breed), do not currently involve the use of genomics.

In all cases, the management of these breeds is part of a territory. It involves different levels of organization: genetic improvement or genetic variability management programs at the level of the animal population, and herd or flock management at the level of the farming system. Management articulates decisions made by individual farmers and collective decisions (Lauvie et al. 2015). The characteristics of farming territories can play a role, to a greater or lesser extent, in the decisions taken by the farmers and breeder collectives involved in this management. Furthermore, numerous approaches to adding value to these breeds have sometimes been developed through products under quality or origin labels (SIQO, in French), some with success stories and others without (VARAPE 2014; Verrier et al. 2021). Recent summer droughts, due to climate disruption, have nevertheless highlighted a new issue of adaptability in situations where this triptych is formalized in a specification: the possibility of derogations, or even the modification of these specifications, thus become issues to be investigated.

Emblematic of the region–breed–product triptych, among the breeds with a breeding program are the Abondance and Tarentaise cattle breeds in the Northern Alps, included in the specifications of four Savoy PDO cheeses (Verrier et al. 2005); the Aubrac cattle breed on the ancient volcanic massif of the same name, which supports a flourishing export sector of Charolais crossbred calves and a local Label Rouge (Borrès et al. 2019, Chapter 4); and the Gasconne cattle breed, which also has a Label Rouge. However, this triptych is not a "universal model", and the relationships between breeds, their farm products and territories can take a variety of forms, as illustrated by the contrasting examples developed below.

Among the breeds under conservation is the Bretonne Pie Noir breed, the "cow of happy farmers" (Bourgault and Quéméré 2016). In addition to a product from this breed benefiting from a collective brand, Gwell®, there are many other products, resulting from on-farm processing practices leading to varied ranges depending on the farmer (Morinière 2017), and sometimes mobilizing processing know-how from other territories (Couix et al. 2016). Lambert-Derkimba et al. (2013) provide a comparison of several cases of local cattle breeds and the impact on their genetic management of the dynamics of economic valorization in their territory.

In terms of volume, the Pyrénées-Atlantiques departemental administrative district is France's second-largest ewe's milk production basin (after the Rayon de

Roquefort and ahead of Corsica). Three local breeds are found in the district, following a west–east gradient from the plains to the high mountains (figures are for 2022, source: Institut de l'Élevage): Manech Tête Rousse (290,000 ewes), Manech Tête Noire (71,000 ewes) and Basco-Béarnaise (89,000 ewes). The milk is processed into various types of cheese, including Ossau-Iraty PDO (5,000 tons produced in 2020 according to the INAO), whose specifications have required the use of milk from ewes belonging to one of these three breeds since 1996. An SO common to the three breeds implements a breeding program for each of them, with a marked dairy orientation (yield and content) inspired by the approach that previously prevailed for the Lacaune breed and the "Roquefort system" (Casabianca and Millet 2018). This orientation mobilizes scientific and technical tools with a "universal" vocation, which Labatut et al. (2012) point out can come into tension with their objectives and practices for some farmers, as is the case for Manech Tête Noire farmers who base their systems on the valorization of altitude pastures and do not mobilize these tools. The authors report on the questioning expressed by some of these farmers as to the hardiness of the animals selected collectively and emphasize that such tensions can undermine cooperation in the collective management of a breed.

The Chèvre des Pyrénées is a local goat breed with a small actual population size found mainly, as its name suggests, throughout the Pyrenees. The association that manages it plays a coordinating role between farmers who may be geographically far apart. Most of the farmers sell products from on-farm milk processing, or meat products, via short distribution channels. The breed, which has remained mixed, is used either in dairy or suckling systems. While food product development is more the result of individual initiatives by these farmers, this does not preclude collective reflection on the breed's development, supported by the association. Among other things, the association has produced common communication tools: leaflets presenting the breed, wrapping paper in the breed's colors, stickers, promotional panels for farms, etc. (VARAPE 2015). On a collective scale, a reflection has also been carried out on the breed's ability to make the most of overgrown environments. The association produced a report entitled "*L'aptitude de la chèvre des Pyrénées à valoriser la broussaille : une opportunité pour les éleveurs et pour le territoire*" ("The ability of pyrenean goats to make the most of scrubland: an opportunity for breeders and the region"), based on a variety of testimonials from farmers (La Chèvre des Pyrénées 2023). In this way, the breed helps to bring together a variety of regional components: dairy and suckler farmers, environments enhanced by grazing, food products sold in circuits that are often local, etc.

The case of the Blanc Bleu Belge cattle breed illustrates the mechanisms by which a breed becomes part of a territory, through dimensions other than the latter's biophysical characteristics. This breed is hyperspecialized for meat production, with most animals displaying a muscular hypertrophy commonly referred to as the "*double-muscle trait*" due to the action of a major gene. In Belgium, this breed is in the majority on beef production farms. Stassart and Jamar (2009) point out that it is associated with growth-finishing systems that rely on concentrates and confined spaces. The relationship with the land is therefore not built on its biophysical dimension. The same authors emphasize that the breed is the focus of a wide range of actors in the Belgian territory: farmers, butchers, veterinarians, consumers, etc. A model is thus constructed, bringing into play standards (addition of the super-double-muscle category to the carcass grading standard, the quest for lean, tender meat, etc.), and the knowledge and know-how of these actors in the territory (Belgian cutting of meat, known as anatomical cutting, widespread use of Cesarean sections, etc.). The authors show how the construction of such a frame of reference around the breed can lead to lock-in, making it more difficult for local breeders to implement other systems (they illustrate this through the example of organic farming systems) and the eventual use of another breed.

4.5. Individual farmers' decisions and how they relate to breeding collectives' decisions

It is well established (see, for example, Elsen 2020) that it is decisions taken within a collective framework that have the greatest impact on the genetic evolution of a breed: definition of breeding goals, recruitment of young male breeding candidates, nature of the selection tests through which they must pass, choice of breeding males, especially those to be used via artificial insemination. In this context, what levers do farmers have at their disposal to make their mark on the genetic make-up of their herd, and how do their decisions depend on the territory in which their farm is located? As is the case on a collective scale, individual farmers can take advantage of the genetic differences that exist between populations or within a given population.

On an inter-population scale (which breed, which crossbreeding plan?), individual choices only come into play at a few key moments in a farmer's career: setting up as a young farmer, starting up a new breeding operation, possibly changing breed along the way. In the case of a family inheritance, the question of choosing a breed may not arise, and continuity is often the order of the day. However, setting up a business can be an opportunity to reconsider the choices made by the previous generation. The question of changing breed along the way generally arises when a farmer perceives major limitations in the herd management and attributes them to the breed.

Whether it is an installation with a new breed or a breed replacement, Couix et al. (2016) interpret this choice as resulting from an "inquiry" carried out by the farmer, an "inquiry" in the sense of Dewey (1938) and considered as one of the figures of adaptation by Zask (2008). These authors illustrate this point with a study carried out in the dairy basin of western France. This vast group of three regions (Brittany, Normandy, Pays-de-la-Loire) accounts for around half of France's dairy cattle herd, three-quarters of which is made up of Holstein cows. Interviews were conducted with two categories of farmers who had made a different breed choice. On the one hand, some 40 farmers who had replaced their Holstein cows with cows from the Montbéliarde breed or the French Simmental breed explained their choice as (i) an alternative to a deteriorating situation (notably declining cow fertility), (ii) a quest for a greater milk–meat mix and (iii) an evolution of their farming system towards more grazing and fewer inputs. On the other hand, some 30 breeders who had settled with an endangered local breed, the Bretonne Pie Noir breed, put forward (i) heritage considerations (conservation of a breed), (ii) the hardiness and ease of handling of the animals within the framework of extensive, low-cost farming systems, as an integral part of a life project and (iii) the quality of the breed's products, valorized through short circuits.

Within a population, individual choices are expressed by a farmer through the culling of certain females and in the self-renewal of females in the herd. As culls are often sudden events (reduced fertility, drop in production, accident, etc.), the room for maneuver lies mainly in renewal, with two questions for the farmer to answer. First, at the herd level: which of the females' present should be chosen to produce offspring that will be kept in the herd? This question can be completed, still at the herd level, by the question of individual choice among new-born females, for which early genotyping represents a new tool. Another question relates to the collective scale: which of the breeding stock available externally, particularly insemination males, should be selected as the fathers of future breeding females?

Depending on their idea of the "ideal" animal and the territorial, social and economic context of their farm, farmers may base their choices on their own observations and/or on published genetic value indexes (for evaluated traits and animals with the necessary information). Perucho et al. (2020) illustrate the diversity of selection criteria used to qualify a "good ewe" for the renewal of the flocks, through survey results from eight farmers of Corsican ewes. These decisions may combine (i) criteria assessed using collective evaluation tools (dairy production evaluated by milk recording and index from a genetic evaluation system), and (ii) criteria chosen and evaluated on a more individual basis (e.g. conformity to breed standard or milking behavior). These authors also analyze the links between individual choices of breeding stock, whether or not farmers are part of the selection scheme, and the characteristics of their breeding systems. In another context, Bérodier et al. (2022) analyze the possibilities offered by the diversity of genetic profiles of Montbéliarde insemination sires for each farmer to build their own sorting grid beyond the objectives defined for the breed as a whole.

In addition to field surveys, national zootechnical databases can shed light on the choice of breeding stock made by farmers. For example, Lambert-Derkimba et al. (2010) analyzed variations in the genetic value indexes of cows from the three main dairy breeds present in the Savoie and Haute-Savoie departmental administrative districts: the Montbéliarde breed mentioned above, and the Abondance and Tarentaise local breeds. At the level of the communal district, the two variation factors considered were classification according to the European mountain policy nomenclature (in ascending order of natural constraints: plain, piedmont, mountain, high mountain) and the existence of a PDO cheese specification (no PDO, PDO without limit on the average milk yield of cows, PDO with a milk yield limit). No effect was found for the Tarentaise breed, which had the lowest numbers of cows and the most unbalanced distribution of these numbers between the different combinations of factors. On the contrary, a significant decrease in the average genetic index of cows for milk yield was observed for the Abondance and Montbéliarde breeds as a function of the degree of constraint linked to a PDO and, only for the Montbéliarde breed, as a function of the degree of constraint linked to the mountain. The differences observed in these cases can be interpreted as a consequence of the farmers' previous choices of breeding animals, especially artificial insemination bulls.

4.6. Conclusion

Genetic management of livestock populations involves decisions made at both farm and population levels. Both the farm and breed territory scales of analysis need to be taken into account, as well as the interrelationships between the dynamics at these two scales. The human, biophysical and technical dimensions of the territories where animals are raised potentially play a part in collective genetic choices as well as those of individual breeders.

Defining collective orientations in terms of genetic management of animal populations that are suitable for a diversity of farmers implies considering the articulation between scales and levels of organization. Recent developments in selection tools and methods, the diversification of the issues to which livestock farming is subject, and the dynamics of global change impacting the territories in which livestock farming is deployed are all important contextual elements and challenges to be taken into account in a forward-looking vision of the articulation between levels of organization in animal genetic management practices. From this point of view, in contrast to the tendencies towards homogenization and standardization of production that have prevailed over the last decades, diversity remains the keyword: diversity of territories, practices and breeding systems, genetic types of animals, products, actors, etc.

4.7. References

Audiot, A. (1995). *Races d'hier pour l'élevage de demain.* INRA Editions, Paris.

Bérodier Gomès, M., Brochard, M., Ducrocq, V. (2022). On farm specific breeding goals for mating lead to higher genetic gains at herd level. In *World Congress on Genetics Applied to Livestock Production*, July, Rotterdam, 1995–1998. doi: 10.3920/978-90-8686-940-4.

Borrès, J.B., Béranger, C., Bonnemaire, J., Devienne, S., Lacombe, P., Rouquette, J.L. (2019). *L'Aubrac. 50 ans de développement. L'élevage à la rencontre des enjeux de territoire.* Éditions de l'Aube, PNR de l'Aubrac.

Bourgault, P. and Quéméré, P. (2016). *Bretonne Pie-Noir. La vache des paysans heureux.* Ouest France Éditions, Paris.

Casabianca, F. and Millet, M. (2018). La ressource génétique locale sous tensions : trajectoires d'évolution des races ovines laitières Corse et Pyrénéennes, entre modèles d'élevage et valorisation collective des fromages de terroir. *Géocarrefour*, 92. doi: 10.4000/geocarrefour.12137.

Couix, N., Gaillard, C., Lauvie, A., Mugnier, S., Verrier, E. (2016) Des races localement adaptées et aussi adoptées, une condition de la durabilité des activités d'élevage. *Cahiers d'Agriculture*, 25, 650009.

Dewey, J. (1938). *Logic: The Theory of Inquiry.* Holt, Rinehart and Winston, New York.

Elsen, J.M. (2020). Le choix des reproducteurs et l'efficacité de la sélection. In *Amélioration génétique des animaux d'élevage*, Verrier, E., Milan, D., Rogel-Gaillard, C. (eds). Éditions Quæ, Versailles.

Fugeray-Scarbel, A., Labatut, J., Lemarié, S., Trommetter, M. (2020). Modèles économiques et formes d'organisation de l'amélioration génétique animale. In *Amélioration génétique des animaux d'élevage*, Verrier, E., Milan, D., Rogel-Gaillard, C. (eds). Éditions Quæ, Versailles.

La Chèvre des Pyrénées (2023). La Chèvre des Pyrénées, page de recherche par mots-clés [Online]. Available at: https://www.chevredespyrenees.org/wordpress/wp-content/uploads/2021/02/Publi-Aptitudes-Pyr%C3%A9n%C3%A9ennes-compress%C3%A9.pdf [Accessed 20 April 2023].

Labatut, J., Bibé, B., Aggeri, F., Girard, N. (2012). Coopérer pour gérer des races locales : conception, rôles et usages des instruments scientifiques de sélection. *Natures Sciences Sociétés*, 20, 143–156.

Lambert-Derkimba, A., Minéry, S., Barbat, A., Casabianca, F., Verrier, E. (2010). Consequences of the inscription of local breeds in protected designation of origin cow cheese specifications for the genetic management of the herds. *Animal*, 12, 1976–1986.

Lambert-Derkimba, A., Lauvie, A., Verrier, E. (2013). How the development of products valorizing local breeds changes breeding goals: Examples from French cattle breeds. *Animal Gentic Resources*, 53, 135–140.

Lauvie, A., Paoli, J.C., Moulin, C.H. (2015). Managing local breeds: A dynamic connected to livestock farming systems that concerns different levels of organization. *Animal Genetic Resources*, 59. doi:10.1017/S2078633614000502.

Lauvie, A., Alexandre, G., Couix, N., Markey, L., Meuret, M., Nozières-Petit, M.O., Perucho, L., Sorba, J.M. (2017). Comment les diverses formes de valorisation des races locales interagissent avec leur conservation ? *Ethnozootechnie*, 103, 7–12.

Lush, J.L. (1994). *The Genetics of Populations*. Iowa State University, Ames [Original published 1948].

Morinière, C. (2017). Quand la valorisation alimentaire encourage la conservation d'une race : le cas de la race Bretonne Pie Noir. *Ethnozootechnie*, 103, 13–18.

Ovaska, U. and Soini, K. (2016). Native breeds as providers of ecosystem services: The stakeholders' perspective. *Finnish Journal of Human-Animal Studies*, 2 [Online]. Available at: https://trace.journal.fi/article/view/52739.

Perucho, L., Paoli, J.-C., Ligda, C., Moulin, C.-H., Hadjigeorgiou, I., Lauvie, A. (2020). Diversity of breeding practices is linked to the use of collective tools for the genetic management of the Corsican sheep breed. *Italian Journal of Animal Science*, 19, 158–172. doi: 10.1080/1828051X.2020.1713027.

Phocas, F., Brochard, M., Larroque, H., Lagriffoul, G., Labattut, J. (2013). Etat actuel et perspectives d'évolution des objectifs de sélection chez les ruminants. *Renc Rech Ruminants*, 20, 129–132.

Sailley, M., Seegers, J., Rouillé, B., Pavi, J. (2022). Autonomie protéique : où en sont la France et l'Europe ? Dossiers techniques de l'élevage – Vers l'autonomie protéique en élevages de ruminants, Institut de l'Élevage, 7–9.

Société d'Ethnozootechnie (1982). Le concept de race en zootechnie. *Ethnozootechnie*, 29, 72.

Stassart, P. and Jamar, D. (2009). Agriculture biologique et verrouillage des systèmes de connaissances. Conventionnalisation des filières agro-alimentaire bio. *Innovations Agronomiques*, 4, 313–328.

VARAPE (2014). Le guide VARAPE : la via ferrata de vos projets collectifs [Online]. Available at: https://idele.fr/detaildossier/le-guide-varape-la-via-ferrata-de-vos-projets-collectifs.

VARAPE (2015). La chèvre des Pyrénées. In *Compilation des synthèses 4 pages. Races partenaires du projet VARAPE*, 42–45 [Online]. Available at: https://idele.fr/fileadmin/medias/Documents/2015_Compilation__2014pages_2013races.pdf [Accessed 20 April 2023].

Verrier, E. (2020). Les différentes étapes d'un programme d'amélioration génétique. In *Amélioration génétique des animaux d'élevage*, Verrier, E., Milan, D., Rogel-Gaillard, C. (eds). Éditions Quæ, Versailles.

Verrier, E., Tixier-Boichard, M., Bernigaud, R., Naves, M. (2005). Conservation and values of local livestock breeds: Usefulness of niche products and/or adaptation to specific environments. *Animal Genetic Resources Information*, 36, 21–31.

Verrier, E., Markey, L., Lauvie, A. (2021). Valorisation des races animales locales par des produits liés à leur territoire : cinq leçons tirées de cas variés en France. *Essais Hors-Série*, 6, 35–45 [Online]. Available at: https://journals.openedition.org/essais/7166.

Verrier, E., Badeau F., Basso B., Bertrand C., Boudry P., Bouffartigue B., Charvolin-Lemaire E., Danvy S., Decourtye A., Duclos D. et al. (2023). Les races locales menacées d'abandon en France : actualisation des listes et extension de la démarche à de nouvelles espèces. *Ethnozootechnie*, 112, 93–110.

Vissac, B. (1993). Société, race animale et territoire : entre les théories et l'histoire, réflexion sur une crise. *Natures Sciences Sociétés*, 1, 282–297.

Zask, J. (2008). Situation ou contexte ? Une lecture de Dewey. *Revue Internationale de Philosophie*, 3, 313–328.

5

Town Microfarms, Country Microfarms in Fields: Innovative Systems in Contrasting Territories

Agnès LELIÈVRE and Kevin MOREL
Paris Saclay University, AgroParisTech, INRAE, UMR SADAPT, Palaiseau, France

5.1. Innovative projects in response to a variety of socio-ecological territorial challenges

Over the past 10 years or so, the microfarm has become a fast-growing concept both in back-to-the-land movements in rural areas (Allens and Leclair 2016; Bedouet et al. 2017; Hervé-Gruyer and Hervé-Gruyer 2019; Morel 2019) and in intra- or peri-urban agricultural dynamics (Aubry et al. 2022).

5.1.1. *Emerging concept of alternative agricultural establishments in back-to-the-land farming dynamics in rural areas*

In France, the number of farms and agricultural jobs is steadily declining (Agreste 2020). The disappearance of farmers goes hand in hand with the expansion of farms, reinforcing industrialization dynamics that are hardly conducive to the sustainable agriculture transition. The renewal of agricultural generations is therefore a key challenge if we are to contribute to the socio-economic dynamism of

Farms and Territory,
coordinated by Philippe MARTIN and Solène PISSONNIER.

rural areas and support a transition towards greater sustainability. In this context, agricultural establishment carried out by new entrants with no agricultural background can play a major role. At present, such projects account for around a third of all agricultural establishments in France (Agreste 2020). Among them, market gardening, organic farming and sales through shorter delivery routes (one intermediary at most between producer and consumer) are the most frequent orientations.

Among these new establishments, microfarms are gaining in popularity. Initially popularized by pioneers (Fortier 2012; Hervé-Gruyer and Hervé-Gruyer 2014), microfarms are alternative farms characterized by:

– *organic vegetable production (market gardening)* as the main income-generating activity;

– *small, cultivated areas* compared to what was commonly accepted as viable in organic market gardening: less than 0.7–1 ha of cultivated area per worker;

– *a high level of diversification*: more than 20–30 types of vegetables, often sold in short circuits;

– *low motorization*: practices based on lighter motorization (rototiller, small tractor) or manual labor (Morel 2016).

These projects are most often led by new entrants with no agricultural background, for whom a return to the land is motivated by strong ecological and social aspirations. In this respect, microfarms are seen as a response to a number of challenges facing rural areas:

– create jobs in the countryside by enabling low-investment start-ups on very small plots of land; this is suitable for new entrants to farming for whom access to land remains a major difficulty, with little start-up capital and/or who do not wish to take on the debt burden of loans needed to take over farms, requiring ever-increasing investment;

– revitalizing the countryside and providing healthy local food via short organic farming circuits;

– experimentation with alternative practices aimed at preserving biodiversity and resources, drawing on sources of inspiration such as permaculture, biointensive market gardening and living soil market gardening (see section 5.2.2).

Initially, microfarms were *alternative farms in rural areas* whose main source of income was production. However, they are increasingly seen as ideal candidates for urban agriculture, due to their small size and claimed ecological dimension, which has led to an evolution of the original concept.

5.1.2. *The evolution of the microfarm concept in the urban context*

The constraints and opportunities of the urban environment have led to an evolution in the concept of the urban microfarm (Daniel 2017; Chang and Morel 2018). In the city, agriculture can be set up in interstices on spatially or temporarily limited scales: the foot of buildings, rooftops, playgrounds, for example, or wasteland awaiting construction. It is also a way of adding value to unused areas. These specific features, together with the surrounding population, mean that microfarms are multifunctional, responding to the needs of the city. Urban microfarms claim to contribute to the sustainability of the city. This can take the form of various environmental services, such as recycling city waste (compost, organic waste, building waste, etc.), retaining rainwater to limit the overloading of stormwater systems and promoting biodiversity. Microfarms also make a strong contribution to cultural and socio-economic needs: agriculture as a means of integration or socialization (conviviality, events), well-being (landscape) and access to healthy food. The educational objective of farms is also often a major component of their construction, to raise awareness of the environment and food for all publics. One of the major concerns and constraints of urban microfarms remains the question of pollution (soil and air) and its impact on harvests and possible production methods (above ground or high up, for example). A possible negative consequence of setting up these microfarms is gentrification of the neighborhood and exclusion of vulnerable populations, which has been observed in New York in some cases (Cohen and Reynolds 2016).

These projects integrate a multitude of social, cultural and ecological activities, in addition to the sale of their food production, which represents a variable part of their income. While they pay at least one salaried employee, these urban microfarms are participatory and rely heavily on volunteer labor, whether for setting up the farm (e.g. building farm facilities) or crop maintenance (Daniel 2017). This can be both an objective of the project and a limitation to its operation, with the need for technical supervision or training time. Often in the form of associations, these structures depend heavily on partnerships (public or private) for access to resources (land, subsidies). Unlike rural microfarms, which are exclusively open-ground, urban microfarms can integrate open-ground, rooftop, container and basement spaces to adapt to a dense urban environment (Grard and Ughetti 2020).

Therefore, the same term "microfarm" can cover different realities in different areas:

– the microfarm as a market garden on a small plot of land, usually in a rural environment;

– the participatory microfarm, whose economic model is based on a variety of activities in addition to agricultural production.

We now explore how these initiatives, responding to the challenges of contrasting territories, are accompanied by changes or innovations at different levels.

5.2. Multiple innovations driven by new farmer profiles

5.2.1. *Profile and agricultural establishment trajectories*

In the EU, a recent report noted the growing proportion of new entrants to agriculture and their higher level of education, a more balanced male–female ratio and a choice to set up on small areas and with alternative farming systems (EIP-Agri FG 2016). Whether on rural or urban microfarms, it is worth noting the high proportion of professional reconversion in farm establishment projects, often with a highly qualified profile (often three to five years of study after the baccalaureate in themes sometimes related but often far removed from agriculture). For example, people taking a BPREA[1] course in agriculture at the Ecole du Breuil (Paris), training in market gardening and urban agriculture, are mostly from backgrounds unrelated to agriculture and the environment and may have worked for almost 10 years before taking this course.

On rural microfarms, we can see forms of gradual establishment (with income reinvested in equipment in the first few years), recovery and self-building to limit borrowing (Morel 2016). One point that remains uncertain is the long-term future of urban farmers. Some see this type of farming only as a halfway house before moving on to peri-urban or even rural areas, but always in microfarms or with strong environmental and social objectives in small-scale farms. All are keen to help recreate a link between urban dwellers and farmers outside the city.

For all these new farmers, it is necessary to acquire training and experience before setting up. For urban agriculture and market gardening, training can be short or long, with or without a diploma (AFAUP 2022). After initial or continuing training, the path to setting up an urban or rural microfarm is fairly diversified (SOL 2020) but involves several stages of project construction and maturation prior to agricultural establishment.

There are several ways to mature and test your project. Support can be provided at various stages by agricultural development organizations such as the Chambers of Agriculture (a public advisory and development agency) via dedicated services

1. Brevet Professionnel Responsable d'Entreprise Agricole: Professional Certificate of Agricultural Business Management.

supporting new entrants with agricultural establishment projects, by the FNAB[2], FNCIVAM[3] or FADEAR[4] networks, or via the Cerfrance network agencies, which provide support to all farmers. As the majority of microfarms are involved in market gardening, they can also mobilize the test areas of the Reneta national network (https://reneta.fr/), which can provide a legal structure, means of production and administrative and technical support so that project leaders can experiment with an agricultural activity before setting up in business. In some regions, those involved in setting up alternative agriculture have formed associations to facilitate the process, such as the INPACT[5] clusters or, for example, the Abiosol network in Île-de-France, which brings together the Community Supported Agriculture (CSA) network and Terres de Lien, enabling farmers to test their project in a test-space cooperative, Champs des Possibles, which belongs to the Reneta network. Abiosol organizes welcome meetings, training courses, farm visits and project support in the form of a business incubator, which enables farmers to test their project over a one-year period, renewable twice, within a secure legal and material framework, with established farmers mentoring new test farmers. The test is taking place on the Toussacq farm in the south of the Seine et Marne region (77). Some 10 entrepreneurs are running their own farming and food businesses there (plant and animal production, primary processing).

In Besançon, a market gardening incubator has been set up by the city council, in partnership with the Chamber of Agriculture, the SAFER[6] and the Jardins de Cocagne network.

Another possibility is for farmers to work as salaried employees before setting up in business. Public policies, with the launch of Territorial Food Projects (Projets Alimentaires Territoriaux, PATs), have helped to make local authorities aware of the importance of agriculture in their area, which can lead to the development of alternative farming initiatives.

---

2. Fédération Nationale de l'Agriculture Biologique: National Federation of Organic Agriculture.

3. Fédération Nationale des Centres d'initiatives pour valoriser l'agriculture et le milieu rural: National Federation of Initiative Centers to Promote Agriculture and the Rural Environment.

4. Fédération Associative pour le Développement de l'Emploi Agricole et Rural: Associative Federation for the Development of Agricultural and Rural Employment.

5. See: https://www.agricultures-alternatives.org/.

6 Société d'Aménagement Foncier et d'Etablissement Rural: Land Development and Rural Development Company.

5.2.2. *Innovations in farm design and management*

From a technical point of view, microfarms are developing innovative practices aimed at intensifying production on small areas. However, these alternative projects are generally strongly marked by ecological or social objectives, which translate into agroecological approaches aimed at enhancing biological regulation by promoting biodiversity and relying as much as possible on local resources to renew fertility. These local resources may come from the farm itself and also from the surrounding area (organic matter, green waste compost, manure, etc.). Compromises are sometimes necessary. For example, for a minority of microfarms, intensification per unit area may be based on the use of hybrid plants fertilized with commercial fertilizers, which is in tension with the quest for autonomy on a local scale.

Microfarms are often inspired by one or more of the following approaches:

– *Permaculture*: a global approach developed during the late 1970s in Australia that proposes design principles for consciously conceiving sustainable human landscapes that mimic natural ecosystems (Holmgren 2014); in the context of a microfarm, this approach may translate, for example, into the interweaving of an intensely cultivated market garden core with a diversity of spaces: orchards, small livestock, ponds, natural areas, etc., to enhance ecological synergies (Hervé-Gruyer and Hervé-Gruyer 2014).

– *Biointensive market gardening*: this approach, inspired by 19th-century Parisian market gardeners, aims to (i) achieve intensification on a small surface area, based on rapid crop successions with a varying proportion of short-cycle, high value-added crops, (ii) achieve spatial densification through manual labor and possibly intercropping, (iii) pay attention to winter cultivation through the use of sheltered areas (greenhouses, canopies, tunnels) and adapted varieties, (iv) reduce costs thanks to low-mechanization systems, (v) increase efficiency for manual labor thanks to ergonomic low-tech tools and (vi) limit tillage on permanent beds (Fortier 2012).

– *Market gardening with a "living soil"*[7] *approach*: an approach that pays close attention to the soil, combining various principles: (i) reduced tillage or no-till, (ii) significant application of organic matter, (iii) plant cover crops, (iv) weed management and soil cover using plastic sheeting and (v) integrated, preventive plant health management.

In addition to these low-tech practices, some urban microfarm projects can be sites for the development of more high-tech techniques, from the use of drones to take photos of sites and communicate, particularly for sites whose economic model

7. Known as *Maraîchage sur Sol Vivant* in French.

relies heavily on communication (events, marketing to attract new customers, etc.), to the development of more specific tools such as harvesting robots or weeding robots.

The constraints of urban agriculture also mean that practices have to be adapted, whether it is the use of containers to avoid soil pollution or the elevation of growing areas to produce table-top products for certain audiences (e.g. the elderly). The type of substrate can also be modified according to available inputs (coffee grounds, local mushroom or brewery waste, green waste and wood shavings from the town, etc.) or the load-bearing capacity of roofs (search for lighter substrates containing expanded concrete or clay balls, for example).

On urban microfarms, with limited surface areas and manual harvesting, it is easier to grow associated crops to take advantage of the beneficial interactions between crops, as well as the temporality of production. However, other constraints can limit the choice of crops: the aesthetic aspect of the farm, the educational aspect and the choice of "unprofitable" crops such as peas, which are very labor-intensive but highly appreciated and expected by consumers. Crop diversity can also be more limited, given the depth of substrate available in tubs or on rooftops. Nevertheless, more "exotic" crops may find their place, such as okra or "greens" (leaf vegetables), which will have a connoisseur audience in the city and can take advantage of the urban heat island. There are also niche products with a high added value, such as edible flowers and microgreens. Alongside these urban microfarms, there are also more high-tech systems, such as indoor vertical farms, aeroponics (substrate-free production with fertilization of roots) and hydroponics (substrate-free or inert substrate production with fertigation). Some microfarms are also adding an aquaponics component (a combination of aquaculture and hydroponics production) to their systems for educational purposes (Aubry et al. 2022).

Figure 5.1. *Intercropping on no-till soil in the greenhouse of a microfarm in Brittany. Photo credit: Kevin Morel*

Figure 5.2. *Permanent beds (involving fixed paths to limit compaction and shallow tillage to improve soil structure) on a microfarm in Lorraine. The dimensions of all the beds are identical, adapted to the width of the hand tools or those towed by the rototiller and to the length of the plastic sheeting and aiming to simplify the organization of rotations. Photo credit: Kevin Morel*

5.2.3. *Business models*

For microfarms whose income comes mainly from production (in rural areas, for example), the search for economic viability is generally based on a combination of several technical and commercial leverages:

– increase production per unit area by intensifying cultivation in time and space (often involving a larger share of greenhouses than in conventional diversified market gardening);

– increase added value by marketing through short distribution channels and possibly by excluding long-cycle, low-added-value crops from their range (e.g. ware potatoes);

– limit input costs by relying on biological regulation and local resources to renew fertility;

– limit investments with low-tech practices and limited motorization (N.B. not all microfarms combine all these aspects exhaustively; Morel 2016).

In urban environments, the search for added value can lead to specialization in an even more restricted range of vegetables or herbs (e.g. greens aromatics) or in niche markets (catering, delicatessen, etc.). For more participative or associative farms, the use of volunteer labor is also an important economic lever, combined with the fact that the microfarm's income comes from a variety of activities in addition to production (Saint-Ges 2021), such as educational activities (welcoming schoolchildren, training, internships), social activities (integration, hospitality, team-building), cultural activities (events) or services (e.g. expertise in urban garden design). The combination of these different activities in urban microfarm business models and the share of production in profitability vary from case to case. This combination can lead to significant management complexity, requiring a wide range of skills and adapted organizational modes (Chang and Morel 2018). In some cases, multi-activity microfarms with a social vocation can be part of a "third place" approach (see Box 5.1).

In terms of technical-economic references (yields, costs, economic models), we need to be careful when considering the "microfarm" object. Indeed, depending on the context and type of microfarm (farm or multi-activity project), technical practices and objectives may vary. Extrapolating references from one type of microfarm to another is a perilous task. It is easy to understand, for example, that a farm that makes its living solely from production is likely to have higher technical efficiency and production levels than an associative project where production represents only part of the income and is primarily a support for educational activities involving schoolchildren or social integration.

Third places is the term described by Oldenburg in 1989 in a book entitled *The Great Good Glace: Cafés, Coffee Shops, Bookstores, Bars, Hair Salons, and Other Hangouts at the Heart of a Community*, which distinguishes between the first two spaces, such as the domestic space of the home and the professional space (school, office, etc.), and this third place, which is a socializing space where people from different backgrounds rub shoulders and exchange ideas, regardless of their origins and cultures.

A recent study (Fab'lim et al. 2020) identified over 100 third-party food centers in France, grouping them into three types.

One type focuses on food justice and seeks to promote social mixing and public integration through food. These third places seek to promote quality food at affordable prices and to change people's behavior towards local, seasonal food. These third places often have a more fragile economic model, dependent on public or private financial support. They were the most fragile during the Covid-19 confinement of 2020. They are, however, real places for social mixing, provided that the association consistently addresses the issue.

A second type is oriented towards citizen food and seeks to rethink the relationship between producer and consumer by moving towards shorter distances to offer alternatives to supermarkets. Agricultural production is not carried out by professionals but by members of the association. The financial and, above all, human equilibrium can be complex, relying as it does on the presence and persistence of members, which is not always guaranteed, and on the need to train new members. These places can be vectors of social innovation. Their survival and resilience vary from one region to another and from one form of governance to another. One example is Ecolieu Jeanot, which has existed since 2011 and is designed as an educational and experimental tool for moving towards a fair, united and sustainable world (see: https://jeanot.fr/le-projet/), combining production and the artistic dimension. Revenues are derived from educational activities and subsidies.

Finally, the third type is that of cooperative work, which seeks to demonstrate the benefits of alternative production and marketing by making farmers more autonomous. These places are more or less open to the public, the production is professional, and they are therefore more productive. Some of these sites are also test areas for setting up. The majority of income comes from agricultural production.

The third place is generally the object around which a network of cooperation and discussion is created between actors in the local food system.

Box 5.1. *Third places: diverse spaces for exchange and training*

5.2.4. *The microfarm as part of a collective farm*

As in the case of third places or test areas, the microfarm can be conceived from the outset as part of a collective project integrating other animal or plant production, processing or even craft activities. Over the last 10 years or so, we have seen an upsurge in collective agricultural establishment run by new entrants in back-to-the-land dynamics who combine a diversity of agronomically and economically complementary activities (e.g. farmer-bakers, small-scale livestock farming, beekeeping, market gardening, orchards, processing, crafts, etc.). Collective agricultural establishments enable several people to take over a medium-sized farm (between 15 and 50 ha), pooling certain equipment and promoting mutual aid and synergies between activities. Located in the Île-de-France region, the Ferme de l'Envol is a case in point. Created in 2020 as an SCOP[8] and SCIC, bringing together producers, local authorities and associations (12 founding partners), its ultimate aim is to provide 13.5 full-time jobs on 75 ha. To date, it includes 12 ha of market gardening, 1 ha of greenhouses, 1 ha of orchards and 3 ha dedicated to biodiversity. In association with a bakery and laying hens, the farm sells through a short distribution network. Eventually, the farm will also include a cheese production and arboriculture workshop[9]. These collective projects are based on a variety of organization, governance, resource-sharing (equipment, buildings, water and electricity networks) and legal forms (Morel 2018a). The land for these farms can in some cases be acquired

8. Société Coopérative et Participative: Cooperative and Participatory Society.

9. https://www.coeuressonne.fr/wp-content/uploads/2022/07/DP-Ferme-de-lEnvol.pdf).

by a landholding company such as Terre de Liens or support networks whose members can become shareholders (Groupement Foncier Agricole, Société Civile Immobilière[10]). Many of these diversified projects incorporate a market-gardening microfarm as a workshop in this mosaic of activities (see Box 5.2).

However, collective agricultural establishments raise specific questions about expectations, governance, financing, and the sharing and use of equipment and buildings. The CASDAR Agri'Coll project (https://www.agri-coll.xyz) was launched in 2020 to identify the obstacles and levers to collective agricultural establishment and to capitalize on existing experience. It offers resource sheets on various themes, such as the notion of entering and leaving the collective, as some farmers do not envisage this activity as a lifelong career but only as a stage in their career.

In Haute-Garonne, Alban of the Collectif de l'an 01 is helping to renovate a grand farmhouse where a dozen people live in a shared habitat and where a few members are developing an agricultural project with two farmer-bakers, two organic market gardeners (one working and one in trial), pigs, chickens, meat sheep and 1 ha of forest. In this collective, Alban's microfarm covers 1 ha with 2,000 m^2 cultivated, including 372 m^2 of greenhouse (2018 figures). He works with a Living Soil Approach, with limited tillage, green manure and the use of local organic matter as fertilizers: grass, shavings, Ramial chipped wood, which two local landscaping companies provide free of charge, and manure from the farm's animal production. He sells his vegetables at an on-site market, which sells all the collective's products, and through complementary bike deliveries in the nearby market town.

Figure 5.3. *Alban's microfarm interacting with the other agricultural activities of the collective project. Photo credit: Kevin Morel*

Box 5.2. *A market-gardening microfarm as part of the puzzle of collective establishments in rural areas*

10. An agricultural landholding group (Groupement Foncier Agricole) and non-trading property company (Société Civile Immobilière).

5.2.5. *Support from local authorities*

In cities, local authorities' support for microfarms and other forms of urban agriculture responds to a wide range of motivations, such as the greening of neighborhoods, their role in the green network (biodiversity) and social and intergenerational links, with a strong educational or professional integration dimension. The productive dimension and participation in food self-sufficiency are mentioned less and less (Bouillier-Oudot et al. 2020) and seem more relevant to rural microfarms than urban ones.

Local authorities can use several planning and urban development documents to support the establishment and maintenance of urban and rural microfarms: regional master plans, Climate, Air and Energy plans, the regional ecological coherence plan at the regional level and, more locally, SCOTs[11] and PLU[12], which enable natural or agricultural areas to be safeguarded. Some local authorities have chosen to set up a department or employ a person dedicated to agricultural issues, to facilitate the process for new entrants.

Local authorities can make land available, as in Besançon with its business incubator, or the City of Paris via its "Parisculteurs" calls for projects, listing available public and private land. Local authorities can also call on the SAFER to pre-empt agricultural land and create a land reserve, sometimes with the creation of microfarms run by public authorities, as in Romainville or Mouans-Sartoux, which employs three municipal employees on 3.5 ha of organically farmed land. The provision of land is often conditional on environmentally friendly practices, most often organic farming (see Box 5.3). In urban areas, local authorities can also help with the supply of inputs (compost, green waste), equipment (loans) or the initial analysis of soil pollution, which determines production methods. In this way, local authorities are increasingly taking the lead, even though there are still obstacles. The case of La Prévalaye in Rennes (35) is a case in point: the land belongs to the city, but after an ambitious program to install farmers, competing uses have arisen, such as sports facilities. There were also concerns about the type of lease to be used, and the town considered new options: different types of call for expressions of interest to find project promoters or going over to municipal management. Today, the project covers 7.2 ha and includes seven agricultural and gardening initiatives. As for the ANRU[13] "Quartiers Fertiles" ("Fertile Neighborhoods") call for projects, it has supported the financing of 100 urban farms between 2020 and 2022.

11. Schémas de Cohérence Territoriaux: Territorial Cohesion Schemes.
12. Plan Locaux d'Urbanisme: Local Urban Planning.
13. Agence Nationale de Renouvellement Urbain: National Agency for Urban Renewal.

This microfarm project was supported by the municipality to diversify the local food supply while enhancing the landscape of a listed site. The microfarm is being developed by Norbert and Nils, both undergoing professional retraining, on 2.2 ha, with 1 ha cultivated in the open field and 2,500 m^2 under shelter. The two market gardeners plan to grow approximately 30 vegetable crops using a "Living Soil" approach. Production will be marketed mainly in partnership with a local network of solidarity grocery stores (possibly supplemented by farm-gate sales and baskets).

The municipality's support was demonstrated by the provision of land (through an 18-year rural lease) and the construction of the farm building, borehole and fencing. In line with the municipality's expectations, ecological infrastructures were installed: hedge, pond, flower strips and wild plant areas.

As the microfarm is located on a listed site, the possibility of installing greenhouses was strongly questioned by the Architecte des Bâtiments de France, and it was also thanks to the mediation of the town hall that the shelters were able to see the light of day. As is often the case in this kind of context, market gardeners and local authorities need to discuss their specific constraints and expectations.

For example, local authorities are often reluctant to use plastic tarpaulins, which are perceived as unsightly, even though they are often crucial to market gardeners' strategies. Intelligent communication is essential to foster mutual understanding and reach satisfactory compromises.

Figure 5.4. *Les Loges microfarm project. Image used with permission of the farm's market gardeners*

Box 5.3. *The microfarm at Loges-en-Josas (Yvelines): an example of a project supported by a town hall*

5.3. Discussion and perspectives

5.3.1. *Microfarms challenge urban–rural boundaries*

In this chapter, our comments may suggest that we distinguish between two main types of microfarm: on the one hand, the "country microfarm", a farm where production is the main source of income, and, on the other hand, the "town microfarm", which is more participatory and whose economy is based on a diversity of activities. While this dichotomy may be helpful in finding one's bearings, it is simplistic. In fact, more and more urban microfarms (especially in the suburbs) may be seeking to develop business models based primarily on production. On the other hand, participatory microfarms with a multitude of activities are developing in rural areas, as part of a third-place or eco-environment dynamic. So, the opposition between "town microfarm" and "country microfarm" has real limits. To approach a microfarm, whatever its territory, it therefore seems more relevant to look at the type of activities it develops (or not) in addition to production, and the share of production in income generation. It is also worth noting that some microfarm projects include productive areas in the peri-urban area (operating as a farm) and areas within the city with a multitude of activities, such as the Potager de la Cantine in Nantes, which was created by a market gardener originally from the peri-urban area. After cultivating 3,000m^2 of land 10 km from Nantes (44) for 10 years, he set up and managed the Potager de la Cantine on the Ile de Nantes in the heart of the city from 2016 to 2022, in addition to his peri-urban microfarm, selling his produce to the adjoining restaurant.

We also find the opposite story, with urban farmers who, after a few years, decide to set up in the suburbs or the countryside, either as a market-gardening microfarm or as part of a collective grouping together diversified productions. Some clearly see these years in urban agriculture as opportunities to test their motivation and acquire skills before setting up in rural areas.

5.3.2. *Integrating microfarms into regional food systems*

Individually, microfarm projects aim to contribute to the sustainable agriculture transition of urban and rural areas by producing local organic food, creating activity on small surfaces and potentially providing a diversity of ecosystem services (which have yet to be objectified by research). However, given the areas involved, these initiatives may seem anecdotal. It is therefore necessary to see what role they can play in the broader transition of the food system. More research is needed on this subject, but several hypotheses can be formulated based on our field experience and past or current research (Box 5.4).

Since the 2018 publication of a PhD thesis on the viability of microfarms (Morel 2016; for dissemination report, see Morel 2018b), several research projects have been conducted on microfarms:

– MicroMa: https://formationcivamgard.fr/?MicromaResultat1;

– MMBIO (results to come): http://qfq.itab.asso.fr/action.php?id=2495;

– MIPS: https://www.mips-aura.fr/;

– MIMABIO: https://www.bio-provence.org/Projet-MiMaBio-Maraichage-Bio-sur-petites-surfaces-en-PACA;

– AURA Chamber of Agriculture study: https://aura.chambres-agriculture.fr/publications/toutes-les-publications/la-publication-en-detail/actualites/maraichage-sur-petites-surfaces/.

On urban microfarms:

– a research paper on the sustainability of microfarms in London (Chang and Morel 2018);

– SEMOIRS (2018–2020) on the evaluation of ecosystem services provided by urban microfarms and their soils. https://hal.archives-ouvertes.fr/hal-03624235/document;

– FEWmeter (2018–2021) on the effects of different forms of urban agriculture on food, energy and water resources. http://www.fewmeter.org/wp-content/uploads/2022/06/FEW-meter_final-report_May_2022.pdf.

Box 5.4. *Past and current research projects on microfarms (non-exhaustive list)*

In rural and, above all, urban areas, microfarms could play a major educational and pedagogical role, helping to raise public awareness of agricultural and food issues and thus influence consumption patterns. In urban areas, a scattering of microfarms (among other urban farming initiatives) could play an important role in greening cities, biodiversity, local recycling of organic matter, rainwater retention, the creation of meeting places between people, etc. Microfarms can also act as intermediaries or points of reconnection between agricultural actors and citizens. For example, some intra-urban microfarms work in collaboration with rural or peri-urban farmers and act as platforms where urban dwellers can find a wide variety of products.

Some activist networks advocate a large-scale multiplication and expansion of microfarms throughout the world, combined with a return to farming by a certain proportion of the population. This is an interesting hypothesis to consider, but it raises a number of questions: to what extent would a large proportion of the population really have the desire and stamina to become farmers? How can they be trained or supported? Can microfarms be generalized beyond the niches they currently occupy? How can microfarms be linked to historic farms?

In any case, microfarms have the advantage of being places where agricultural alternatives are being tested, both in terms of practices and also marketing methods and new business models based on services other than food production alone. Perhaps one of their major contributions could be to fuel debate and renew imaginations about agriculture.

5.4. References

AFAUP (2022). Annuaire formations Agriculture urbaine 2021–2022. Online catalogue of training provided by the French Association for Professional Urban Agriculture [Online]. Available at: http://www.afaup.org/wp-content/uploads/ANNUAIRE-FORMATIONS-AFAUP-2021-pages.pdf.

Agreste (2020). Recensement agricole 2020 [Online]. Available at: https://agreste.agriculture.gouv.fr/.

d'Allens, G. and Leclair, L. (2016). *Les néo-paysans*. Éditions du Seuil, Paris.

Aubry, C., Giacchè, G., Maxime, F., Toussaint Soulard, C. (2022). *Les agricultures urbaines en France. Comprendre les dynamiques, accompagner les acteurs*. Éditions Quæ, Versailles.

Bedouet, L., Rabhi, P., Hulot, N. (2017). *Créer sa micro-ferme : permaculture et agroécologie*. Rustica, Paris.

Bouillier-Oudot, M.H., Rocq, S., Vedeau, F. (2020). Quelle politique du ministère de l'agriculture et de l'alimentation en matière d'agriculture urbaine? Report 19052, CGAAER [Online]. Available at: https://agriculture.gouv.fr/la-politique-du-ministere-en-matiere-dagriculture-urbaine.

Chang, M. and Morel, K. (2018). Reconciling economic viability and socio-ecological aspirations in London urban microfarms. *Agronomy for Sustainable Development*, 38(1), 9.

Cohen, N. and Reynolds, K. (2016). *Beyond the Kale: Urban Agriculture and Social Justice Activism in New York City*. University of Georgia Press, Athens, GA.

Daniel, A.-C. (2017). Fonctionnement et durabilité des micro-fermes urbaines. Une observation participant sur le cas des fermes franciliennes [Online]. Available at: http://www.cityfarmer.org/2017DanielACD.pdf.

EIP-Agri FG (2016). New entrants into farming: Lessons to foster innovation and entrepreneurship. Report, Brussels.

FAB'LIM, UMR Innovation INRAE, Chaire AgroSYS Institut Agro, InCitu, DRAAF Occitanie (2020). Les tiers-lieux nourriciers engagés pour la transition agroécologique et alimentaire, Passer à l'action [Online]. Available at: http://fablim.org/wp-content/uploads/2020/12/FABLIM_Brochure_tiers_lieux_nourriciers_HD.pdf.

Fortier, J.-M. (2012). *Le jardinier-maraîcher : manuel d'agriculture biologique sur petite surface*. Ecosociété, Montreal, Quebec.

Grard, B. and Ughetti, M. (2020). Pourquoi mettre des fermes dans les villes ? *The Conversation*, 16 July [Online]. Available at: https://theconversation.com/bd-pourquoi-mettre-des-fermes-dans-les-villes-142107.

Hervé-Gruyer, P. and Hervé-Gruyer, C. (2014). *Permaculture*. Actes Sud, Arles.

Hervé-Gruyer, C. and Hervé-Gruyer, P. (2019). *Vivre avec la terre. Méthode de la ferme du Bec Hellouin*. Actes Sud, Paris.

Holmgren, D. (2014). *Permaculture : Principes et pistes d'action pour un mode de vie soutenable*. Rue de l'Échiquier, Paris.

Morel, K. (2016). Viabilité des microfermes maraîchères biologiques. Une étude inductive combinant méthodes qualitatives et modélisation. Thesis, INRA, AgroParisTech, Paris Saclay University.

Morel, K. (2018a). Installation collective néo-paysanne. *Pour*, 234–235(2), 153–161.

Morel, K. (2018b). Viabilité des microfermes maraîchères biologiques. Diffusion des principaux résultats de thèse. Thesis, INRA, AgroParisTech, Paris Saclay University [Online]. Available at: https://hal.archives-ouvertes.fr/hal-01930607.

Morel, K. (2019). Les projets néo-paysans de microfermes en France : quelles visions et quelles pratiques agricoles alternatives ? *Bulletin de l'association de géographes français. Géographies*, 96(4), 659–675. doi: 10.4000/bagf.6072.

Saint-Ges, V. (2021). Business models des organisations marchandes et productives de l'agriculture urbaine. *Innovations*, 64(1), 91–118. doi: 10.3917/inno.pr2.0r100.

SOL (2020). Toutes paysannes, tous paysans ! *SOL* [Online]. Available at: https://www.sol-asso.fr/wp-content/uploads/2020/10/BD-SOL-Version-digitale.pdf.

6

The Rise of Professional Gathering in Rural Areas

Valentin ASSELAIN and Florence PINTON
Paris Saclay University, AgroParisTech, INRAE, UMR SADAPT, Palaiseau, France

6.1. Introduction

Gathering wild plants immediately conjures up images of nature's abundant resources and/or natural environments that have undergone little transformation and are in good health. It also conjures up images of an archaic activity, even predatory in its contemporary form, which the processes of cultivation and domestication of species should have put an end to. It conjures up vague and contradictory imaginaries, oscillating between a fantasized perspective of nature, a precarious and marginal activity or one imbued with an unusual esoteric charge. However, today's wilderness is no longer the wilderness of the Paleolithic era. It encompasses spontaneous plants that colonize more or less anthropized environments and are the object of self-consumption, domestic or commercial exchanges. With the emergence of discourse on the economic value of biodiversity and, more generally, on the values attributed to nature, the way we look at them has been renewed. Finally, the resources collected are often not the private property of the gatherers but are seen as common property in the same way as land and water. The species collected may be considered by naturalists as heritage to be preserved as elements of biodiversity. In the field, they are governed by the rural code and environmental law.

In mainland France, while gathering remains a "family" practice, it is a fast-growing professional activity, the fruit of a return to the land and also the consequence of a

Farms and Territory,
coordinated by Philippe MARTIN and Solène PISSONNIER.

strong social demand for natural products that we are seeing on a global scale. In recent years, there has been an unprecedented increase in the number and variety of species sought (Lescure et al. 2018)[1]. Initiated in the 1970s by a few “outsiders”, this return to the wild has been fueled by popular knowledge and practices abandoned by farmers during agricultural modernization. The socio-environmental turnaround of the 1990s has seen this trend intensify. For the most part, gatherers are young and not-so-young, eager to return to the land and/or to re-establish themselves in a rural area. Given their atypical trajectory, the ways in which land is passed on, and their dependence on other practices, access to land and the conditions of appropriation of these plants raise particular questions. Their great heterogeneity and the multiplicity of issues they face today make them a complex object of study.

The question arises as to the place of professional gathering in a rural environment that is increasingly attractive to certain segments of the population, and how it relates to contemporary agricultural systems. If we accept that we are living in a new era, in which we have moved from the application of the performativity[2] of agricultural modernity on living things to its contestation, then the rise of commercial gathering demonstrates a more or less assertive desire to reconnect economic, social and environmental issues through the mediation of materiality and living things, and to break away from a strictly productivist approach. The many ways in which our modes of production are being called into question have become conducive to the requalification of the wild and its inclusion in productive systems. In fact, the practice of gathering is not based on antagonism with nature but can follow in its wake to take advantage of natural dynamics. It involves gathering spontaneous plants outside a production cycle that has been mastered as a whole.

Because it has remained in the shadows for so long and has no statutory recognition as such, there has been little supervision of this profession. It may be perceived as offering greater freedom of action and distance from authoritative bodies, attracting individuals seduced by the limits of autonomy, where they feel they will find enjoyment in regaining control over their lives. However, it also poses difficulties in terms of collective dynamics, whether in terms of its impact on biodiversity, the sustainability of the wild plant resources, the conditions of setting up or training and support.

1. The NGO TRAFFIC claims that 60–90% of aromatic and medicinal plants marketed worldwide come from the wild, that this globalized trade represents millions of tons of plant ingredients, and that it has increased threefold since 1999 (Timoshyna et al. 2020).

2. The notion of performativity refers to the capacity of certain deliberate statements to change reality and the practices they designate.

In this chapter, we return to these different aspects by asking the question of the influence of gathering on territories and natural environments, focusing more specifically on the world of professional gatherers, the collectives that represent them, and the interactions they develop with plants and other users of the countryside. Their practices raise questions about their relationship to work, to plants and to the construction of knowledge. How can we define the socio-professional universe of gatherers, and to what extent do their trajectories appear to counterpoint what has been forged by modernity? Are they engaged in this movement of return to rurality in the image of the neo-rurals of the 1970s (Hervieu-Léger and Hervieu 1979; Deléage 2018), are they the "symbol of resistance in a world that has become precarious" (Lowenhaupt Tsing 2017, authors' translation) or are they the bearers of a conception of humanity's relationship with its environment that breaks with modernity (Latour 1991) because it would show new ways of collaborating and cohabiting with the land (Centemeri 2019)? We do not claim to answer these questions directly, but we do seek to understand how the transformations they bring with them are part and parcel of a process of agroecological transition.

This work is based on data from previous projects[3] and on several years of surveys of gatherers and the various structures they work with or that support them[4]. Numerous research collaborations and the building of partnerships with stakeholder institutions have contributed to a better understanding of professional gathering and the industries that support it. Finally, this work benefits from the findings of a thesis on the professionalization process of gatherers[5].

Our research approach is to follow the gatherers in their life experience, in order to describe a socio-professional field in the making. The prism we have chosen to use is their attachment to the rural environment and the links they maintain with other users (farmers, foresters, breeders, hunters, landowners, etc.) and their usual contacts (managers, chambers of agriculture, inspectors from certification and training bodies, local authorities, etc.). Our qualitative, ethnographic data are at the

3. At the initiative of the SIMPLES syndicate, a reflexive survey marked the start of F. Pinton's collaboration in 2010 with the Association Française des Professionnels de la Cueillette de plantes sauvages (AFC, French Association of Professional Wild Plant Gatherers), followed in 2014 by action research involving several institutions (Flores 2019), the completion of several internships under the aegis of AgroParisTech and ongoing work around the feasibility of a gathering observatory.

4. We would like to thank all the organizations we worked with (AFC, CBNPMP, CNPMAI, CPPARM, University of Lausanne) as well as all the gatherers we met. Finally, we benefited from the advice of Jean Paul Lescure and his proofreading of the document.

5. The thesis, directed by F. Pinton, has been supported by an Ademe grant (2021–2024). Between 2021 and 2023, the recipient conducted over 40 in-depth interviews with the gatherers, from whom the verbatim quotes are taken.

crossroads of environmental sociology, the sociology of professions and pragmatic sociology. Our hypothesis is that these different actors are more or less directly involved in transforming the ways of producing and acting rural, despite a socio-technical and institutional environment that is rather unfavorable to them.

After tracing the genealogy of gathering and its singular character in terms of practices and inclusion in agroecosystems, we discuss the heterogeneity of gatherers, their motivation for gathering and their relationship to the agricultural and rural world. In the third part, we look at the creation of gathering territories and the initiatives we have observed in terms of regulating gathering.

6.2. A centuries-old business embedded in agroecosystems

Gathering is an uncommon way of producing. The recent rise in the profile of the profession by associations, the professionalization initiatives that have given rise to various positions and the rationalization that is sometimes at work in this ill-defined socio-professional field prompt us to briefly characterize this profession, both in terms of its history and its contemporary specificities.

6.2.1. *The social history and renewal of a practice*

Wild plant gatherers have long been associated with folk knowledge and confined to the figure of the *paysan*[6], as subject to the whims of nature and the market (Bromberger and Lenclud 1982; Larrère and de la Soudière 2010). Gathering was done for domestic use or to sell and earn extra income. After a brief period of growth in commercial gathering to supply emerging industries, the recomposition of rural spaces and living patterns observed from the 1980s onwards (Hervieu and Viard 1996) put an end to the complementarity between farming and gathering practices. With the professionalization of agriculture came the end of a certain relationship with nature and its reinvention on a small scale by new actors in search of a new art of living (Hervieu and Léger 1983) and interested in the popular uses of plants. They saw the production and processing of aromatic and medicinal plants as an economic activity compatible with their values and the safeguarding of a disappearing cultural heritage (Lieutaghi 1983). Agricultural modernization, the rural exodus and the development of environmental policies were to confine first-generation gatherers to peripheral territories marked by a high level of abandonment

6. Translation note: in French, the word “peasant”, that is, *paysan*, derives from the word for “country”, *pays*, and as such connotes an attachment to the territory and the land.

(mid-mountain areas) and considered more natural and less polluted. Later, the expected greening of agriculture and consumers' attention to product traceability gave a new impetus to gathering activities.

6.2.1.1. *Towards professionalization*

Due to its history and the ways in which it has been redeployed, the gatherer profession has no specific status that takes into account the singularity and multiplicity of its activities, combining natural gathering practices, agricultural production and processing[7]. With an economy that has long been perceived as informal, underground or anecdotal, these activities have long been considered marginal by government services, whether they work for agricultural supervision or nature protection. Gatherers, who transform their plants by hand, prefer production methods that require little agricultural land and market their produce themselves.

There are few statistics available on a national scale to put a figure on this revival of gathering, which can be described in two waves: that of the 1980s, which coincided with the first neo-rural movement, and that which succeeded it in the 2000s. We propose to use as indicators the increase in the number of professional gatherers, the emergence of numerous forms of collective organization that bear witness to the reality of these mobilizations and the innovations that underpin them, the renewal of knowledge production around gathering and also the recurring testimonies in our interviews.

As a sign of converging interests or a desire to rationalize practices, gatherer organizations are the ideal places to discuss and develop their social identities. The best known of these is undoubtedly the Intermountain Syndicate for the Production and Economy of Medicinal Plants (SIMPLES[8]), created in 1982 at the initiative of a group of medicinal plant producers intent on structuring and professionalizing their practices. Holders of the SIMPLES brand conform to exacting specifications and the syndicate is set up as a network for the exchange of practices and the defense of "the professionalization of the gathering activities, respect for good gathering practices and sound management of the resource, in partnership with the [AFC]" (authors' translation). This syndicate has taken up the fight for recognition of the herbalist profession[9]. In 2016, following multiple internal discussions and favorable

7. Sorting out, drying, sometimes making home-made cosmetics, food products, etc.

8. Syndicat Inter-Massifs pour la Production et L'Economie des Simples.

9. The herbalist diploma was abolished in 1941 in favor of pharmacists, who retain a monopoly on the sale of medicinal plants listed in the French Pharmacopoeia, with the exception of 148 liberated plants (2008 decree) (Bost 2016). Recognition of this profession would give producers access to more plants and enable them to process and sell medicinal plants in complete transparency.

political support, the Fédération des Paysannes Herboristes (FPH) was set up[10] to put pressure on the authorities. Five years later, Senator Joël Labbé introduced a bill on the creation of the *paysan-herboriste*[11] profession, and a Senate commission was initiated to work on the issue (Labbé and Imbert 2018). In 2023, the syndicate federates 162 holders of the brand, 386 non-certified members and 400 supporters[12].

The creation of the AFC in 2011 opened a period of promotion for the profession of professional wild plant gatherers in all its forms. Its aim is to unite wild plant gatherers around the issue of resource management while also defending their interests in a field that has become competitive. The organization has 240 members, with nearly 200 expected to join in 2023. This organization must make gatherers credible interlocutors for all economic and/or institutional actors in the perfume, alimentary and medicinal plants (PPAM[13]) sector and establish collective rules in consultation with environmental managers (Julliand and Garreta 2017). The creation of a gathering charter backed up by a book of good practices was one of the first actions taken to gain recognition for the profession and set themselves apart from gatherers with little respect for the environment. A charter for the plant-using industries that are members of the AFC has also been drafted.

The Aromatic and Medicinal Plants Committee (CPPARM[14]) is an inter-professional body that renewed its statutes in 2009 to acquire a Protected Designation of Origin (PDO). It supports the development of the industry through various actions coordinated with its partners. Initially focused solely on agricultural production, it gradually opened up to gathering issues, with a gatherers' cooperative joining its structure in 1987, followed by actions developed in conjunction with the AFC, and finally the creation in 2014 of "Gentiana Lutea", an associative structure whose main aim is to safeguard yellow gentian (*Gentiana lutea* L.), an emblematic wild plant of the Auvergne region much sought-after in the manufacture of liqueurs and considered vulnerable. The association has created the collective trademark

10. The FPH, created on their initiative, brings together the SIMPLES, the National Federation of Organic Farming (FNAB, Fédération nationale de l'agriculture biologique), the Biodynamic Agriculture Movement (MABD, Mouvement d'agriculture biodynamique) and producer groups practicing direct sales.
11. The majority of the gatherers refer to themselves as peasants to emphasize their link to the territory (the "*pays*") in which they live, and the mode of production far from the conventional model of contemporary agriculture.
12. See: https://www.syndicat-simples.org/presentation-du-syndicat/.
13. Plantes à Parfum, alimentaires et Médicinales.
14. Comité des Plantes à Parfum Aromatiques et Médicinales.

"Gentiane – Filière développement durable", which will be registered with the National Institute of Intellectual Property (INPI[15]) in 2021.

6.2.1.2. *The diversity of knowledge circulating within the gatherers*

Because it does not belong to a professional field, the practice of gathering is not based on a stabilized training repository (Pinton et al. 2015) but rather on empirical, experiential and localized knowledge (Compagnone et al. 2018). Access to knowledge represents a real field of activity, with acquisition processes taking place in frameworks that are not often very institutionalized. First-generation *paysan* gatherers are strongly committed to the reappropriation of popular knowledge, where a good knowledge of plants and their medicinal properties are fundamental to their processing and sales activities. The private schools that meet these requirements are renowned, and many producers have passed through them[16]. We are also struck by the intense circulation of knowledge in the field, through informal exchanges, visits and internships, self-training, etc. Mentoring and companionship therefore play an important role in the acquisition of knowledge. Our surveys reveal a number of areas of naturalist, agronomic and therapeutic knowledge, while scientific training (botany, ethnobotany, plant chemistry, ecology, etc.) is unevenly used by individuals, depending on their original level of training and career path.

There is a lack of specialized agricultural vocational training in PPAM for anyone wanting to set up as a farmer and obtain aid (Labbé and Imbert 2018). Producers have to go through public establishments such as the Agricultural Vocational Training and Promotion Centers (CFPPAs[17]), which deliver a Vocational Diploma for Farm Managers (BPREA[18]), most of them ill-suited to the expectations of gatherer-producers. Well aware of this obstacle to their installation, the most committed are organizing and campaigning to offer training courses more closely linked to gathering, PPAM production or processing, and with a view to retaining control over training[19].

Know-how is an essential feature of gathering. It relates to the gesture, the tool, the plant concerned and the gathering environment. All field surveys have shown

15. Institut national de la propriété intellectuelle.
16. There are also several university courses in phytotherapy and aromatherapy (Diplôme Universitaire).
17. Centres de Formation Professionnelle et de Promotion Agricole.
18. Brevet Professionnel de Responsable d'Exploitation Agricole.
19. The SIMPLES syndicate is calling for "*paysan-herboriste*" training courses "designed by the profession" (SIMPLES AGM 2021).

that the technical act is imbued with a sensitive relationship with the plant and incorporates affective and relational dimensions. While the definition of good practices is not perceived in the same way by all gatherers, depending on whether their activities are part of an artisanal approach or supply the industrial sector, what unites producers is their particular relationship with plants. Because their values are more firmly rooted in ecology, some immediately place respect for living things at the heart of their practices and their discourse. Others are more pragmatic, with sustainability objectives aimed more at guaranteeing access to resources (Julliand et al. 2019).

6.2.2. *One foot in, one foot out: how gatherers belong in the rural community*

Gathering is an integral part of the rural environment, but it is rooted in a paradigm that distinguishes it from the agricultural profession and blurs our understanding of the various components of the rural environment.

Following on from work on "rural utopias" (Hervieu-Léger and Hervieu 1979; Stuppia 2016) or the settlement of relatively precarious urban populations in the countryside (Ponchelet et al. 1998), contemporary gatherers share some similarities with "neo-farmers", an emerging category at the turn of the century. In phase with societal expectations, they form collectives alien to the agricultural development apparatus that accompanied agricultural modernization in France, where nature was relegated to the role of threat. Their practices distance themselves from the principles that forged the productivist model, believing that sustainable, responsible gathering requires regenerative cooperation with living things.

In this way, gathering can evolve from a simple one-off collection to numerous forms of management that encourage the presence and renewal of desired species or maintain the environment. The cultivation of plants is another mode of action, whether this involves seeds or plants taken from the wild, exchanges between producers or purchases from local nurseries or structures such as the National Conservatory of Perfume, Medicinal, Food and Industrial Plants (CNPMAI[20]), which has its own "catalog". The principle of "semi-culture", claimed by some and expressed in the quote below, is to let plants grow in a biotope that is favorable to them and whose evolution can be controlled. Note that the "local plant" label created

20. Conservatoire National des Plantes à Parfum, Médicinales, Alimentaires et Industrielles.

in 2015 to certify plants and seeds produced from material sourced from local "wild" populations tends to enhance these principles[21]:

> In my opinion, if there's a future in gathering, it's specifically in a cultural-gathering approach, I don't know how to put it. In other words, to … either, reseed places, you know for example lavender areas that don't grow back very well, somewhere in the mountains, work them roughly, sow them, see what happens … Or to control the sites, not let them become overgrown. I think that's one future for gathering (Gatherer-cultivator, Hautes-Alpes 2022, authors' translation).

References to the wild and the domesticated find a new place, no longer in opposition or along a gradient, but rather as a form of cooperation. Generally speaking, it is not a question of working plants but of working with plants. Plants orient the behavior and practices of the gatherers, with the latter lending them intentions, ways of acting or ideas (Kazic 2022; Pouteau 2022). The two categories would thus be more closely intertwined so as to respect as closely as possible the "aspirations" of the cultivated plant. It is on this capacity to cooperate fruitfully with the living, on this subjectivity specific to each individual, that the objective of productivity and the evaluation of the value accorded to nature are anchored (Julliand et al. 2019).

Most of the gathered species also depend on agro-sylvo-pastoral environments such as natural meadows, heaths, wastelands, hedgerows, forest edges and forests, whose characteristics guarantee their renewal.

Arnica, for example, is dependent on the management of acid meadows, and its presence is incompatible with liming and fertilization. The production of gentian, whose rosettes disappear under the teeth of sheep, requires a good balance between grazing quality for cattle and gentian density. Lavender and thyme sites depend on the maintenance of sheep pastures; hedgerow and forest edge management are favorable to hawthorn; Aubrac tea is favored by forest clearings, whereas bear's garlic (*Allium Ursinum L.*) struggles in these environments.

21. A simple collective brand created on the initiative of three networks: the National Botanical Conservatories (Conservatoires Botaniques Nationaux), Afac-Agroforesteries and Plante et Cité. It is the property of the French Office for Biodiversity (Office Français de la Biodiversité).

Figure 6.1. *Gathering sea buckthorn (Hippophae Rhamnoides L.) by the river, Hautes-Alpes (2022)*

While the sustainability of resources depends on gathering pressure, their abundance is largely dependent on agricultural and forestry practices whose permanence is not always self-evident, given the prevailing management models. The French Environment Code seeks to remedy this by applying to species considered endangered (species protection). However, it should be noted that the bans on destroying, cutting, mutilating and uprooting plants do not apply to routine farming operations on plots of land that are usually cultivated, which limits their scope[22], while the presence of certain plants depends on owners or managers who endeavor to define what should be valued or protected in their territory according to their own logic. Encouraging the agroecological transition, reaching agreements and developing cooperation with the various actors in rural areas is therefore an essential dimension of the permanence of gathering and the anchoring of gathering professionals in rural areas.

22. Book II of the Rural Code on Nature Protection.

Another particularity, supported by the members of the SIMPLES syndicate historically organized according to mountainous regions, known as “massifs”[23], was to militate and open up to more anthropized areas and attract new generations. Discussions focused on the nature of the project to be defended and the contours of more demanding specifications than the organic brand, giving the SIMPLES brand its specificity. “Either we focus on a mainly technical project (to protect the brand that guarantees organic quality), thereby excluding those located on the plains or near urban centers deemed too polluted, or we adhere to a more ambitious project to reclaim territories” (SIMPLES AGM 2015, authors’ translation), to influence conventional models and regional planning from small enclaves. In the early 2000s, and despite the persistent reluctance of some members, the syndicate opened up to producers based in low-lying regions such as Burgundy, then Brittany and Normandy, and more recently Pays de la Loire. To adapt to these new environments, members had to revise the specifications that set out the conditions required to label farms[24]. Since then, their numbers have continued to grow, as evidenced by the increase in the number of producers and members of all producer-cultivator professional organizations and associations.

6.3. Make a living by gathering

Today, gatherers are stepping out of the shadows and moving into new territories. In this movement, the gatherers’ entire economy and the national dimension of their activities are being brought to the fore. How can we give a global representation of this profession? We have already emphasized the special relationship with plants and living beings in general. We are going to spell out the particularities of a profession that does not fit neatly into established frameworks.

6.3.1. *Characterizing today’s gatherers*

It is possible to identify different gatherers’ profiles and to tell the story of a little-known profession. Gatherer typologies have been proposed on the basis of field surveys (Laucoin 2012; Muraz 2018; Pascal 2018; Pinton et al. 2015). They are all based on variables that seek to objectify the professional gatherers, highlighting their practices and production systems. It is worth noting that women are over-represented in these professions, where attention to plants and the care they

23. The Alpes, Auvergne, Bourgogne, Bretagne, Cévennes, Grands-Causses, Jura-Vosges, Limousin and Pyrénées.

24. For example, as part of a restoration approach, taking into account indicator plants for soil management, hedge planting, distance from major traffic routes, etc.

provide certainly play a role in their greater involvement. For example, the survey of producers and "sympathizers" for the SIMPLES syndicate in 2021 shows a ratio in their favor (71% women). Producing and gathering by PPAM is not a random choice, and it is likely that gender contributes to it. These "ecofeminist neo-ruralities" (Rimlinger 2022) remain to be put into perspective. Nevertheless, we feel it is important to return to two fundamental aspects of the gatherers' profession. First, we look at the links that gatherers forge with other stakeholders in the spaces where they work. Second, we look at the personal motivations of gatherers in practicing their chosen profession.

6.3.1.1. *Connections and roots with the farming community*

This singular way of producing by resorting in part to the "wild" no longer makes access to agricultural land, controlled by the farming profession since the introduction of co-management in France, indispensable (Sencébé et al. 2013). The majority of gatherers encountered are, in fact, *paysans* with little land, and sometimes even "landless *paysans*". Available data on the size of their land holdings show that most of them are settled on areas of less than 5 hectares, while the investment in time required for gathering is offset by the manufacture of processed products. From an economic point of view, it is not possible to speak uniformly of the professional practice of gathering. For example, there are gatherers whose gathering volume is counted in tons and others in a few hundred grams. Some will value their plants in their fresh state, while a good proportion will gather modest quantities but add value by processing them. Generally speaking, the use of wild plants and PPAM requires modest investments, compared with other agricultural activities: on the order of a few tens of thousands of euros, necessary for the purchase of a vehicle, the construction of a dryer and containers. Income from this activity varies greatly according to profile, with some being dependent on Active Income Welfare (RSA[25]) to live on, while others, organized in professional groups or directly linked to laboratories or manufacturers, can earn net incomes of €1,500 to €2,000 a month.

The interviews conducted between 2021 and 2023 reveal three ways of relating to the agricultural and rural world:

Some gatherers claim a social identity attached to "farmers" or "*paysans*" and defend principles and values close to "agrarian individualism" (Bloch 1931; Beau 2017): respect for private property, mutual recognition by demonstrating the work accomplished and the productivity of the activity carried out, importance of mutual

25. Revenu de Solidarité Active.

respect and good neighborly relations. For these profiles, the plant represents the basis of an economic model that enables a certain income to be earned. It is not necessarily correlated with a lifestyle and discourses drawing on an imaginary of autonomy, militancy or the promotion of alternatives to the dominant modes of production, agricultural or otherwise.

At the interface between this incorporation into the farming/*paysan* world and a more militant stance, the second profile of gatherers openly criticizes certain agricultural practices deemed harmful and responsible for the disappearance of natural environments, while at the same time maintaining good neighborly relations, or even cooperation, with other farmers or users. The values mentioned above are also shared, but a certain distance can be observed. For these gatherers, who most often own land, the relationship with plants and natural environments is rich in meaning, and plants represent the foundation or an important part of a way of life that goes beyond mere professional activity. The type of commitment varies, from sharing knowledge on botanical walks, to involvement in nature protection associations, while others are involved in trade unions such as the *Confédération Paysanne*.

Finally, the last profile, which is in the minority, appears to be at odds with the farming world, sometimes even in opposition, or more simply out of mutual ignorance, and involved in alternative networks of shared knowledge. As before, the basis of the professional activity is the relationship with the plant and the environment in which it grows, or the social role of intermediary that the gatherers represent between the consumer and the natural environment.

In all cases, gathering is based on a specific relationship with the countryside, which is crossed and crisscrossed as the plants sought are located. In the same way, the social origins of the gatherers, often non-agricultural, with life trajectories characterized by reconversions and specific motivations, reveal a discrepancy with other users of rural spaces. This invites us to better understand the motivations behind wild plant gathering.

6.3.1.2. *Making a living off of one's passions*

The interviews reveal a number of common features in the life paths of the gatherers. Professional reconversions are often the result of questioning their previous activity, deemed out of step with certain values, or the source of more or less profound trauma, or of a lifestyle that no longer suits them. They reorient themselves towards a project that ensures continuity between professional experience and proximity to natural environments, through "practical experience of the living" (Ginelli 2015).

6.3.1.2.1. Discovering a vocation

Being a professional gatherer means investing passionately in the activity. Some people talk about the sense of freedom they feel when they gather, others about a sense of calm in relation to the rest of the world at the time, or about access to a new form of perception, not to mention a certain vertigo in the face of the countless discoveries made among plants. These areas, where the influence of man is less pronounced, are desired as much for the resources they generate, as for the sensitive and aesthetic experiences they enable, and also for the values they apply. What is important to seek out, experience and pass on, then, is not so much the land as a way of living on it, a mode of operation with its own particular knowledge (Julliand et al. 2019):

> [Question] What do you enjoy about gathering?
>
> For me it is not possible to … Why do you love someone? We don't know. I'm in my place when I do that, I don't ask myself questions. Lots of people ask themselves questions in their work: "Why am I doing this?" And in fact, when I'm doing this, I don't ask myself any questions! […]" (Gatherer-cultivator, Puy-de-Dôme 2022, authors' translation).

Entering the profession of plant gatherers is also often a synthesis of different interests, passions and aptitudes, whether physical or derived from previous personal, professional or academic experience. The need to earn an income is combined with the need to make a living off of one's passion:

> I find that it is a bit of a synthesis of agricultural activities, a production thing, starting from the living to produce something, and bringing that to others and earning a living from it, that is something I really liked about the agricultural world, and this nature side … A big word for "nature", but a bit of "the wild", that really spoke to me, and then the adventurous side of my job as a vegetation census taker, and then, beyond that, for me, driving around in a truck, looking for jobs left and right, that is something I really liked. And so, gathering is a bit of a synthesis of all that (Gatherer-cultivator, Loire 2022, authors' translation).

Finally, gathering, often associated with cultivation, is also practiced as part of a pluriactivity: the practice of so-called alternative medicine, seasonal work, activities, etc. It can be a facet of a "holistic" or "decompartmentalized" lifestyle that goes beyond the work-leisure dichotomy. The latter is linked to specific values, which often emphasize notions of autonomy, freedom, the need to be outdoors, "artisan" work (Sennett 2010) and the rejection of certain practices or imperatives from other lifestyles.

6.3.1.2.2. Reinvesting in a bygone profession: herbalists and "*paysan-herboristes*"

A particularity of wild plant gathering is the fact that it can be a component of a vanished profession, now undergoing a strong revival, that of the herbalist (Bost 2016; Labbé and Imbert 2018). While not all gatherers claim to be herbalists, which implies a relationship to care and a specific social role, it is still part of this passionate component of the profession. Our interviews show that the desire to invest in the profession of herbalist is often justified by the desire to treat or prevent ailments (most often "everyday aches and pains", but sometimes real pathologies as well), by investing in a vacant social role: "pharmacies have the monopoly, but they don't use it!" says a gatherer-naturopath.

The FPH's raison d'être is to gain recognition for the existence of this profession, by working towards the creation of a professional title and state recognition of training provided in CFPPAs (Labbé and Imbert 2018). A job specification has been drawn up for inclusion in the National Vocational Certification Register (RNCP[26]). The title was awarded in September 2023 for a renewable three-year period, making the FPH the only body authorized to offer training course.

By joining the movement to promote *paysan* agriculture and its values, the herbalist profession and its representatives are being reshaped. The FPH seeks to reshape the profession of herbalist. The *paysan-herboriste*, who masters the medicinal uses of plants as well as the ways of gathering or cultivating and processing them, is now joining the "salesman-herbalist" who only buys and sells plants. As a result, we can see that under a profession name that often combines different professions (farmer-herbalist, gatherer-distiller, gatherer-cultivator, etc.), there is a wide range of know-how. Many gatherers have mastered not only gathering, cultivation and processing techniques but also sales channels and techniques, web page management and knowledge of the regulations governing all these activities.

In conclusion, deciding to become a gatherer means at the very least to be in daily connection with the "natural" environments that are sought after and loved. More broadly, it is a way of life in which plants are an end in themselves, and not

26. Répertoire National des Certifications Professionnelles.

just a professional object. In this way, we believe that practicing this profession enables us to perpetuate a "logic of subjectivation" (Dubet 2009), distancing ourselves from anything that prevents us from appealing to our convictions, creativity or freedom. It is also an opportunity to hybridize the knowledge and practices acquired throughout one's life and to combine a "cognitive passion" (Roux et al. 2009) with specific gestures and know-how. From this perspective, the trajectories and relationships to work of gatherers are closer to professions such as beekeeping (Fortier et al. 2019), fishermen such as seaweed or shellfish gatherers (Rétière 1998), or transhumant shepherds.

6.3.2. *"Bricolage" overflow and adaptation*

6.3.2.1. *One profession, many practices, many statuses*

Now that gathering is becoming established in all parts of the countryside, is gathering considered by legislators to be equivalent to an agricultural activity in France? Article L 722-1 of the Rural Code and Maritime Fisheries defines it as an activity that is an "extension of the act of production". However, in practice, it is clear that rural organizations do not see it so clearly. This is a recurring question at general meetings, whether at SIMPLES or at the AFC (Lescure and Pinton 2019), where in 2013, it was put by the MP of Puy-de-Dôme to the Minister of Agriculture Stéphane Le Foll. The latter replied that gathering was indeed part of agricultural activity "even if these activities are limited to part of the production cycle" (authors' translation). As a result, wild plant gatherers are covered by the agricultural social protection scheme. In fact, as a self-employed worker, a gatherer is considered a non-salaried agricultural worker as long as they earn, and their activity exceeds certain thresholds.

Many register themselves as artisans with the Chamber of Trades and Professions (CMA[27]). Some have commercial status and are attached to the Chambers of Commerce and Industry (CCI[28]). Some of them, however, gain access to the status of agricultural worker defined by the 2014 Law on the Future of Agriculture, Food and Forestry[29], provided they have access to a minimum area of land or sufficient income or provide a set number of hours of work. For those under the age of 40 years, the challenge may be to obtain the Young Farmer's Assistance

27. Chambre des Métiers et de l'Artisanat.
28. Chambres de commerce et d'industrie.
29. Loi d'avenir pour l'agriculture, l'alimentation et la forêt de 2014: Wherein a farmer is considered to be a farm worker if he or she is registered with the Register of agricultural Assets (Registre des actifs agricoles), managed by the Permanent Assembly of Chambers of Agriculture (APCA, Assemblée permanente des chambres d'agriculture).

(AJA[30]). Those with too little land or low incomes register with the Agricultural Social Security (MSA[31]) to benefit from the agricultural social protection scheme as joint contributors:

> The fact remains that I have the same status as the biggest cereal grower in Beauce, and that I'm the head of a farm. [...] I'm not sure there are many cereal growers in the Beauce who know that a gatherer has the same agricultural status as them. We are not all farm managers, there are the joint contributors, but today that is the status I have, and I make a point of going to meet them and saying that we are farmers in the same way, in the same capacity (Gatherer-cultivator, Hautes-Alpes 2022, authors' translation).

Nevertheless, several gatherers express doubts as to the relevance of obtaining the status of farm manager, due to the fact that expected turnover will be too high to qualify. In all cases, these statuses are ill-suited to the reality of gathering: no French Nomenclature of Activity (NAF[32]) code comes close to gathering wild plants, and many gatherers discover that they are attached to activities that are more or less distant from their profession: gathering berries, gathering wild-growing non-timber forest products... More recently, INPI's introduction of a one-stop shop for all business start-ups is confronting applicants with even greater difficulties, according to the AFC, which is much in demand on this subject (AFC 2022). Be that as it may, most of the gatherers interviewed underline the fact that the administrations with whom they have contact have little knowledge of gathering. Bounced from department to department, this legal and administrative vagueness can lead to delays in setting up, inadequate training and the maintenance of the precarious status of joint contributor offers little social protection:

> [...] I often hear from trainees who want to set up in business: "I called to get information from the MSA", and I say: "Call three times, you'll get three different pieces of information, you take the average and that is it! "[...] Then, not long ago, I had a friend come over, I gave her a legal, tax, etc. course, I told her: "you do that, you don't ask anyone for anything, anyway if you don't ask for aid, if you don't ask for CAP, subsidies, there are no controls! As long as you pay your dues, what is important in the end? You pay your taxes, you pay your contributions, the rest is paper. Knowing that you've got so many surfaces, so many

30. Aide aux Jeunes Agriculteurs.
31. Mutualité Sociale Agricole.
32. Nomenclature d'Activité Française.

things, so much stuff ... who cares!" (Gatherer-cultivator, Puy-de-Dôme 2022, authors' translation).

The absence of a status specifically linked to the gathering activity and the difficult recognition of the gatherers' world by the agricultural profession invite us to draw a parallel with the difficulty of public policies to respond to the challenges of setting up in agriculture in view of the changing profiles of project carriers (Maresca 1988), in this case those "outside the family circle" (HCF[33]). Some of them are insufficiently supported by public policies to gain access to land (Barral and Pinaud 2017) and to the profession (Rémy 1987, 1997; Coly 2020), as they lack the expected attributes. If we accept that gathering is agricultural in nature, it is urgent to renew the ways of thinking about entry into these professions in the prism of a process of re-establishing agricultural professions and extending them to people eager to work with nature, whatever their origin. Finally, while the lack of a framework for gathering, particularly in terms of status, may penalize gatherers in relation to other professionals, it is necessary to question the very notion of "framework", which may prove ambiguous.

6.3.2.2. *Framing an ancient practice*

In France, the agricultural framework was built by bringing together agricultural actors who, in agreement with the State, decided on technical and productivist orientations to bring about the emergence of the modern farmer, integrated into the market and freed from natural constraints (Alphandéry and Billaud 2009). Gathering is not subject to the same challenges, given that it is a recent profession, subject in part to the rural code and in part to the environmental code. The prospect of greater supervision relates to the motivations of gatherers and their projects, some of which are similar to those of neo-farmers. As we have already seen, lifestyle is a determining factor in the choice of gathering as a profession. Moving towards stricter regulations and standards means overturning the perception of work and the identity forged within it. It is a way of calling into question the broad "operational autonomy" (Bouleau and Gramaglia 2015; Julliand and Garreta 2017) that gatherers have in relation to their environment. Conversely, the desire to regulate the profession, recurrent during interviews when discussing competition, endangering plant populations, poor gathering practices and conflicts, also stems from the question of economic survival. In short, gathering and its professionalization, in the broadest sense of the term, seems like a challenge: is professional gathering always the kind of gathering that is sought after and experienced with passion? In this case, the doubts – and even outright hostility – about the risks of standardizing

33. *Hors Cadres Familiaux*: "outside the family circle for three generations".

knowledge, increasing controls or the imperatives of transparency reveal the difficult standardization of a highly individual and solitary practice.

Finally, gathering has become a marginal activity in the economy. Human subsistence has long been based, for the most part, on the gathering of spontaneous flora. The narrative of "domestication" and the transition to agriculture, which relegated the hunter-gatherer to a semblance of oblivion or true archaism, is more complex than we think: "low-density food production" (Scott 2019), between wild and domesticated, has been the norm for many societies for centuries and remains so for some peoples. In a complementary way, the "rewilding" practices analyzed today suggest a need to return to naturalness (Dalgalarrondo and Fournier 2020), after several centuries of repression (Moscovici 1990). There is also a growing interest in wild edible plants and the knowledge associated with them, as described in numerous articles. Finally, should gathering these days be considered a taking of a "common good" in the sense of a shared environment that needs to be collectively protected, or does it remain an infringement of the law, referring to an "exclusivist" conception of property rights (Girard 2016)? The following quotation illustrates the tensions between these various positions:

> Don't recognize the gatherers. I don't know why. Because we don't have land, because we don't own property? It is like people, nomads, whether they are travelling or not, are still people, in some countries we don't even recognize their territory … It is biblical, it is the nomad against the sedentary, it is the farmer against the hunter-gatherer. On the whole, we are a society of belonging, of property. We go and pick other people's things! It is wrong! […] It has always been wrong. Because we've always had a problem with nomads, with people who hunt, with people who are a bit out of the ordinary, who don't belong to anything, with hermits in the woods, because at some point you've got to fit in, you've got to farm, you've got to work! You have to work. That is it, isn't it? We, the *paysan*-farmers, are aligned with the land a little more, but when they see us leave on the truck, to them it seems like we didn't work, that it wasn't work that we were doing. I think I work a lot more than them, they sit in their tractor all day, I don't think they do what I do! (Gatherer-cultivator, Puy-de-Dôme 2022, authors' translation).

6.4. Gathering territories and rural areas: separate paths or mutual learning?

For those who gather large quantities in widely scattered locations, identifying species and establishing perennial territories are strategic activities (Pascal 2018).

For others, these territories are just as important, but they are created more through acquaintance relationships. In all cases, cooperation with rural actors facilitates access to coveted plants. We discuss below the two social components of gathering territories: the "spaces experienced" (Frémont 1974) by gatherers and other users, and the institutional spaces governed by regulations, uses and practices[34].

6.4.1. *The making of gathering territories*

In the course of our surveys, we found that anyone who regularly forages knows their territory, defined as all the areas where they forage. Most often, it consists of a mosaic of environments that form "sites" or "stations", as they put it. This territory is evolving in the sense that it is not the property of the gatherers, even though they have exclusive rights to it. The owner may change its use. It may evolve, plants sought may disappear, the environment may be transformed and the seasons and the ecological successions must be taken into account (AFC 2022; Pascal 2018).

This making of the territory is based on an almost permanent process of prospecting and scouting. The gatherers' range of action varies according to their marketing methods. For many gatherers, the combination of cultivation and gathering activities limits their geographical gathering area to the neighborhood, which can be up to 60 km or one or two hours away by car from their home. However, for others, or for specific plants, they can go much further on an occasional yet regular basis, depending on seasonality and their needs. Prospecting is therefore not only limited to the possibility of accessing a single gathering site, once and for all, but also to the deployment of methods to secure access to other gathering spots. This perspective of sustainability places the gatherers in a situation where they have to come to terms with the various stakeholders involved in the gathering site, whether human (owner, user, contractual, manager, other gatherers, etc.) or non-human (animals, other plant populations, etc.).

It has been shown that gathering "requires a significant a global overview of the territory, its trails, plants, ecosystems and seasons", derived from experience, from which arises "the desire to take care" (Doyon 2019, authors' translation), and it can be said that the gatherers develops "care" (Brugère 2021) applied to these environments, what others will call "management strategies". There are many different types of knowledge involved, and testimonials reveal a wide variety of

34. The socially constructed dimension of territories is based on practices, representations and actions that contribute to their identification and delimitation. The forms of appropriation at work, whether cultural, social, economic or political, are closely dependent on the actors involved (individuals, social groups, collectives, public authorities) (Chery et al., 2018).

ways of approaching plant populations and their environments. The gradient of these different approaches ranges from a sensitive, aesthetic perception, based on affinities, cross-referencing of factors or clues, to research based on scientific indications concerning soil composition, topographical characteristics or phytosociology. Many gatherers testify to oscillating between these two approaches[35]. In all cases, the notion of a "sixth sense", a relationship with knowledge that is intuitive and internalized, comes up regularly (Pascal 2018), and it is sometimes difficult for gatherers to explain how they manage to find sites.

Figure 6.2. *Gathering elderberry (*Sambucus nigra *L.), Hautes-Alpes (2023)*

Legally, all gatherings must be authorized by the landowner, who has a "right of exclusion" (Girard 2016). Owners are most often private individuals but may also be institutions such as the National Forestry Office (ONF[36]) or a commune within its territory. For example, the ONF issues written authorizations for gathering, theoretically in return for payment of a fee. In the case of a commune, we are not aware of any agreements involving financial compensation. The majority of gatherers say they agree with the need to ask for permission to gather, in the search

35. The particular case of salaried seasonal gatherers, hired by a "boss" to gather on authorized sites, is not covered by this chapter.

36. Office National des Forêts.

for a relationship that can be sustainable on a human level and not have to hide or make themselves dubious in the eyes of other users. However, the reality in the field regularly involves unauthorized gathering, for example, when a series of trees is gathered from a multitude of owners who are often absent or unidentified, or when the gathering is minimal and takes place over a short period of time on the site:

> Well … things that have a common market value, like arnica and gentian, a lot of people gather them, everyone knows that it is a resource that can be monetized, and I wouldn't even consider gathering them without asking. After that, there's a bit of … I'd say there's a lot of time spent. If I stay in the same place for three days, yeah, I'll ask. On the other hand, the fig tree […], you stop gathering for a quarter of an hour, half an hour, you stop gathering, you get back in the truck, you leave again. If I had to ask permission for all the fig trees, it would be an unmitigated mess […] (Gatherer-cultivator, Loire 2022, authors' translation).

Generally speaking, the question of authorization is a matter of debate. Some will make it a sine qua non condition, others criticize the obligation to seek authorization to extract a resource that interests no one, and most will "oscillate" between these two poles. The dominant legal imaginary, centered on "absolute" private property, thus deserves to be questioned, notably through the prism of a new way of seeing this property as a "bundle of rights", shared between different users (Girard 2016).

In any case, even though authorization is granted to a gatherer who practices "responsible" gathering, there is no guarantee that the activity will continue. "Thefts" can occur, and situations of competition can arise when owners give their authorization to several people at the same time, sometimes fostering a culture of secrecy similar to that found in mushroom gathering, or at other times situations of open conflict, with, on the margins, threats or violence between gatherers. Over and above these issues of authorization, top-down initiatives are being put in place to regulate gathering in order to prevent conflict and promote sustainability.

6.4.2. *Frameworks and governance for gathering activities*

Concerns about the state of the resource for plants considered vulnerable have emerged, notably through the AFC. Its ambition is to build a gathering ethic in consultation with environmental managers and to transform the image of gatherers, making them their partners. A change of vision has also made itself felt within the public bodies responsible for protecting flora: the French Ministry of Ecological

Transition (MTE[37]), the French Biodiversity Office (OFB), the National Botanical Conservatories, regional and national parks, and nature reserves. Having once been viewed as a damaging activity, gathering is now seen in a more nuanced light. Gatherers are professionals who are able to take a specific look at the environments they work in and help to improve understanding of their condition and how to protect them.

With this in mind, the MTE provided financial support to the AFC, enabling it to publish, after several years of in-house work, its "Guide to Good Practice" (AFC 2022; Chaber et al. 2013), enacting numerous recommendations on how to work with plants. This approach constitutes a form of "ecologization of work" in the sense of taking into account the active and fragile nature of natural entities (Mouret and Lainé 2021), in this case plants. Technical booklets on gathering particular species have also been published at regular intervals. For each species treated, they detail the procedures to be followed to ensure the plant's survival, and the actions to be avoided (e.g. felling the plant, gathering all the reproductive parts of a site, etc.). These results could become the basis of a shared standardization in which the gatherers would have embraced the role claimed by the AFC of "nature sentinel" in line with the naturalist logics of environment managers (Julliand et al. 2019). This would recognize the ability of these "plant workers", in direct contact with nature, to provide expertise on environments[38].

However, the recognition of gatherers as legitimate actors in the management of environments is not unanimously shared, particularly by certain botanists and managers we interviewed. In their view, gathering remains an activity that is essentially harmful to plant populations, and a conservation approach would be antagonistic to a commercial objective. However, gathering is regulated by environmental policies, the most recent of which is the 2016 Law for the Reconquest of Biodiversity, Nature and Landscapes[39], from which a few articles aim to adapt gathering regulations to biodiversity issues. At present, plant species are protected at the national level (1989 decree), the departmental level (by various prefectorial decrees) or within protected areas (national parks, nature reserves, Natura 2000 areas) (Cambacedes and Garreta 2018). These policies are binding on gatherers, who must respect their conditions: not gathering or no longer gathering, requesting

37. Ministère de la Transition Ecologique.
38. Granjou uses the term "nature professionals" (2010) to refer to all professional naturalists (associations, government managers, etc.). It is necessary to distinguish them from "nature workers" as defined by Salmona (1986), and plant workers, which has a more restrictive meaning and today refers to those who work in proximity to plants.
39. La loi pour la reconquête de la biodiversité de la nature et des paysages.

authorization from France's Departmental Authority for Territories (DDT[40]), indicating gathering locations and quantities gathered, etc.

Is it desirable and possible to regulate gathering by the stakeholders concerned? Several recent initiatives seem to point in this direction. For example, the Parc National des Cévennes has authorized professional gathering of some 50 plants in the "heart zone", after two years of consultation with local users. In return, anyone gathering more than two kilograms per day is required to submit a request for authorization to the park, indicating the volumes desired and those actually gathered, as well as their sites, with a view to improving botanists' knowledge of plant populations. Recognized as one of the human activities present in the protected area, alongside other agricultural and livestock practices, gathering has acquired a new status.

Another example is arnica, an emblematic Vosges plant that is currently under threat and has given rise to concerted management initiatives in a number of locations. The best known of these is the multi-party agreement (departmental council[41], communes, mixed development syndicate, farmers, companies using arnica) on the Markstein plateau, set up following liming practices by farmers that led to the disappearance of arnica. Following conflicts of use, the Catalan Pyrenees PNR[42] has also proposed accompanying measures based on field experiments to better understand vegetation dynamics and the balances that are established with pastoral activities or climate change. These management methods are likely to become more widespread, but the "cautious" nature (Champy 2012) of the profession, faced with unpredictable dynamics such as climatic disruptions and increasing gathering pressure on certain sites, must always be borne in mind.

In short, the management of gathering territories reflects the ambiguities of the profession's framework, discussed above. The change in the way gathering is viewed seems to benefit gatherers and, at the very least, integrate them as stakeholders in the territory. The development of agroecology, which refers to a change in the agricultural paradigm, could also benefit them. However, two limiting factors remain. The first relates to the occupation of rural areas, which are still predominantly dominated by a productivist imagination, leading to the concentration of land and the disappearance of natural areas, as well as the artificialization of soils and the spread of pollution in areas that were previously relatively unspoiled[43].

40. Direction Départementale des Territoires.

41. Conseil départemental, communes.

42. Parc Naturel Régional.

43. A large proportion of the wild plants marketed must be certified organic to meet market demand.

Climatic disturbances also need to be taken into account. They reinforce the specific nature of the tasks carried out by "nature workers", of which gatherers are a significant part.

6.5. Conclusion

Making a living from gathering plants is not limited to a stereotypical activity, cut off from the rest of society in favor of a direct relationship with natural spaces and the wild. Nevertheless, such a relationship is sought after by a majority of professional plant gatherers, whose practices are rooted in a vision of the world and a relationship with plants perceived as living beings to be cared for.

Combining two ways of producing plants that modernity had artificially separated, that of cultivating and that of picking in the wild, gatherers are struggling to find their place in the agricultural environment of the 21st century, itself in crisis. Inclusion of the gathering activities in agricultural and forestry systems, on a territorial scale, is subject to the precise contexts in which it is deployed (agricultural model, environmental management, landowner willingness) and suggests cooperative relationships. In contrast to the process of agricultural modernization based on "encapsulated" knowledge (Compagnone et al. 2018) and "top-down" techniques, the circulation of knowledge among gatherers largely frees itself from a standardized training system and draws on the experiences of each individual. Spurred on by the AFC and defended by several professional organizations, the regulation of gathering and the professionalization of its actors seems to be an unavoidable path, but it must not fall into the many shortcomings that professionalization can engender (Rétière 1998; Parent 2014), particularly within the agricultural profession or among nature professionals (Granjou et al. 2010; Alphandéry and Fortier 2013; Fortier et al. 2022).

As seen, the gathering profession is multi-faceted and cannot be reduced to a simple act of production. The ways in which it can be recognized are currently being hotly debated, as the approach presents a number of risks, and there are dividing lines within the gatherer community itself. From this point of view, the profiles of gatherers are similar to those of agricultural project initiators struggling to set up in business. While it is important to organize the profession so as to gain access to subsidies, disseminate best practices and regulate gathering, it is also necessary to decompartmentalize public policies and revisit our conception of work, here in contact with nature. We also need to rethink the way in which rural areas are occupied today and integrate the agroecological transition with the rural transition, with all its utopian implications for the social and institutional transformation of the agricultural, environmental and forestry sectors. The ecological crisis is also contributing to these transformations, overturning the dominant representations of rural space, which is becoming a place of emancipation (Gazo 2017) for some, a "land of opportunity" (Rimlinger 2022) for others. It justifies this renewed interest in the modes of production and organization of work of alternative spheres insofar as they can be bearers of innovations and "solutions" moving in the direction of the ecological transition as a project for society. Nor can we ignore the fact that our consumption-based social model will be called into question by the availability of wild plants (Mercan 2021).

6.6. References

AFC (2022). Guide de bonnes pratiques de cueillettes de plantes sauvages : Partie Générale. AFC, Milly la Forêt.

Alphandéry, P. and Billaud, J.-P. (2009). Retour sur la sociologie rurale. *Etudes rurales*, 1(183), 9–22, doi: 10.4000/etudesrurales.8895.

Alphandéry, P. and Fortier, A. (2013). Les données sur la nature entre rationalisation et passion. *Revue du MAUSS*, 2(42), 202–220. doi: 10.3917/rdm.042.0202.

Barral, S. and Pinaud, S. (2017). Accès à la terre et reproduction de la profession agricole. Influence des circuits d'échange sur la transformation des modes de production. *Revue française de socio-économie*, 1(18), 77–99.

Beau, R. (2017). *Éthique de la nature ordinaire. Recherches philosophiques dans les champs, les friches et les jardins*. Éditions de la Sorbonne, Paris.

Bernard, B. (2005). Le "cadrage-débordement" : un concept pour comprendre le changement. Le cas de la gestion des forêts publiques. *Revue du centre d'études et de recherches en administration publique*, 10.

Bloch, M. (1931). Les caractères originaux de l'histoire rurale française. *Bulletin de l'Association Guillaume Budé*, 56, 62 [Online]. Available at: http://www.persee.fr.

Bost, I. (2016). Ethnologie des herboristes en France, de l'instauration du certificat en 1803 à aujourd'hui. PhD Thesis, Paris Nanterre University.

Bouleau, G. and Gramaglia, C. (2015). De la police de la pêche à celle de l'environnement : l'évolution d'une activité professionnelle dédiée à la surveillance des milieux aquatiques. In *Activités professionnelles à l'épreuve de l'environnement*. Octares.

Bromberger, C. and Lenclud, G. (1982). La chasse et la cueillette aujourd'hui. Un champ de recherche anthropologique ? *Études rurales*, 7/35 [Online]. Available at: http://www.persee.fr.

Brugère, F. (2021). *L'éthique du "care"*. Presses Universitaires de France, Paris.

Cambacedes, J. and Garreta, R. (2018). The gathering of wild plants: Exploitation of naturel resources and conservation of a natural heritage. *Journal of New Sciences, Agriculture and Biotechnologie*, 13, 3355–3365.

Centemeri, L. (2019). *La permaculture ou l'art de réhabiter*. Éditions Quæ, Paris.

Chaber, L., Julliand, C., Moreau, D. (2013). Pré-étude du projet de charte nationale de cueillette professionnelle et de plantes. Premiers éléments de diagnostic. MEDD Report, FranceAgriMer.

Champy, F. (2012). *La sociologie des professions*. Presses Universitaires de France, Paris [Online]. Available at: https://www.cairn.info/la-sociologie-des-professions--9782130594697.htm.

Chery, J.-P., Pinton, F., Berthome, K. (2018). Redéfinir le territoire pour envisager l'avenir. Working Group coordinated by P. Martin and F. Pinton. AgroParisTech.

Coly, B. (2020). Entre transmettre et s'installer, l'avenir de l'agriculture ! *Les avis du CESE*, 10.

Compagnone, C., Lamine, C., Dupré, L. (2018). La production et la circulation des connaissances en agriculture interrogées par l'agro-écologie : de l'ancien et du nouveau. *Revue d'anthropologie des connaissances*, 12(2), 2. doi: 10.3917/rac.039.0111.

Dalgalarrondo, S. and Fournier, T. (2020). *L'utopie sauvage. Enquête sur notre irrépressible besoin de nature*. Les Arènes, Paris.

Deléage, E. (2018). Retour à la terre : entre promesses et contradictions. *Ecologie & Politique*, 2(57), 39–49.

Doyon, S. (2019). Les cueillettes commerciales au Québec : capter la diversité socio-environnementale. *EchoGéo*, 47. doi: 10.4000/echogeo.16873.

Dubet, F. (2009). *Le travail des sociétés*. Le Seuil, Paris.

Durand, J.M. (2020). Apprenons à accepter la part de végétal qui est en nous ! *L'ADN* [Online]. Available at: https://www.ladn.eu/nouveaux-usages/usages-et-style-de-vie/apprenons-accepter-part-vegetal-en-nous/.

Flores (2019). Document de synthèse des travaux du projet (2014–2018). UNIL, AgroParisTech, CBNPMP and FEH.

Fortier, A., Dupré, L., Alphandéry, P. (2019). L'autonomie entre marché, rapport à la nature et production de soi. Approche sociologique des pratiques apicoles. *Développement durable et territoires*, 10(2), 8. doi: 10.4000/developpement-durable.14580.

Fortier, A., Alphandéry, P., Dupré, L. (2022). Résister à la professionnalisation. Socio-histoire des catégories "amateurs" et "professionnels" en apiculture [Online]. Available at: https://hal.inrae.fr/hal-03781148v1/file/ResisterProfessionnalisationFortierAlphanDuprPosprint2022.pdf.

Frémont, A. (1974). Recherches sur l'espace vécu. *L'Espace géographique*, 3(3), 231–238. doi: 10.3406/spgeo.1974.1491.

Gazo, C. (2017). Emancipation et nouvelle paysannerie. Appréhension du fondement matériel de l'autonomie politique et de la liberté dans l'imaginaire des nouveaux paysans. Master's Thesis, Paris Diderot University.

Ginelli, L. (2015). Loisirs de nature et environnement—Liens avec les professions, spécificités des épreuves. In *Activités professionnelles à l'épreuve de l'environnement*, Arpin, I., Bouleau, G., Candau, J., Richard-Ferroudji (eds). Octares, Toulouse.

Girard, F. (2016). La propriété inclusive au service des biens environnementaux Repenser la propriété à partir du bundle of rights. *Cahiers Droit, Sciences & Technologies*, 6, 185–236. doi: 10.4000/cdst.505.

Granjou, C., Mauz, I., Cosson, A. (2010). Les travailleurs de la nature : une professionnalisation en tension. *SociologieS*. doi: 10.4000/sociologies.3296.

Hervieu, B. and Léger, D. (1983). *Des communautés pour les temps difficiles. Néo-ruraux ou nouveaux moines*. Le Centurion, Paris.

Hervieu, B. and Viard, J. (1996). *Au bonheur des campagnes (et des provinces)*. Editions de l'Aube, Paris.

Hervieu-Léger, D. and Hervieu, B. (1979). *Le retour à la nature. "Au fond de la forêt... L'État"*. Le Seuil, Paris [Online]. Available at: https://www.cairn.info/le-retour-a-la-nature--9782020051798.htm.

Julliand, C. and Garreta, R. (2017). Pour un autre regard sur la cueillette commerciale de plantes sauvages : les pratiques gestionnaires des cueilleuses et cueilleurs professionnels. In *3èmes rencontres végétales du Massif central. Découvrir, comprendre et protéger la flore et la végétation du massif central*.

Julliand, C., Pinton, F., Garreta, R., Lescure, J.-P. (2019). Normaliser le sauvage : l'expérience française des cueilleurs professionnels. *EchoGéo*, 47. doi: 10.4000/echogeo.16987.

Kazic, D. (2022). *Quand les plantes n'en font qu'à leur tête. Concevoir un monde sans production ni économie*. La Découverte, Paris.

Labbé, J. and Imbert, C. (2018). Rapport d'information fait au nom de la mission d'information sur le développement de l'herboristerie et des plantes médicinales, des filières et métiers d'avenir. Report No 727, French Senate.

Larrère, R. and de la Soudière, M. (2010). *Cueillir la montagne*. Ibis Press, Paris.

Latour, B. (1991). *Nous n'avons jamais été modernes. Essai d'anthropologie symétrique*. La Découverte, Paris.

Laucoin, V. (2012). La cueillette de plantes sauvages sur le territoire d'agrément du CBN Massif central : état des lieux et perspectives. Master's Thesis, University of Lille II and University of Lorraine.

Lescure, J.-P. and Pinton, F. (2019). La cueillette au XXIème siècle, l'invention d'un métier, communication à la fête des SIMPLES. Cornimont.

Lescure, J.-P., Thévenin, T., Garreta, R., Morisson, B. (2018). Les plantes faisant l'objet de cueillettes commerciales sur le territoire métropolitain. Une liste documentée. *Le Monde des Plantes*, 517, 19–39.

Lieutaghi, P. (1983). *Les simples entre nature et société. Histoire naturelle et thérapeutique, traditionnelle et actuelle, des plantes médicinales françaises*. EPI, Mane.

Lowenhaupt Tsing, A. (2017). *Le champignon de la fin du monde. Sur la possibilité de vivre dans les ruines du capitalisme*. La Découverte, Paris.

Maresca, S. (1988). La réussite d'une politique. Interrogations sur les effets sociaux des incitations publiques. *Cahiers d'Economie et sociologie rurales*, 6(1), 5–40.

Mercan, A. (2021). *Manuel de phytothérapie écoresponsable. Se soigner sans piller la planète*. Terre Vivante, Mens.

Moscovici, S. (1990). *Hommes domestiques et hommes sauvages. Le marxisme et la question naturelle*. Bourgeois, Paris.

Mouret, S. and Lainé, N. (2021). AAC "Natures au travail". *Revue d'anthropologie des connaissances*, 1.

Muraz, M. (2018). La cueillette commerciale de plantes sauvages sur les Monts d'Ardèche PNR des Monts d'Ardèche. Master's Thesis, Montpellier SupAgro.

Ostrom, E. (2010). *Gouvernance des biens communs, pour une nouvelle approche des ressources naturelles*. De Boeck, Brussels.

Parent, P. (2014). "Seuls les médecins se piquent d'acupuncture ?" Le rôle des associations Professionnelles de praticiens dans la régulation de pratiques professionnelles de médecine chinoise en France. *Terrains & travaux*, 2014/2(25), 21–38.

Pascal, E. (2018). Récoltes de pratiques et de savoirs chez les cueilleurs de plantes sauvages : permanences et évolutions d'un groupe professionnel autour de la problématique de la gestion de la ressource. Master's Thesis, Muséum National d'Histoire Naturelle, AgroParisTech, Projet Flores.

Pinton, F., Julliand, C., Lescure, J.-P. (2015). Le producteur-cueilleur, un acteur de l'interstice ? *Anthropology of Food*, S11. doi: 10.4000/aof.7902.

Ponchelet, D., Mundler, P., Cartier, S., Rémy, J., Fabre, C., Laurent, C. (1998). L'activité agricole des ménages ruraux et la cohésion économique et sociale. *Économie rurale*, 12, 21 [Online]. Available at: http://www.persee.fr.

Pouteau, S. (2022). La condition de créature agricole : dignification des plantes et agroécologie. *La pensée écologique*, 9(2), 112–123. doi: 10.3917/lpe.009.0112.

Poux, X. and Aubert, P.-M. (2021). *Demain, une Europe agroécologique. Se nourrir sans pesticides, faire revivre la biodiversité*. Actes Sud, Arles.

Rémy, J. (1987). La crise de professionnalisation en agriculture : les enjeux de la lutte pour le contrôle du titre d'agriculteur. *Sociologie du travail*, 415–441 [Online]. Available at: http://www.persee.fr.

Rémy, J. (1997). Les sans-dot de l'agriculture : faut-il aider les installations sans aide ? *Économie rurale*, 33–37 [Online]. Available at: http://www.persee.fr.

Rétière, J.-N. (1998). Une indépendance sous contrôle, la professionnalisation de la pêche à pied. *Annuaire de droit maritime et océanique*, 233–260.

Rimlinger, C. (2022). Mise en perspective des néo ruralités écoféministes. *Colloque Néo ruralités à l'ombre des catastrophes*, 9–11 November, University of Nanterre.

Roux, J., Charvolin, F., Dumain, A. (2009). Les "passions cognitives" ou la dimension rebelle du connaître en régime de passion. Premiers résultats d'un programme en cours. *Revue d'anthropologie des connaissances*, 3(3), 369–385. doi: 10.3917/rac.008.0369.

Salmona, M. (1986). Questions d'épistémologie, d'éthique et d'action/formation sur le travail avec la nature et les travailleurs de la nature et du vivant. *Colloque "diversification des modèles de développements ruraux"*, 17–18 April.

Scott, J.C. (2019). *Homo Domesticus. Une histoire profonde des premiers Etats*. La Découverte, Paris.

Sencébé, Y., Pinton, F., Alphandéry, P. (2013). Le contrôle des terres agricoles en France. Du gouvernement par les pairs à l'action des experts. *Sociologie*, 4, 251–268.

Sennett, R. (2010). *Ce que sait la main. La culture de l'artisanat*. Albin Michel.

Stuppia, P. (2016). La révolution dans le jardin. Utopies communautaires et expériences néo-rurales françaises après Mai 68. *Éducation et sociétés*, 37(1), 49–64. doi: 10.3917/es.037.0049.

Timoshyna, A., Zhang Ke, Z., Yang, Y., Ling, X., Leaman, D. (2020). The invisible trade: Wild plants and you in the times of COVID-19 and the essential journey towards sustainability. *Traffic* [Online]. Available at: https://www.traffic.org/publications/reports/the-invisible-trade-wild-plants-and-you-in-the-time-of-covid-19.

7

How Can Biodiversity Be Taken into Account on Farm Territories?

Aude BARBOTTIN[1], Daphnée DURANT[2], Raymond REAU[3] and Hélène GROSS[4]

[1] *Paris Saclay University, AgroParisTech, INRAE, UMR SADAPT, Palaiseau, France*
[2] *Domaine expérimental de Saint-Laurent-de-la-Prée, INRAE, UE0057, Saint-Laurent-de-la-Prée, France*
[3] *Paris Saclay University, AgroParisTech, INRAE, UMR Agronomie, Palaiseau, France*
[4] *ACTA – Les Instituts Techniques Agricoles, Paris, France*

7.1. What is biodiversity in agricultural landscapes?

Growing awareness of the need to combat biodiversity loss, coupled with political and social demands, is prompting rural stakeholders to build and manage agroecological farming systems in multifunctional landscapes. Biodiversity is a key component of these areas, and farmers in particular, are called upon to preserve and manage it sustainably. While biodiversity lies at the heart of the relationship between humans and nature, its protection, conservation and management are hampered by the multiplicity of attitudes and viewpoints adopted by those involved in agricultural landscapes.

Agriculture has a close relationship with the biodiversity it uses (crop beneficials, soil, flora and fauna, etc.), impacts, helps to preserve or suffers (crop pests, wildlife damage, etc.). The relationship between agriculture and biodiversity is often posed in terms of compromise or cohabitation (Barbault et al. 2009). This

Farms and Territory,
coordinated by Philippe MARTIN and Solène PISSONNIER. © ISTE Ltd 2025.

dichotomy often overlooks the fact that agriculture has beneficial effects on biodiversity at different scales and levels of organization, generating habitats and food resources necessary for certain components of biodiversity. At the same time, the benefits of biodiversity for agriculture are also numerous for agricultural production in the broadest sense, through the ecosystem services it provides.

The notion of biodiversity is polysemous, evolutionary and has a multitude of meanings. From a formal point of view, biodiversity can be defined as the variability of the ecosystems and the organisms included (Convention on Biological Diversity 1992, Article 2). It can then be broken down into four organizational levels: genetic diversity, species diversity, ecosystem diversity and, more recently, functional diversity. To these four dimensions can be added the evolutionary dimension (phylogenetic for the past) and the networks of interactions between organisms, which go beyond the description of species and the functions they carry. These different levels of biodiversity are not fixed in time, and studies focus on conserving what already exists, as well as analyzing the dynamics of adaptation to change (i.e. species appearance/extinction ratio).

With the rise of the concept of ecosystem services, the concept of "biodiversity" has been enriched by a utilitarian vision, which has profoundly changed the way we understand biodiversity. The coexistence of these different schools of thought has meant that biodiversity has been seen either in terms of the services or benefits it provides to humankind or in terms of its own existence value (Le Roux et al. 2008). Today, this notion of value has been broadened to include values associated with nature that go beyond utilitarian and existence values (Intergovernmental Science-Policy Platform on Biodiversity and Ecosystem Services 2022).

7.2. Biodiversity as seen by farmers

Whether they are interested in biodiversity or not, successful projects to conserve or promote biodiversity in rural landscapes cannot be achieved without the involvement of farmers.

When asked about biodiversity, the majority of farmers state they are interested in it (i.e. Dephy[1] network surveys; Herzon and Mikk 2007). However, this interest in biodiversity varies according to the taxa considered. When it comes to birds and mammals, the majority of farmers would like to know more about all species. An exception applies to species they know to be likely to cause damage to crops or

1. See: https://ecophytopic.fr/pic/concevoir-son-systeme/la-perception-de-la-biodiversite-par-les-agriculteurs-du-reseau-dephy.

livestock: corvids and pigeons that consume seedlings, rodents that degrade soils, consume and stain stored feed and so on. When it comes to plants and insects, which are often better known, a distinction is made between "harmful" species and species of interest or those that are tolerated. This perception of biodiversity also varies according to the type of cropping systems. The work of Kelemen et al. (2013) shows, for example, that organic farmers adopt a more philosophical and holistic vision of biodiversity, linking it to the notion of ecosystems and the interrelationships between the organisms that make them up, compared to farmers considered to be conventional. The latter more spontaneously perceive biodiversity in terms of species richness or associated functions (pollination, soil structuring, etc.). The authors note, however, that there is considerable heterogeneity in the perception of biodiversity within this latter group. Beyond the type of farming practiced, the perception of biodiversity depends on the personal background of each farmer and their interests outside the workplace (initial training, family sensitivity, hunting activities, etc.), as well as the more or less biodiversity-rich environment in which the farm is located.

In their report, Maljean and Peeters (2001) define two categories of biodiversity from the farmer's point of view:

– Biodiversity that is planned, implemented and managed by the farmer for production (with no preconceived ideas here about the product, level, or mode of production; i.e. a crop for its grain, and also a hedge planted to generate a microclimate). Biodiversity is defined, at the gene level or in terms of the plant varieties and animal breeds, or even the strains of micro-organisms voluntarily used by the farmer.

– Associated biodiversity, made up of ecotypes[2], wild species and communities[3] that occur spontaneously in farming systems. Some of this biodiversity has functionalities that have a positive effect on production, such as wild plant species holding a forage interest, decomposing or nitrogen-fixing micro-organisms, parasitoids and predators of crop pests, pollinators, soil workers, etc. Conversely, some of this associated biodiversity can have negative effects on production, such as fungi causing plant diseases, crop pests, etc.

Finally, some of this associated biodiversity plays a role in the functioning of agroecosystems and is not perceived in terms of its impact on production (neutral

2. Ecotype: sometimes called subspecies, an ecotype is a genetically distinct variety, individual or population (morphe) of a given species that displays characteristics adapted to different habitats.

3. A community is a system in which living organisms share a common environment and interact.

perception), even when it contributes to the provision of ecosystem services. This is the case, for example, with birds, butterflies, amphibians, chiropterans, etc.

This classification is not absolute. Some species may be planned and then become associated with negative effects (e.g. volunteers of a previous crop with the next crop in the rotation) or change "status" according to the number or period of presence in agricultural plots (e.g. some arable weed species may be useful in providing resources for pollinators but have negative impacts if they are too numerous or limit crop growth at certain key stages of their development). Because of the temporal dynamics of agroecosystems, "associated" biodiversity evolves. This evolution depends on short- and long-term adaptation processes, from the individual to ecosystems.

For this chapter, we focus primarily on associated biodiversity, i.e. not established by the farmer for production.

7.3. Can agriculture and biodiversity conciliation occur at the farm territory level?

The biodiversity in actual agricultural landscapes is a co-construction by humans and nature. The development of agriculture from antiquity to the beginning of the 20th century led to the opening of landscapes and, in livestock farming areas, to their structuration by a network of hedges to enclose plots of land. Agricultural landscapes have become heterogeneous landscapes in which an adapted biodiversity has developed, the species most sensitive to these changes at the landscape scale having disappeared.

In the second half of the 20th century, the modernization of agriculture was supported by the development of the chemical industry (synthetic fertilizers and pesticides) combined with the reorganization of landscapes (land consolidation, drainage, irrigation and opening of the countryside). The aim of this modernization, which was to increase food production and thus farm productivity, had negative effects on the biodiversity of agricultural landscape. These effects were reported early (e.g. *The Silent Spring* (Carlson 1962)) and were subsequently confirmed by the abundant scientific literature on the subject. While the negative effects of agriculture on biodiversity dominate the debate, it should not be forgotten that agriculture also supports certain forms of biodiversity through the habitats and resources it generates.

Numerous factors have been highlighted in the scientific literature to explain the negative effects of agriculture on biodiversity: the intensification of farming practices (e.g. for birds: Benton et al. 2002; Boatman et al. 2004), the simplification of land use at the landscape scale (crop rotation), time (e.g. for birds: Siriwardena 2001) and the loss of landscape heterogeneity (semi-natural vs. cultivated elements) (Fahrig et al. 2011).

The issues of agricultural production (in quantity and quality) and biodiversity conservation raise the question of land use by opposing two strategies: (i) separating areas dedicated to agricultural production, where surfaces are almost totally occupied by the cultivated species with high productivity (high yield per hectare), and areas dedicated to biodiversity conservation, which will be protected (land sparing); and (ii) pursuing both objectives in the same place (land sharing) through an agriculture in which biodiversity would be conserved (Green et al. 2005). The partitioning of space according to a "land sparing" or "land sharing" strategy is the subject of much debate, in particular, because its validity seems to depend on the level of specialization of the species considered. For example, forest species will disappear as soon as this kind of habitat is cleared and cultivated. A key point in the debate also concerns the level of volume production targets set. If the latter are low, then they can easily be achieved with areas managed using little or no synthetic inputs or small areas managed more intensively with high yields; the opposite is much more complex.

This debate on the division of space between areas dedicated to nature and cultivated areas cannot be envisaged without discussing production levels per hectare volumes produced, diets and the economic strategies associated with them. It also calls into question the roles, objectives and ambitions that society ascribes to agriculture: food production alone or also land management, the provision of ecosystem services, etc.

In French landscapes, the proposed division between "untouched by human activity" natural areas and productive areas seems difficult, if only because there are no large areas that are effectively free of human activity. Agricultural landscapes must therefore enable both objectives to be pursued simultaneously: conserving biodiversity and producing food and non-food resources. This dual objective can be achieved through the simultaneous presence of cultivated areas and semi-natural habitats (hedges, copses, grass strips, fallow land, etc.). We are returning here to the notion of landscape heterogeneity, whose effects on biodiversity are well documented (Fahrig et al. 2011; Kremen et al. 2012; Hass et al. 2018; Sirami et al. 2019).

Given the impact of agricultural transformations in the second half of the 20th century on landscapes, two major challenges are (1) to recreate heterogeneity (enhance the diversity of land use at the landscape scale) and diversity (of habitats) in agricultural areas where these are lacking and (2) to improve the habitat quality for the species that live there (reducing the pressures and disturbances linked to agricultural practices).

7.4. How can we support biodiversity conservation at farm level?

The issue of biodiversity conservation lies at the interface between at least two sectoral policies, the agriculture policy and the environmental one, and occupies three levels: international, European and national.

7.4.1. *Environmental policies for biodiversity conservation*

At international and national levels, these initiatives are in line with states' ratification of the Convention on Biological Diversity (CBD) (Secretariat of the Convention on Biological Diversity, Handbook of the Convention on Biological Diversity[4]), recently updated at the Kunming-Montreal meeting (Kunming-Montreal Global Biodiversity Framework, 2022). The CBD aims to achieve the conservation of biodiversity; the sustainable use of natural habitats and species; and the equitable and sustainable sharing of the benefits arising from the utilization of genetic resources. Following France's signature of the CBD in 1994, National Biodiversity Strategy was implemented. This strategy is currently being implemented in a new phase covering the period 2022–2030 (Stratégie Nationale pour la Biodiversité 2030[5]). Actions relating to the agricultural sector include, among others:

– regulatory aspects, in particular the use and the toxicity of plant protection products (Measures 3.2 and 3.5 of Objective 3);

– restricting soil artificialization (Objective 4);

– measures to promote the agroecological transition of agricultural production and food systems (Objective 6), in line with the European ambitions for the development of organic farming;

– the promotion of biodiversity in business strategies, in particular the creation of added value linked to labels or standards explicitly integrating biodiversity.

In addition to the national biodiversity strategy, which aims to bring coherence to all biodiversity-related policies, other initiatives aimed at reducing agricultural pressures on the environment have also been implemented at France's national level,

4. See: https://www.cbd.int/history/.
5. See: https://www.ecologie.gouv.fr/strategie-nationale-biodiversite.

such as the Grenelle de l'Environnement from 2007 to 2012, which led to the implementation of the Ecophyto program; the Stratégie de Création d'Aires Protégées or the Trames Vertes et Bleues scheme, which led to hedgerow preservation and planting programs (including the recent Plantons des haies[6] program as part of the National Agroforestry Plan) and the Conférences Environnementales, including the one held in 2012, which announced the drafting of a framework law on biodiversity and the creation of the French Biodiversity Agency[7] (resulting from the fusion of ONEMA[8], the National Parks Public Establishment[9], the Marine Protected Areas Agency[10] and the public interest group ATEN), which became the French Office for Biodiversity in 2019[11].

At the European level, there have historically been two directives aimed at protecting biodiversity: the Birds Directive (1979) and the Habitats Directive (1992). These established the regulatory basis for the European Natura 2000 network of natural land and marine sites identified for the rarity or fragility of their wildlife (animal or plant species) and their habitats. Following these directives, two successive action plans were set up to halt the loss of biodiversity and ecosystem services. The first, drawn up in 2001, aimed to halt the loss of biodiversity by 2010. This target was not met and was extended to 2020. The new biodiversity strategy aims to put biodiversity "on the road to recovery by 2030".

The integration of agricultural areas into the European strategy is based on the Common Agricultural Policy (CAP) as part of the "farm to fork" strategy, developing programs to promote biodiversity and a subside system based on performance targets. For example, member states are to take measures to reduce the use of chemical pesticides in general and the risks associated with them by 50% by 2030 and to reduce the use of the highest-risk pesticides by 50% by 2030. This approach is supported by the European Pollinator Initiative[12].

The EU strategy also calls for at least 10% of agricultural land composed of topographical features (hedges, ponds, isolated or aligned trees, copses, ditches, etc.) with high biological diversity (Vallé et al. 2023) and for at least 25% of land to be converted to organic farming by 2023. These objectives are confirmed in the

6. See: https://agriculture.gouv.fr/francerelance-45-meu-pour-planter-7-000-km-de-haies-en-2-ans.
7. Agence Française de la Biodiversité.
8. Office National de l'Eau et des Milieux Aquatiques: National Office for Water and Aquatic Environments.
9. Établissement Public des Parcs Nationaux.
10. Agence des Aires Marines Protégées.
11. Office Français pour la Biodiversité: https://www.ofb.gouv.fr/.
12. See: https://environment.ec.europa.eu/topics/nature-and-biodiversity/pollinators_en.

European agricultural strategy (CAP 2023–2027) with the implementation of "Eco-regimes"[13], which aim to support the agroecological transition and were expected to represent 25% of the first pilar subsides.

In addition to the financing of "Eco-regimes", regulatory actions to conserve biodiversity largely mobilize funding associated with the CAP. Actions to preserve and enhance biodiversity at the farm level are financed through voluntary agri-environmental schemes, of which the Agro-Environment-Climate Measures (AECMs) are the translation. These AECMs have established themselves as the instrument of choice for promoting biodiversity with "action-oriented" subsides, i.e. paying farmers for the implementation of "good agricultural practices" and land management.

AECMs are financial subsides paid annually to voluntary farms, which commit for five years to comply with a set of specifications. These farms are remunerated based on an additional cost/loss of profit compared with the practice usually implemented and are classified into three categories: localized AECMs, which corresponds to commitments made at the field or topographical unit level (e.g. creation of an intercrop favorable to pollination), system AECMs (also known as flat-rate AECMs, e.g. transition of phytosanitary practices, farm carbon balance, etc.), which apply to the entire farm system, and AECMs for the conservation of genetic biodiversity. To be eligible for AECMs, agricultural fields must be located in an area covered by an Agro-Environment-Climate Project (AECP) validated by the region. These approaches mainly involve committing farmers to implement the "good practices" laid down during the AECM development phase.

While AECMs appear to be promising tools for conserving biodiversity in agricultural areas[14], there is some controversy about their effectiveness, some believing that they have partly failed to achieve the desired results (Kleijn et al. 2006, 2001; Kleijn and Sutherland 2003; Boatman et al. 2010) – either because the "good practices" chosen were not really adopted beyond the five years of the commitment or because the effect of these "good practices" did not translate positively into results in terms of biodiversity. An analysis of their implementation in Natura 2000 sites shows that, while promising (Princé et al. 2021), this system is hampered by administrative difficulties and regulatory changes that limit farmer take-up of AECM (report on AECM in Natura 2000 areas). The need to improve the financial sustainability of the schemes over the long term, to finance the coordination of these schemes with farmers and to monitor the results of the

13. Not all of these measures are explicitly aimed at biodiversity, but they do help to reduce pressure on ecosystems, cut down pollution, etc.
14. See: https://www.conservationevidence.com/index.php.

measures implemented are all factors put forward to improve the effectiveness and uptake of these measures by farmers.

Some authors suggest that they could largely be improved by defining targeted ecological outcomes and making payments conditional on the delivery of results ("payments for environmental services" principle proposed by the EU) (Burton and Schwarz 2013; Börner et al. 2017) or by improving the adaptability of measures to local conditions (Concepción et al. 2020).

Another subject of debate concerns the foundations on which these measures are based. They are based on the principle that protecting biodiversity generates costs that can be offset by the AECMs. The "agroecology" position is that biodiversity is necessary for the proper functioning of the agroecosystem, and that it is therefore in the farmer's interest to protect biodiversity in order to optimize his system.

Beyond the production activity, some see agriculture as a source of environmental services, in this case the preservation of biodiversity. They consider that it is more appropriate to remunerate farmers for these services (see Payments for Environmental Services) than for any losses or additional costs generated by changes in their practices in favor of biodiversity (Kleijn et al. 2015).

7.4.2. *Examples of initiatives to support farmers in conserving biodiversity*

In addition to the actions carried out by the public authorities, initiatives supported by farmers' associations and groups have also been implemented. For example, the five-year national Agriculture and Biodiversity program (2004–2009) was launched by the League for the Protection of Birds[15] and some agricultural networks, providing an opportunity for naturalists and farmers to engage in in-depth dialogue to define actions to promote biodiversity on around 130 farms in 18 French départements (Clément et al. 2009).

The aim of these initiatives is to develop a greater account of biodiversity on farms. They are carried out by companies, naturalist associations, farmers' groups and agricultural structures or by institutional structures. Here are a few (non-exhaustive) examples from each of these categories.

15. Ligue pour la Protection des Oiseaux (LPO).

Des Terres et des Ailes[16] *(https://www.desterresetdesailes.fr/)*

Des Terres et des Ailes is a program supported by the LPO association and the French Chamber of Agriculture[17]. Launched in 2018, it is explicitly aimed at farmers who wish to implement one or more actions to preserve biodiversity on their farm territory. Through this program, volunteer farmers can access feedback from their peers, resource documents and personalized support. This program has led to the completion of 5,224 semi-natural elements, the creation of 931 ha of grassland and the planting of 57,137 ml of hedges since 2018.

Paysans de Nature[18] *(https://www.paysansdenature.fr/paysans-de-nature-presentation/)*

Paysans de Nature is an association that supports farmers in protecting biodiversity in agricultural landscapes. The association was born of a partnership between a network of farmers and the LPO Vendée. Paysans de Nature is also a registered trademark (2017). The association participates in the dissemination and implementation of agricultural systems for which the defense of wild biodiversity is a central activity. Its main objective is to create and manage natural areas enhanced by agriculture or to protect them within the framework of contracts or laws with farmers.

Observatoire Agricole de la Biodiversité (OAB)[19] *(https://www.observatoire-agricole-biodiversite.fr)*

The OAB is a participatory science program launched in 2009 by the French Ministry of Agriculture and coordinated by the French "National Museum of Natural History"[20], the University of Rennes 1 and the LADYSS laboratory (CNRS). It proposes observation protocols for ordinary biodiversity[21] in order to raise awareness of farmers to biodiversity and gain a better understanding of it. The aim is to respond to an acknowledged lack of indicators for monitoring the state of biodiversity in agricultural environments, according to farming practices. This program is based on a network of volunteer farmers and demonstration sites. Five groups of species are monitored, according to the functions they perform in agricultural landscapes (Table 7.1 sourced from the OAB).

---

16. Land and Wings.
17. Chambres d'agriculture, formerly Assemblée permanente des Chambres d'Agriçulture.
18. Nature's Peasants.
19. Agricultural Observatory of Biodiversity.
20. Muséum National d'Histoires Naturelles.
21. Ordinary biodiversity: as opposed to remarkable biodiversity, this represents the biodiversity that surrounds us, i.e. all the species that are abundant in a given ecosystem.

Protocols	Related agricultural themes
Solitary beehives	Pollination, landscape quality
Butterfly transects	Pollination, landscape quality
Earthworm plots	Soil fertility
Terrestrial invertebrate plots	Pest monitoring, crop auxiliaries
Bat recorder	Crop auxiliaries, landscape quality

Table 7.1. *Available and implemented OAB protocols and associated ecosystem services for farmers (from https://www.observatoire-agricole-biodiversite.fr)*

The results of this monitoring are published annually in scientific journals (e.g. Billaud et al. 2021).

The Agrifaune network (https://www.agrifaune.fr/)

Set up in 2006 following the approval of an Agriculture – Wildlife – Hunting[22] agreement, this network brings together stakeholders from the agricultural and hunting worlds. It contributes to the development of agricultural practices that reconcile economics, agronomy, environment and wildlife objectives. The overall objectives of the program are to:

– promote the integration of wildlife into productive and economically efficient agriculture;

– carry out actions to preserve biodiversity;

– promote the references acquired through the network;

– contribute to the sustainable development of rural areas.

Voluntary farmers can join the network if their farm is located in a project area supported by the Agrifaune program. Through this network, they can obtain information, carry out experiments on their farms and host or organize farm visits.

Six technical groups contribute to the production of knowledge on the contribution of one or more elements to the preservation of biodiversity: field edge management; intercropping; equipment; viticulture; pastoralism and small mountain fauna and grove.

In 2019, 68 départements were involved in the program, representing 150 partner structures and 30 agricultural high schools.

---

22. Agrifaune, agriculture – faune sauvage – chasse.

The Harmony charter (https://www.harmony.info/fr-fr/index.html)

Launched in 2008 by the LU brand (Mondelez Group), the Harmony chart was designed to reduce the environmental footprint associated with the production of the cereals needed to make their cookies. It involves a partnership with volunteer farmers whose fields are located as close as possible to the wheat processing units. Farmers who sign up to the chart receive a premium for implementing a set of good farming practices and the creation and management of biodiversity-friendly facilities. The chart covers 30 measures aimed at:

– reducing the use of pesticides and fertilizers (choice of wheat varieties for resistance and cookie quality; recommendations on crop rotation and soil preservation; use of decision-making tools for implementing treatments; recording and justification of each treatment);

– reducing carbon emissions, in particular by reducing the use of fertilizers;

– preserving water resources;

– preserving and promoting local biodiversity. Farmers who are partners in the program allocate 3% of each of their Harmony wheat fields to the planting of flowered strips for pollinating insects, as well as for auxiliary insects. All phytosanitary products are prohibited on flower strips.

The charter evolves every year and is adapted to local conditions.

In Europe, 1,600 farmers have signed up (figures for 2020, including around 1,300 in France), with a 20% reduction in the Treatment Frequency Index (TFI) on committed fields between 2008 and 2020 compared with the national average, and 1,450 ha of flowered areas attracting 15 million bees and 25 species of butterfly. In France, some 90,000 tons of wheat are produced under this chart.

Club AGATA (https://noe.org/qu-est-ce-que-le-club-agata)

Club AGATA is a program run by the Noé association, which aims to develop tools and methods to support the agroecological transition in agriculture. It is based on the results obtained by the Fermes Noé pilot farms. Its ambition is to better characterize the links between farming practices and biodiversity so as to better integrate its preservation into the management of agri-food chains, from suppliers to consumers. The club is not aimed directly at farmers but mobilizes players in the agri-food chain (agricultural cooperatives, retailers, producer groups, processors, distributors, label holders, etc.). The objectives, indicators and methods to be developed are drawn up collectively by participants and members of the Noé association.

AgriBEST® (https://agribest.fr/)

AgriBEST® is a diagnostic tool developed by Coopération Agricole de l'Ouest and CDC Biodiversité (Biodiversity subsidiary of Caisse des Dépôts et Consignations[23]). It aims to support farmers wishing to implement biodiversity conservation actions on their farms through a self-diagnosis of the biodiversity performance of their practices. The aim is to mobilize farmers around biodiversity conservation, trigger voluntary action and facilitate dialogue between actors. AgriBEST® is not a prescriptive tool but provides elements to help farmers choose freely, on their own or with support, how to improve their farming practices in terms of biodiversity, by integrating all the dimensions (economic, organizational, structural, etc.) and sectoral and territorial issues of the project they wish to achieve. Since its launch, 181 diagnoses have been carried out.

7.5. Examples of integrating biodiversity into farm projects: challenges, results and difficulties

Integrating biodiversity into farm projects requires farmers to change their point of view so that biodiversity says something about the agroecosystem they manage. Whether to improve the expression of ecosystem services (regulatory processes, for example) or to conserve "wild" biodiversity on the farm, agronomic decisions need to be adapted to time scales longer than the growing season and to spatial scales larger than the cultivated field. Farmers must also move from a decision-making process based on standardized indicators to a contextualized decision based on knowledge of how agroecosystems function.

Farmers also need indicators to monitor and evaluate the effects of their actions.

7.5.1. *Thinking and acting using result-oriented biodiversity conservation strategy? The case of the Saint-Laurent-de-la-Prée farm*

For several years now, a group of engineers and technicians has been testing an adaptive management approach based on a "results-oriented strategy" on an INRAE experimental farm.

Located on the French Atlantic coast, in the "dried marshes" of Saint-Laurent-de-la-Prée, near Rochefort-sur-Mer (Charente-Maritime), the INRAE experimental farm Ferme de la Prée, has been running a system experiment called Transi'marsh since 2009 with the objective to implement the sustainable agriculture transition of

23. Deposit and Consignment Office.

the farm and to promote wild and domestic biodiversity (Durant et al. 2020). The Ferme de la Prée was converted to organic farming in 2017.

Numerous actions have been – and are being – carried out on the farm: planting hedges, diversifying crop succession, sowing grass and flower strips around and within cultivated fields, biodiversity-friendly management of meadows, extensive grazing, installing nest boxes and bat shelters, digging ponds, etc.). Biodiversity protection is integrated into the farm project, and the aim is to move on from the application of a list of "good practices" (which has been done for several years) to the adoption of a more adaptive, result-driven approach, based on quantified objectives to be achieved in terms of the desired biodiversity on the farm.

As part of a research program financed by INRAE (ADORE[24] project), Ferme de la Prée is applying this result-based approach. This is based on an action strategy involving expected results and planned actions, all clearly explained and shared by the farm's research team. The action plan is thus goal-oriented (Cerf and Sebillotte 1988) and driven by the principles of adaptive management (Prost et al. 2018). This is a four-step iterative process (Ferrané et al. 2020) based on a step-by-step design logic (Meynard et al. 2023).

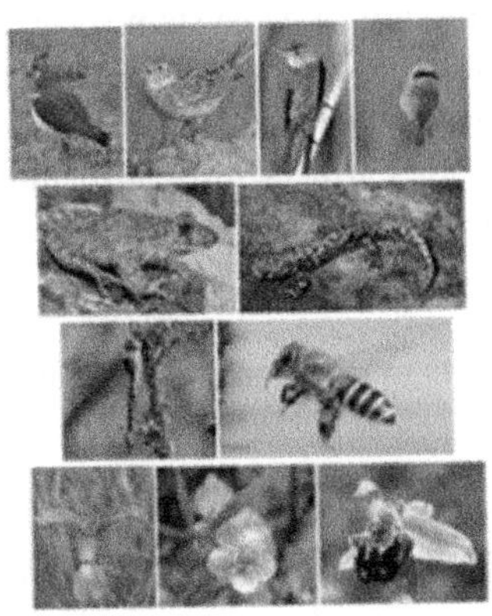

Figure 7.1. *The 11 selected target species. From top to bottom and left to right: lapwing (*Vanellus vanellus*), skylark (*Alauda arvensis*), red warbler (*Acrocephalus scirpaceus*), red-backed shrike (*Lanius collurio*), spotted pelodyte (*Pelodytes punctatus*), marbled newt (*Triturus marmoratus*), Aechne printanière (*Brachytron pratense*), bees (here the honey bee* Apis mellifera*), iris maritime (*Iris reichenbachiana*), buttercup (*Ranunculus ophioglossifolius*) and orchids (here the* Ophrys abeille Ophrys apifera*). INRAE or royalty-free photographs*

24. The ADORE project is funded by the INRAE BIOSEFAIR metaprogram (see: https://www6.inrae.fr/biosefair/Nos-actions/Projets-exploratoires/ADORE). The Transi'marsh program is financially supported by INRAE and the Nouvelle-Aquitaine region.

Before stage 1, 11 target species were selected (Figure 7.1), based on the expertise of members of the collective and local naturalists, and considering that the preservation of these species was an issue for the area in which the farm is located. The aim here is to achieve a rich biodiversity, with the emphasis on the heritage value of this biodiversity.

Stage 1 involves defining the project in its version for the coming year. Abundance objectives to be reached by 2025 are set for each target species, and the actions to be taken as intermediate objectives are formalized in the form of a dashboard as proposed by Girardin et al. (2019) (e.g. for the Skylark, see Figure 7.2). These dashboards make the logic of action explicit through the chain of causal relationships between the expected abundance targets, the habitats considered important to preserve/implement for the target species and the practices and/or semi-natural elements planned in the fields to generate those habitats.

Stage 2 is the project implementation phase, with the concrete realization of the actions decided in stage 1. Figure 7.3 shows some of the actions to be taken in 2021 to promote skylark protection, such as extensive cattle grazing (no more than 1 LU/ha) on natural grasslands, the creation of unmown strips (so-called "refuge strips") and late mowing of grass strips (not before September).

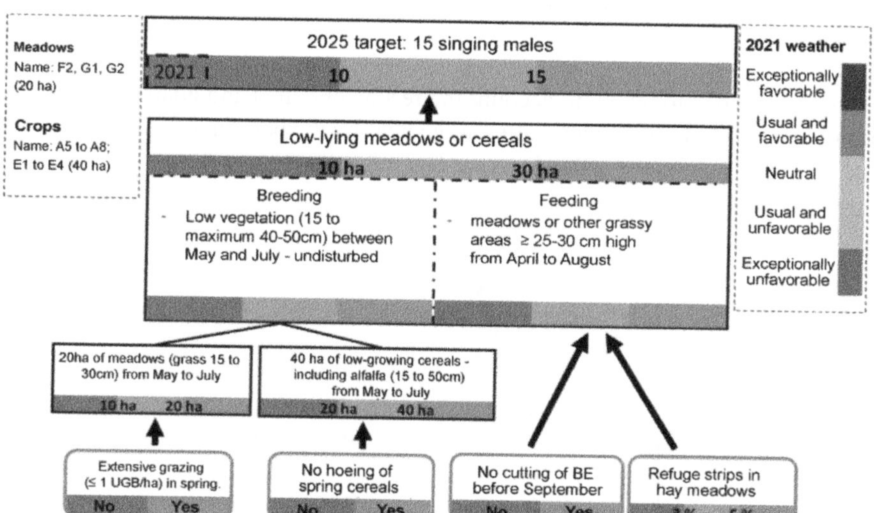

Figure 7.2. *Project dashboard for the skylark (*Alauda arvensis*). For a color version of this figure, see www.iste.co.uk/martin/farmsandterritory.zip*

Figure 7.3. *Examples of practices implemented in 2021 in favor of the skylark: A: extensive cattle grazing; B: 3 m wide refuge strips in natural grasslands; C: late mowing of grass strips. INRAE*

Stage 3 consists of observing or measuring target species and habitats, as well as the practices that have been implemented, and summarizing them in "dashboards of results obtained". This monitoring is carried out by specialists for each species or group of species.

Finally, stage 4 involves an analysis, and an evaluation, of the results obtained according to the expectations, and a diagnosis of the causes of failures, the conditions for success or the gap between the practices carried out and the results obtained. This stage is essential in the learning process of participants, both in terms of knowledge of the local situation and more generic knowledge of how agroecosystems work.

The next step is to take into account the results obtained and adapt the project by proposing a new action plan and, after agreement by the participants, to carry out the same learning cycle in the following year (Figure 7.4).

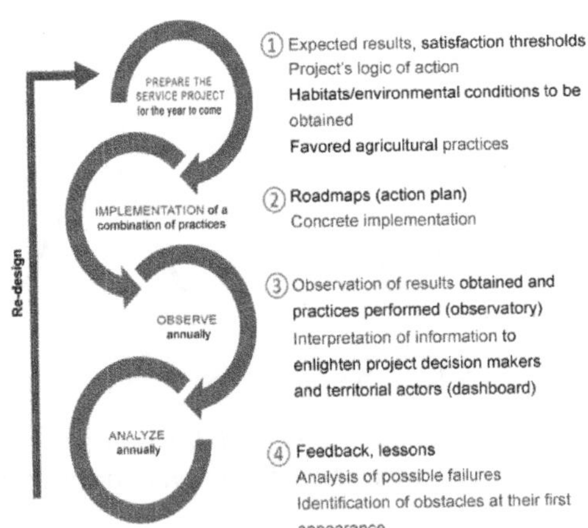

Figure 7.4. *Stages in the process and learning cycles*

Implementation of the first two learning cycles in the Ferme de la Prée (2021 and 2022) shows variable success depending on the species. Some results are promising, such as for the skylark and the maritime iris, with good abundance and habitat results. For other species, the target habitat has been reached, but the expected number of individuals of the species to be conserved has not been observed. In this case, the question arises as to the spatial scale to be taken into account in the process (as in the case of the lapwing) or the time required for the species to respond to the change (as in the case of the marbled newt). For others, despite the conformity of what has been done, the habitat is not satisfactory (as in the case of the red-winged warbler).

The iterative process highlighted (1) the need to translate each of the objectives into an operational "roadmap" for the participants and (2) the development of a "culture of compromise" within the collective involved in the approach, the dashboard being a good dialogue tool for reconciling agricultural production and biodiversity.

The advantage of this approach is that it makes explicit the target biodiversity that the farmer wishes to conserve and formalizes the relationship between their management actions and the results observed for this biodiversity. It is currently being tested on an experimental farm, with fewer constraints on the farming system than on private farms. The approach has been implemented within a collective for which biodiversity conservation is a shared issue. Extending the method to "real farms" would imply that biodiversity becomes an issue for the farmer, whether this biodiversity is perceived as a heritage (Barbottin et al. 2020), as a production factor (ecosystem service) or to meet the expectations of other stakeholders interested in biodiversity (industries, naturalist associations, hunting or fishing federations, public authorities, etc.) (Reboud 2019). The feasibility and conditions for implementing this type of approach on private farms remain to be explored, with questions such as: Who takes charge of observations on biodiversity and habitats? Who analyzes the results? How can we support the farmer in revising their project?

7.5.2. *Enhancing production through bocage and wetlands: the case of the Forgineau farm*

While in the previous case, biodiversity conservation was based on the choice of a few target species, it can also be achieved through the conservation, restoration and maintenance of the landscape. This is the choice made by Antoine Pasquier on his farm.

The Forgineau farm is an organic farm located in the commune of La Petite Boissière in the Deux-Sèvres region of France. Antoine Pasquier, who took over the family business in 2013, is a livestock farmer (beef cattle). The herd comprises approximately 60 mothers and their offspring. Of the 105 hectares of his farm, 80% is grassland. The meadows are managed using dynamic rotational grazing (grazing small 0.3 ha fields over a short period). In addition to the grass and hay produced on the farm, fodder crops are also grown to feed the herd. The cultivated fields have been managed using simplified cultivation techniques since 2000 (no plowing and minimal mechanical intervention).

Antoine Pasquier's objective is to conciliate biodiversity and agricultural activity, without making compromises, i.e. without jeopardizing the farm's economic profitability.

Passionate about bocage, in 2013, he launched the first stage of his agroecological project to restore the bocage on the farm[25]. The bocage is not seen here as a simple landscape element but is integrated into the farm's production objectives. In particular, it enables the production of trees for timber, and the pruning from its upkeep contributes to the production of wood chips for heating (Figure 7.5(a) and (b)).

In addition to its value for wood production, the bocage is particularly rich in biodiversity and limits climate change in agricultural areas. From a biodiversity point of view, the bocage is a mosaic environment combining closed areas (copses, old wide hedges, etc.), open areas (cultivated fields or meadows), interface areas between open and wooded areas and wetlands (ponds, for example).

Since 2013, around 3.5 kilometers of planted hedges and wetlands (ponds and the Ouin riverbanks) have been recreated (Figure 7.5(d)). The planting is carried out taking into account the mechanical constraints for field management (2–3 ha plots). Over the past few years, 6,380 trees and shrubs have been planted over a five-kilometer stretch, and 22 ponds have been created. Around 800 m of bocage remain to be created. In parallel with the restoration of the bocage, 4.5 ha of agroforestry have been planted in 2022, using around 30 species (Figure 7.5(c)).

One of the main difficulties lies in matching the timing of bocage interventions with key stages in the life cycles of biodiversity, particularly birds. Work must be carried out outside the nesting period and must not impact berry production from the hedge, which represents winter foods for the bird communities living on the farm. Added to these constraints, bocage management has to be done before fields are too

25. These various actions carried out on the farm are supported by the Bocage Pays Branché association and the Établissement Public Territorial de la Vilaine, as well as the Nouvelle-Aquitaine region.

wet in winter in a naturally damp environment, and fences have to be installed to limit access to hedges and coppices by cows while at the same time not penalizing maintenance operations and grazed areas.

Figure 7.5. *Examples of work carried out. a: Trimming the hedge to produce wood chips and opening up a footpath. b: Cutting back tall trees in the grove. c: Planting an agroforestry plot with different species of tall trees. d: Restoring a wetland on the farm. © A. Barbottin*

Including biodiversity conservation in the objectives of the farm implies that farmers can monitor the evolution of biodiversity they want to preserve (a landscape, a species or a group of species, etc.) as a result of the action plan that they implement (ecological engineering, changes in practices, etc.). It implies: (1) defining the biodiversity to be conserved; (2) the spatial scale at which actions will be implemented (field, part of the farm territory, the whole farm territory or even beyond) and the objective pursued (restoration/creation of habitats, conservation of a particular species, favoring a support function for an ecosystem service, etc.). Once these elements have been established, it is possible to define indicators "for biodiversity", i.e. that will estimate the impact of actions on biodiversity (Bockstaller et al. 2019).

7.6. Monitoring biodiversity

The literature contains numerous indicators for monitoring biodiversity. These indicators are related to species, taxonomic group descriptors (species richness, total abundance, local abundance, density of individuals, etc.), ecosystem or ecosystemic functions (i.e. relating to species' life traits).

Indicators used to quantify biodiversity can be classified according to two philosophies:

– evaluation and monitoring indicators, generally mobilized ex post for evaluation purposes. They assess the actions implemented, or results, most often in terms of changes in the compliance rate over time (Bockstaller et al. 2019);

– result indicators as part of project management, where action plans are modified whenever deemed necessary by the project's governance, in a design approach and not just one of monitoring progress.

The question of biodiversity indicators for assessment and monitoring purposes in action programs follows the Rio Conference (1992). The need to conserve and measure variations in biodiversity, both at the global level and within the EU, led to the setting of biodiversity targets in 2004. For France, these objectives are set out in the National Biodiversity Strategy (2004). This strategy aims to establish biodiversity indicators to monitor conservation policies. These indicators focus, for example, on genetic diversity, species diversity and habitat diversity (Table 7.2).

Genetic diversity	Number of plant varieties and animal breeds, registered and certified for marketing purposes, in the main crop and livestock categories
Specific diversity	Index of specific diversity of common birds
	Fish species richness
	Status of national red list species
Habitat diversity	Conservation status of habitats of community interest on Natura 2000 sites
Ecological fabric	Map of local land-use diversity with low levels of artificial development
	Dominance of semi-natural elements in the landscape
Ecosystem functioning	Tree defoliation
	Standardized global biological index for watercourses

Table 7.2. *Biodiversity indicators used in the French National Biodiversity Strategy (after Levrel 2007)*

The number of indicators that can be used to monitor the state of biodiversity is vast (over 350). In reality, this diversity of indicators is limited by the scale at which they can be used. Indicators derived from ecological monitoring are numerous and are mobilized at local scales. National indicators, which focus on biodiversity in a generic sense, are based on standardized data and are few and far between (Levrel 2007).

While there are many indicators available, they are developed to analyze temporal variations at scales often larger than that of the farm or field, and the correlation between the level of the indicator and a practice or set of practices is not so easy to demonstrate. Moreover, the biodiversity indicators used for evaluation require naturalist skills and knowledge that farmers do not necessarily possess.

However, the absence of a simple available indicator for farmers to monitor their biodiversity-related actions is not a hindrance to action (see the EU Saint Laurent Ferme de la Prée or AgriBEST® system). The CASDAR AgriBirds project has shown that the implementation of avian biodiversity measurement protocols by farmers, even neophytes, enables them to take ownership of birds as part of their farm's heritage. All participants, regardless of their initial level of ornithology, were able to apprehend a level of avifaunal diversity and observe differences according to season, landscape environment, weather and time of day (Barbottin et al. 2020).

The nature of the indicators and the way they are collected then need to be rethought according to the sensitivity of the taxa considered, the time available and the farmers' objectives (Manneville et al. 2016; Tasser et al. 2019). This approach is difficult for farmers to carry out alone if they do not have a naturalist background. Individual or collective support is needed, and the work of CASDAR AgriBirds has shown that farmers' learning is very often linked to the interactions they had with ornithologists. Farmers then become creators of naturalist data on their territory. These data can be used beyond the farm in participative science programs to analyze biodiversity trends (Billaud et al. 2021).

7.7. Conclusion

Integrating biodiversity as an issue for farms means no longer seeing the environment as a single constraint. The farmer then looks after an agroecosystem in which biodiversity becomes a production tool (ecosystem service) or a companion, witness to a "job well done". This "job well done" is no longer measured by the level of production achieved but by the reproducibility of the agroecosystem over time. This change in point of view calls for a cultural transformation that goes beyond the farmer as an individual.

Since the 19th century, innovation and design in agriculture have historically focused on technical aspects related to production and have led farmers to think of their management actions outside the biophysical functioning of the farm territory. These developments have taken the "agroecosystem" and all the complex relationships between species and their associated functions, including "heritage" biodiversity, out of the scope of agricultural design and innovation. Integrating biodiversity and associated ecosystem services into the design at the farm and the regional level means moving beyond the "one technique = one solution" framework. Given the knowledge available and the validation mechanisms (which fall outside the scope of generic knowledge), the mechanisms to be mobilized go beyond the design process usually used in agriculture (Berthet 2013). The historical developments in knowledge production have also led to a split within agricultural education between teaching to train future farmers and teaching for nature management and protection. Knowledge of how agroecosystems and biodiversity function has taken a back seat, although this is changing under the impetus of the Local Plan on Teaching to Produce Differently[26].

For farmers, the challenge is to reappropriate and recontextualize knowledge to which they have little or no access. Biodiversity then acquires a new meaning and says something about the agroecosystem they manage. The ability to observe biodiversity then becomes an imperative condition for change, an essential quality, not only for the pleasure of observing, but also to appropriate and conserve the biodiversity present on farm territory.

7.8. References

Barbault, R., Baudry, J., Burel, F., Doussan, I., Garnier, É., Herzog, F., Lifran, R., Trommetter, M., Lavorel, S., Roger-Estrade, J. et al. (eds) (2009). *Agriculture et biodiversité : Valoriser les synergies*. Éditions Quæ, Versailles. doi: 10.3917/quae.inra.2009.01.

Barbottin, A., Andrade, C., Omnès, F., Hélène, G. (2020). AGRIBIRDS – Concevoir et diffuser des indicateurs "Oiseaux" pour les acteurs du monde agricole. *Innov. Agron.*, 79, 19–30. doi: 10.15454/kxaw-4b30.

Benton, T.G., Bryant, D.M., Cole, L., Crick, H.Q.P. (2002). Linking agricultural practices to insect and bird populations: A historical study over three decades. *J. Appl. Ecol.*, 673–687.

Berthet, E. (2013). Application des principes de conception innovante à la gestion de conflits. Cas de la préservation de la biodiversité dans une plaine céréalière intensive de Poitou-Charentes. *Réunion du Réseau Recherche Antibiotiques Animal (R2A2).*

26. Plan Local Enseigner à Produire Autrement.

Billaud, O., Vermeersch, R., Porcher, E. (2021). Citizen science involving farmers as a means to document temporal trends in farmland biodiversity and relate them to agricultural practices. *J. Appl. Ecol.*, 58, 261–273. doi: 10.1111/1365-2664.13746.

Boatman, N.D., Brickle, N.W., Hart, J.D., Milsom, T.P., Morris, A.J., Murray, A.W.A., Murray, K.A., Robertson, P.A. (2004). Evidence for the indirect effects of pesticides on farmland birds. *Ibis*, 146, 131–143. doi: 10.1111/j.1474-919X.2004.00347.x.

Boatman, N.D., Pietravalle, S., Parry, H.R., Crocker, J., Irving, P.V., Turley, D.B., Mills, J., Dwyer, J.C. (2010). Agricultural land use and Skylark Alauda arvensis: A case study linking a habitat association model to spatially explicit change scenarios. *Ibis*, 152, 63–76. doi: 10.1111/j.1474-919X.2009.00969.x.

Bockstaller, C., Lasserre-Joulin, F., Meiss, H., Sausse, C., Van Der Werf, H., Denoirjean, T., Ranjard, L., Angevin, F., Manneville, V., Michel, N. et al. (2019). Les indicateurs de biodiversité pour accompagner les agriculteurs : embarras du choix ou pénurie ? *Innov. Agron.*, 75, 73–86. doi: 10.15454/BZMZZF.

Börner, J., Baylis, K., Corbera, E., Ezzine-de-Blas, D., Honey-Rosés, J., Persson, U.M., Wunder, S. (2017). The effectiveness of payments for environmental services. *World Dev.*, 96, 359–374. doi: 10.1016/j.worlddev.2017.03.020.

Burton, R.J.F. and Schwarz, G. (2013). Result-oriented agri-environmental schemes in Europe and their potential for promoting behavioural change. *Land Use Policy*, 30, 628–641. doi: 10.1016/j.landusepol.2012.05.002.

Cerf, M. and Sebillotte, M. (1988). Le concept de modele general et la prise de decision dans la conduite d'une culture. *C. R. Acad. Agric. Fr.*, 74, 71–80.

Clément, O., Faupin, G., Gauduchon, V., Guillou, E., Guyot, D., Jourde, P., Le Roy, E., Liozon, R., Macé, F., Métais, M. et al. (2009). Agriculture et Biodiversité – comment améliorer la biodiversité sur votre exploitation ? Recueil d'expériences. LPO, CIVAM, FARRE, FNAB.

Concepción, E.D., Aneva, I., Jay, M., Lukanov, S., Marsden, K., Moreno, G., Oppermann, R., Pardo, A., Piskol, S., Rolo, V. et al. (2020). Optimizing biodiversity gain of European agriculture through regional targeting and adaptive management of conservation tools. *Biol. Conserv.*, 241, 108384. doi: 10.1016/j.biocon.2019.108384.

Durant, D., Martel, G., Chataigner, C., Farruggia, A., Kernéïs, E., Prieur, M., Roux, P., Tricheur, A. (2020). Comment évoluer vers davantage d'autonomie au sein des systèmes de polyculture-élevage ? : l'expérience d'une ferme expérimentale en marais. *AFPF*.

Fahrig, L., Baudry, J., Brotons, L., Burel, F.G., Crist, T.O., Fuller, R.J., Sirami, C., Siriwardena, G.M., Martin, J. (2011). Functional landscape heterogeneity and animal biodiversity in agricultural landscapes. *Ecol. Lett.*, 14, 101–112. doi: 10.1111/j.1461-0248.2010.01559.x.

Ferrané, C., Reau, R., Prost, L. (2020). Qualité de l'eau en aire d'alimentation de captage : gestion dynamique avec la démarche TRANSIT'EAU. *L'eau en milieu agricole, outils et méthodes pour une gestion intégrée et territoriale.*

Girardin, P., Guichard, L., Bockstaller, C. (2019). *Indicateurs et tableaux de bord. Guide pratique pour l'évaluation.* Tec et Doc – Lavoisier, Paris.

Green, R.E., Cornell, S.J., Scharlemann, J.P.W., Balmford, A. (2005). Farming and the fate of wild nature. *Science*, 307, 550–555. doi: 10.1126/science.1106049.

Hass, A.L., Kormann, U.G., Tscharntke, T., Clough, Y., Baillod, A.B., Sirami, C., Fahrig, L., Martin, J.-L., Baudry, J., Bertrand, C. et al. (2018). Landscape configurational heterogeneity by small-scale agriculture, not crop diversity, maintains pollinators and plant reproduction in western Europe. *Proc. R. Soc. B Biol. Sci.*, 285, 20172242. doi: 10.1098/rspb.2017.2242.

Herzon, I. and Mikk, M. (2007). Farmers' perceptions of biodiversity and their willingness to enhance it through agri-environment schemes: A comparative study from Estonia and Finland. *J. Nat. Conserv.*, 15, 10–25. doi: 10.1016/j.jnc.2006.08.001.

IPBES (2022). Summary for Policymakers of the Methodological Assessment Report on the Diverse Values and Valuation of Nature of the Intergovernmental Science-Policy Platform on Biodiversity and Ecosystem Services. doi: 10.5281/zenodo.6522392.

Kelemen, E., Nguyen, G., Gomiero, T., Kovács, E., Choisis, J.-P., Choisis, N., Paoletti, M.G., Podmaniczky, L., Ryschawy, J., Sarthou, J.-P. et al. (2013). Farmers' perceptions of biodiversity: Lessons from a discourse-based deliberative valuation study. *Land Use Policy*, 35, 318–328. doi: 10.1016/j.landusepol.2013.06.005.

Kleijn, D. and Sutherland, W.J. (2003). How effective are European agri-environment schemes in conserving and promoting biodiversity? *J. Appl. Ecol.*, 40, 947–969. doi: 10.1111/j.1365-2664.2003.00868.x.

Kleijn, D., Berendse, F., Smit, R., Gilissen, N. (2001). Agri-environment schemes do not effectively protect biodiversity in Dutch agricultural landscapes. *Nature*, 413, 723–725. doi: 10.1038/35099540.

Kleijn, D., Baquero, R.A., Clough, Y., Díaz, M., De Esteban, J., Fernández, F., Gabriel, D., Herzog, F., Holzschuh, A., Jöhl, R. et al. (2006). Mixed biodiversity benefits of agri-environment schemes in five European countries. *Ecol. Lett.* 9, 243–254. doi: 10.1111/j.1461-0248.2005.00869.x.

Kleijn, D., Winfree, R., Bartomeus, I., Carvalheiro, L.G., Henry, M., Isaacs, R., Klein, A.-M., Kremen, C., M'Gonigle, L.K., Rader, R. et al. (2015). Delivery of crop pollination services is an insufficient argument for wild pollinator conservation. *Nat. Commun.*, 6, 7414. doi: 10.1038/ncomms8414.

Kremen, C., Iles, A., Bacon, C. (2012). Diversified farming systems: An agroecological, systems-based alternative to modern industrial agriculture. *Ecol. Soc.*, 17, art44. doi: 10.5751/ES-05103-170444.

Le Roux, X., Barbault, R., Baudry, J., Burel, F., Doussan, I., Garnier, E., Herzog, F., Lavorel, S., Lifran, R., Roger-Estrade, J. et al. (eds) (2008). Agriculture et biodiversité. Valoriser les synergies. Expertise scientifique collective. Report Summary, INRA.

Levrel, H. (2007). *Quels indicateurs pour la gestion de la biodiversité ?* Les Cahiers de l'IFB, Paris.

Maljean, J.F. and Peeters, A. (2001). Agriculture intégrée et biodiversité : impacts et mesures politiques. In *Stratégie Paneuropéenne de la diversité biologique et paysagère – Conférence paneuropéenne à haut niveau sur l'agriculture et la biodiversité : vers une intégration de la diversité biologique et paysagère pour une agriculture durable.* La Documentation francaise, Paris

Manneville, V., Michel, N., Amiaud, B. (2016). INDIBIO : élaborer des indicateurs relatifs aux effets des pratiques agricoles sur la biodiversité dans les systèmes d'exploitation d'élevage. *Innov. Agron.*, 49, 83–97. doi: 10.15454/1.4622765655890154E12.

Meynard, J.-M., Cerf, M., Coquil, X., Durant, D., Le Bail, M., Lefèvre, A., Navarrete, M., Pernel, J., Périnelle, A., Perrin, B. et al. (2023). Unravelling the step-by-step process for farming system design to support agroecological transition. *Eur. J. Agron.*, 150, 126948. doi: 10.1016/j.eja.2023.126948.

ONU (2022). Cadre mondial de la biodiversité de Kunming-Montréal.

Princé, K., Rouveyrol, P., Pellissier, V., Touroult, J., Jiguet, F. (2021). Long-term effectiveness of Natura 2000 network to protect biodiversity: A hint of optimism for common birds. *Biol. Conserv.*, 253, 108871. doi: 10.1016/j.biocon.2020.108871.

Prost, L., Reau, R., Paravano, L., Cerf, M., Jeuffroy, M.-H. (2018). Designing agricultural systems from invention to implementation: The contribution of agronomy. Lessons from a case study. *Agric. Syst.*, 164, 122–132. doi: 10.1016/j.agsy.2018.04.009.

Reboud, X. (2019). Le lien entre agriculture et biodiversité : ce qui a bougé sur le plan de la société, des idées, de la réflexion entre 2008 et 2018. *Innov. Agron.*, 75, 1–14. doi: 10.15454/7IEFC5.

Secretariat of the Convention on Biological Diversity (2005). *Handbook of the Convention on Biological Diversity* [Online]. Available at: https://www.cbd.int/doc/handbook/cbd-hb-all-en.pdf.

Sirami, C., Gross, N., Baillod, A.B., Bertrand, C., Carrié, R., Hass, A., Henckel, L., Miguet, P., Vuillot, C., Alignier, A. et al. (2019). Increasing crop heterogeneity enhances multitrophic diversity across agricultural regions. *Proc. Natl. Acad. Sci.*, 116, 16442–16447. doi: 10.1073/pnas.1906419116.

Siriwardena, G. (2001). Changes in agricultural land-use and breeding performance of some granivorous farmland passerines in Britain. *Agric. Ecosyst. Environ.*, 84, 191–206. doi: 10.1016/S0167-8809(00)00210-3.

Tasser, E., Rüdisser, J., Plaikner, M., Wezel, A., Stöckli, S., Vincent, A., Nitsch, H., Dubbert, M., Moos, V., Walde, J. et al. (2019). A simple biodiversity assessment scheme supporting nature-friendly farm management. *Ecol. Indic.*, 107, 105649. doi: 10.1016/j.ecolind.2019.105649.

Vallé, C., Viol, I.L., Kerbiriou, C., Bas, Y., Jiguet, F., Princé, K. (2023). Farmland biodiversity benefits from small woody features. *Biol. Conserv.*, 286, 110262. doi: 10.1016/j.biocon.2023.110262.

8

Sociometabolic Approaches to Conceptualizing and Supporting Agri-Food Transitions

Caroline PETIT[1] and Thierry BONAUDO[2]

[1] *INRAE, UMR SADAPT, Palaiseau, France*

[2] *AgroParisTech, UMR TETIS, Montpellier, France*

8.1. Introduction

Agri-food systems have intricate and profound links with the ecological crisis. The unsustainability of standardized, industrialized agri-food systems, embedded in globalized economic organizations and based on the exploitation and control of living organisms, is widely recognized within the scientific community (Tello and González de Molina 2018). Taken as a whole, from input production to final consumption, agri-food systems account for approximately one-third of global greenhouse gas emissions (Crippa et al. 2021). In addition to climate change, the linear agro-industrial production model contributes to the overexploitation of natural resources, ecosystem destruction and the disruption of key biogeochemical cycles. Despite the spectacular increase in agricultural yields since the mid-20th century (Liu et al. 2015; Domingues et al. 2019), a significant portion of the world's population remains under- or malnourished. Concomitantly, there is a rising prevalence of non-communicable diseases (obesity, diabetes, coronary heart disease) linked to relative overnutrition, where caloric intake exceeds individual energy expenditure. These dysfunctions occur even as, according to recent FAO (2019) and UNEP (2021) reports, food losses and waste along the supply chain amount to one-third of global agricultural production. The environmental, economic, health-related,

Farms and Territory,
coordinated by Philippe MARTIN and Solène PISSONNIER.

ethical and social challenges, including value distribution, animal welfare, and the human–nature relationship, underscore the urgent need for a new "social contract" between farmers, actors in the agri-food sector and society.

In terms of agriculture and food, many initiatives, such as short supply chains, alternative food systems and community and cooperative systems, advocate food re-territorialization with the aim of countering the harmful effects of globalized food chains (Rastoin and Meynard 2020). Beyond these initiatives, which are organized through collectives but remain scattered across territories, several public policy frameworks set general objectives for reconfiguring the links between agriculture and food in the context of ecological and energy transition[1]. The succession of regulations and commitments by public authorities, from the municipal to the regional level, aligns with the trend towards the re-territorialization of agri-food systems. As Bonnefoy and Brand (2014) already pointed out, this process suggests a "potential transversality between agricultural and food-related actions and a possible interconnection with other territorial issues" (authors' translation), such as inequalities in access to quality food, the structuring of local economic sectors, adaptation to climate change, territorial planning, etc.

The ecological, economic and social crises compel us to rethink linear agri-food systems in favor of sobriety, circularity and autonomy. Circularity in particular has become a key objective, both because material recycling is fundamental in agriculture and because the circular economy has emerged as a strategic watchword in the economic sphere[2]. By reshaping the spatial organization of activities and their interactions, thus influencing territorial development models, the goal of circularity is to enable more efficient resource use, greater autonomy (fewer inputs) and increased resilience (Niang et al. 2020). This requires rethinking agriculture in its multiple connections to the territory, by relocating and recoupling resources, production and the management of residual organic materials. However, circularity alone cannot address all challenges. Understanding and managing the complex interrelations within agri-food systems, as well as their interactions with the

1. Since the early 2000s, France has adopted several regulations to support transformations at several levels of the agri-food system: the Grenelle laws in 2009 and 2010, the Methanization-Nitrogen Autonomy Energy Plan (le plan énergie méthanisation autonomie azote) in 2013, the Law for the Future of Agriculture, Food And Forestry (la loi d'avenir pour l'agriculture, l'alimentation et la forêt) in 2014, the Law on Energy Transition for Green Growth (la loi relative à la transition énergétique pour la croissance verte) in 2015 or the EGalim law in 2018.

2. The circular economy can be generically defined as an economic system that, at every stage of the product life cycle, seeks to use resources more efficiently and reduce environmental impacts while contributing to people's well-being. As with other fuzzy concepts, there are lively debates about conceptions of the circular economy (see Arnsperger and Bourg 2016).

environment and other economic sectors, requires a systemic approach. This idea, already implicit in sustainable development and integrated resource management frameworks, is gaining prominence in the context of climate change and, more broadly, planetary boundaries (Rockström et al. 2009). In this regard, the analytical framework of societal metabolism appears particularly well suited to capturing the cross-sectoral dynamics at play and analyzing territories under urban influence through their material characteristics, such as high population density, potential for food relocalization and diverse organic matter deposits.

Our aim is to highlight the strengths and limitations of sociometabolic approaches in understanding and supporting agri-food transitions. First, we clarify the epistemological foundations of sociometabolic approaches and their adaptation across different disciplines. Second, we present a case study of agri-food metabolism, addressing both its diagnostic and prospective dimensions. Finally, we discuss the mutual contribution of sociometabolic approaches and agronomy.

8.2. Society's metabolism: overview of epistemological foundations and current research trends

The notions of urban, social, socio-economic or territorial metabolism are gaining ground in scientific and public spheres. Without claiming to be exhaustive, we provide an overview of the epistemological foundations and current scientific developments that constitute a constellation of scientific and operational approaches to understanding the interactions between human societies and the biosphere.

The application of the concept of metabolism to the functioning of societies is based on a general principle of analogy with living organisms[3]. In disciplines that study the functioning of living organisms, metabolism refers to the set of biochemical reactions that sustain various functions (survival, reproduction, growth, response to environmental stimuli, communication, etc.). This process involves the intake of external resources, their transformation into products and energy, and the release of waste at the end of these reactions. During the 19th century, significant advances were made in chemistry and biology in understanding these biochemical reactions. For ecology, which was emerging as a scientific discipline at the same time, the analysis of material and energy flows became a key method for understanding the functioning and organization of ecosystems. In the social sciences, Karl Marx was one of the first to use this analogy in relation to social

3. This is not to say that companies, socio-systems or societies are, in themselves, living organisms.

processes, in particular to address the co-evolution between nature and society and the relationship between human societies and the biosphere (Haribey 2012). From these reflections emerges the notion of metabolic rift, largely theorized by sociologist John Bellamy Foster, which refers to the transformation, under the effect of the expansion of capitalism, of society's relationship to nature, in particular through the lens of production, through the disconnection between cities and their countryside, embodied in the limited return of nutrients to agricultural soils and the loss of soil fertility. This work has paved the way for an understanding of the material and energetic relationships between societies and natural resources.

The role of cities has been extensively studied from this perspective, particularly because they concentrate large populations with ever-greater supply needs and generating increasing waste. The notion of urban metabolism emerged in the 1960s with the seminal work of Abel Wolman, an engineer specializing in sanitary engineering, who published "The Metabolism of Cities", highlighting urban metabolic problems resulting from inflows (food, water, fuel) and outflows (wastewater, solid residues, atmospheric pollutants) (Wolman 1965). Ecologist Eugene Odum also contributed to this notion in the 1970s, considering the city as a heterotrophic ecosystem totally dependent on other ecosystems and fossil fuels (Odum 1989).

During the same period, discussions on the limits to natural resources and environmental degradation emerged, documented in particular in the 1972 Meadows et al. report, better known as *The Limits to Growth*, commissioned by the Club of Rome. These reflections marked a turning point in the very conception of economic growth. Contradictory debates ensued between two positions: green growth (Allenby 1992) and the bioeconomic approach, supported by the work of N. Georgescu-Roegen (Missemer 2015). The latter particularly influenced the degrowth movement.

Since the late 1980s, industry has been implementing industrial ecology strategies, notably through the creation of industrial symbioses (Kalundborg, a Danish port city, was one of the first to test this concept – Jacobsen 2006), with the aim of closing cycles, i.e. turning the waste of some to the resources of others. By extension, these strategies are to some extent part of the circular economy[4]. It is

4. Historians F. Jarrige and T. Le Roux (2016, pp. 65–66) point out that "industrialists and chemists construct […] fascinating diagrams of the circular economy, in which residues serve as the starting point, contributing to the interdependence of commodity chains and products. This enables them to justify the legitimacy of their products and residues. It is also interesting to note that the theories of recycling correspond to the major phases of waste production, as if the occurrences of debates were ultimately correlated to the insolubility of the problem" (authors' translation).

important to specify that industrial ecology is not understood in a restrictive sense, limited to industrial activities alone, but is applied to "industrial society" as a whole, including all the actors involved in a post-industrial revolution society that consumes and generates material flows. In France, the terms industrial and territorial ecology (ITE) and territorial bioeconomic systems (Buclet 2011; Brullot et al. 2017; Wohlfahrt et al. 2019) have emerged, making it easier to grasp the integration of non-industrial actors in the strict sense and broader territorial issues. These terms underpin action-oriented strategies in territories, through forms of synergy and mutualization between economic actors, similar to industrial symbioses. In France, these initiatives are supported in particular by Ademe and the Institut de l'économie circulaire. Arnsperger and Bourg (2016) summarized the scope of these debates and clarified the contours of a genuinely circular economy, which is not equivalent to green growth, involving not only circularity of material flows but also a net reduction in flows, achieved through a very low growth rate.

Currently, internationally, research that examines the relationships between human societies and the biosphere, considering the biophysical aspects of societies, its impacts on the environment and its adaptation strategies, is referred to as sociometabolic research (Haberl et al. 2019). Pauliuk and Hertwich (2015) define socioeconomic metabolism as "the self-reproduction and evolution of the biophysical structures of human society. It comprises those biophysical transformation processes, distribution processes, and flows, which are controlled by humans for their purposes". To fully understand the interactions between societies and the biosphere, three elements need to be taken into account: (i) metabolism as a process involving flows; (ii) biophysical structures such as buildings, machines, the artifacts used and all associated techniques (together (i) and (ii) constitute the biophysical basis of society); (iii) the cultural system of society including norms, practices and actors governing this biophysical base. Material and energy flow accounting approaches, such as material flow analysis (MFA), pioneered by Ayres and Kneese (1969), are frequently used. They are based on a so-called stock-flow model, which considers a flow as a quantity of matter, substance or energy taken from a stock and transformed during a productive process. Other models introduce the notion of fund, based on the fund-flow theory of Georgescu-Roegen (1971). Funds are the production factors that do not fundamentally change their state during the production process (workers, machines, buildings, livestock, land, etc.). As functional elements of ecosystems, they operate at different time steps from the production process, provide services and therefore need to be preserved and properly managed. The distinction between a stock and a fund is particularly useful in the case of agricultural systems. For example, land can be seen as a resource, a stock, which can be diminished by urbanization, or it can be seen as fund, a life-supporting system with its own functionalities, which can be degraded by inappropriate practices.

In the international academic sphere, two scientific communities dominate sociometabolic research, one based at the Institute of Social Ecology in Vienna (Fischer-Kowalski and Haberl 2007a; Krausmann et al. 2008) and the other at the University of Barcelona (Giampietro et al. 2009). The French community is mainly structured around territorial ecology, the scientific genealogy of which has been outlined in several publications (Barles 2010; Bognon 2014; Jambou et al. 2018). Through the notion of *territorial metabolism*, it emphasizes all human activities within a given territory. Territorial metabolism thus refers to the way in which human societies, embedded in a territory, consume and transform materials and energy, both in quantity (material flows, manufactured products, water, food, energy, etc.) and quality (nature of flows), depending on the socio-technical systems in place (infrastructures, techniques, stakeholder organizations, governance) (Barles 2017). Territorial metabolism offers an integrative perspective of the issues surrounding the consumption and use of resources by societies, with an analysis according to the degree of linearity and circularity of the system. The underlying observation is that industrial society, further subjected to urbanization, is characterized by an open, linear metabolism that is highly intense in materials and energy, with numerous socio-economic and environmental consequences: externalization of the metabolism, dependence on distant territories, and environmental footprint (see, in particular, Barles 2015; Esculier et al. 2019; Tanguy et al. 2020). These studies have a dimension viewed through the lens of territorial biogeochemistry, which focuses on the degree of openness of biogeochemical cycles and their environmental repercussions (Billen et al. 2012).

8.3. Agriculture in the metabolism of societies

As a fundamental need, food is a cornerstone of the metabolism of societies. In sociometabolic research, agriculture is typically regarded as a sector to be integrated into material accounting, as biophysical structures such as land and livestock generate agricultural flows (Fischer-Kowalski 1998).

Sociometabolic research, fundamentally systemic, examines agricultural systems as an element in a nexus that also connects food, energy, materials and the land resources (Fraňková et al. 2018). It also highlights the underlying mechanisms behind the transformation of sociometabolic regimes, defined as the "dynamic equilibrium of a system of society–nature interaction" (Fischer-Kowalski and Haberl 2007b). Agriculture has been studied through its role in supplying cities (Barles 2007), considering the agronomic and environmental performance of large agricultural systems (Le Noë et al. 2017) or even from a historical perspective, as in Güldner and Krausmann (2017) on nutrient management. Territorial biogeochemistry research focuses more directly on agricultural systems, but its scale of analysis is often quite broad, ranging from regional to continental levels, as shown in the works of Lassaletta et al. (2014), Le Noë et al. (2017), Harchaoui and Chatzimpiros (2018), Domingues et al. (2019), Billen et al. (2021) and Bonaudo et al. (2021). Because they necessarily use large-scale statistical data, these studies view agriculture in terms of broad categories (e.g. conventional versus organic farming) or major commodity chains (cereal, dairy, vegetable, etc.). As a result, as noted by Fraňková et al. (2018), the biophysical aspects of local food systems have not been fully addressed within sociometabolic approaches.

The agronomy community is still underrepresented in the field of sociometabolic research. Agronomic research explores topics shared with sociometabolic research, but in a more sectoral manner, such as changes in agricultural practices, involvement in short supply chains, synergies between plant and animal production, and the use of organic waste products. More recently, some contributions from agronomists have more clearly adopted sociometabolic approaches, particularly in the field of nutrient cycle management. Modeling approaches for agro-industrial systems or at territorial scales aimed at promoting organic matter recycling have been developed (Fernandez-Mena et al. 2016, 2020; Nesme and Withers 2016; Moinard et al. 2021). In parallel, research has analyzed the flows of substances such as N, P and K and their efficiency of use in livestock systems (Gameiro et al. 2019). The work of Harchaoui and Chatzimpiros (2018) on the energy footprint of agriculture is also significant. At the territorial and organizational levels, contributions have been made on crop–livestock interactions (Moraine et al. 2017), the notions of food print and feeding basin (Benoit et al. 2011), and the interactions between agricultural sectors and territories (Madelrieux et al. 2017). Apprehending agri-food systems with sociometabolic approaches also obviously concerns other research communities, as demonstrated by Madelrieux and Redlingshöfer (2023).

The notion of *agri-food metabolism* can be derived from the concept of societal metabolism to refer to the set of processes implemented by human societies to provide food. More precisely, it encompasses all processes related to the extraction and transformation of materials and energy for food consumption, including the treatment of resulting waste. These processes are carried out by actors through biophysical structures, as defined by Haberl et al. (2004). The structures involved in agri-food metabolism are highly diverse and refer to the constituent subsets of the food system (Rastoin and Ghersi 2010). They include agro-supply (cooperatives, seed companies, fertilizer manufacturers, etc.), agricultural production (professional farms and forms of domestic production), collection and processing (cooperatives, mills, oil mills, slaughterhouses, agri-food industries, etc.), distribution structures (logistics platforms, wholesalers, food retailers including supermarkets, markets, specialist stores, e-commerce, etc.), consumption sites (home, out-of-home catering, etc.), and structures for sanitation and waste management (sewage treatment plants, incinerators, biowaste composting platforms, biogas plants, etc.) (see Figure 8.1). Understanding agri-food metabolism means focusing, at the scale of the food system, on the material dimension of the biophysical flows generated by the processes involved, as well as their implementation by groups of actors operating in a specific context of techniques, policies and institutions.

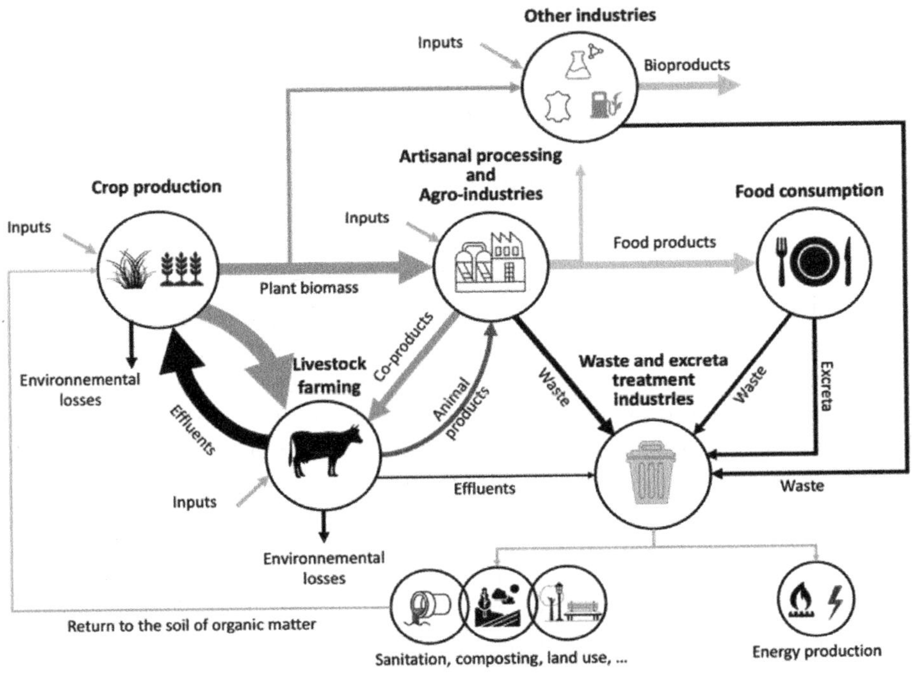

Figure 8.1. *Diagram of an agri-food system. Several key issues can be addressed in the study of flows within agri-food systems, including: the relationship between crop production and livestock farming; competition for biomass use; the connection between agricultural production and food supply; waste and loss management; productive efficiency; the autonomy and circularity of flows; the interdependence and resilience of the flow system; logistics and infrastructure within the flow system; the governance system, etc.*

In recent years, research has emerged at the crossroads of sociometabolic approaches and agri-food issues at the scale of specific territories, notably within the territorial ecology community (with pioneering work at the municipal scale in Aussois – Buclet et al. 2015), as well as in research on the production and use of biomass of agricultural origin (Marty et al. 2022). The following section presents the results of a study conducted in a peri-urban area within the Île-de-France region.

8.4. For agri-food metabolism approaches at the local scale

Complementary to the previously mentioned studies, it is essential to develop approaches at local scales that account for the diversity of existing agricultural and their interactions, both among different production systems and with other key functions of agricultural sectors (agro-supply, collection, storage, processing, distribution networks). As an illustration, this section presents the case of the agri-food metabolism of the Saclay Plateau. Beyond encompassing the entire agri-food system, this study is innovative in that it articulates several approaches: a local-scale diagnosis of flows and the participatory construction of contrasting territorial scenarios to explore potential transition pathways[5].

8.4.1. *Diagnosis of agri-food metabolism at the territorial scale: the case of the Saclay Plateau*

In this study area, we began by diagnosing the flows within the agri-food system, considering all agricultural materials, as well as upstream and downstream flows related to agricultural production. We took into account the diversity of local farming systems by identifying contrasting models of technical systems through in-depth interviews with farmers. Rather than relying on a standard classification, we adopted a typological approach aligned with the methodological proposals of Hubert Cochet (2011), which aim to reconstruct the historical trajectory of production systems.

Three types of farming system were identified. By analyzing the material flows within each type and the material interconnection between them, we assessed their contribution to the local food supply and developed an objective understanding of the relationships between agriculture and the territory. These relationships are described as symbiotic, as they emphasize cooperation between entities, mutual benefits and the reuse of materials (Petit et al. 2018).

Material flows between the three identified farming systems and the local population were analyzed, along with the factors influencing these flows, such as the functioning of farming structures and the production and commercial strategies of

5. This work has been the subject of specific publications (Petit et al. 2018, 2019; Tedesco et al. 2018; Verger et al. 2018; Petit 2021).

the actors involved. The quantification of material flows revealed a significant disconnect between agricultural production and food consumption. Only 13% of local agricultural production is used locally, primarily through short supply chains, while just 8% of the total food consumed originates from the territory (Tedesco et al. 2017). The "conventional large-scale cropping" systems rely on imported mineral inputs and sell their products on national and international markets, making them largely disconnected from their local territory. Only 0.2% of cereal production is consumed locally. "Diversified" systems have more locally embedded flows but remain relatively non-autonomous, with significant incoming flows (external inputs) and dependence on exports to regional and national markets (outgoing flows from the territory). "Small-scale" systems (livestock farming, orchards and market gardening) are strongly rooted locally, but have a marginal contribution to local territorial metabolism. Livestock products and fruits and vegetables are strongly linked to local consumption (with 65% and 100% of these products consumed locally, respectively), but their production remains very low compared to the total consumption in the territory (4% and 5%, respectively). The food needs of the population are considerable in comparison to the productive capacities of local agricultural systems, leading to significant incoming flows (food imports) to meet local demand. Downstream, outgoing flows related to the human excreta and waste are also substantial (see Figure 8.2). This assessment of agri-food system flows suggests that new local production channels could be developed to address local needs.

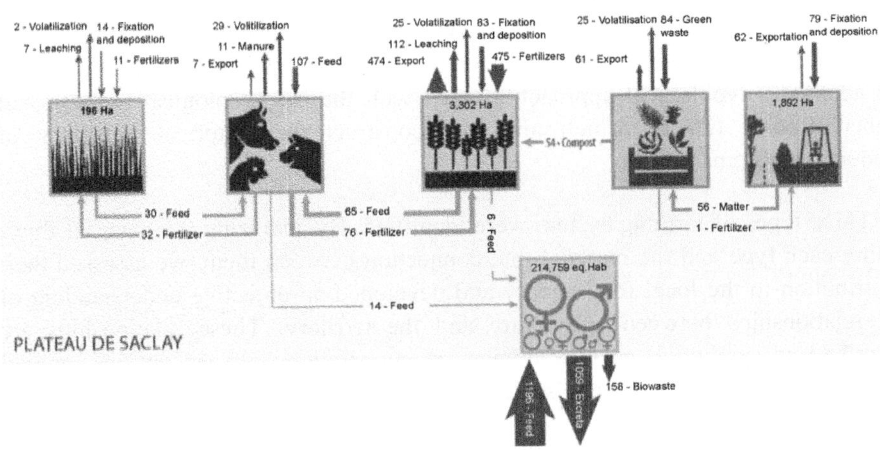

Figure 8.2. *Illustration of the agri-food metabolism of the Saclay Plateau (~ 2016). Material flows are expressed in kilotons of nitrogen. Illustrations by Emilie Bonnet ©*

In summary, the approach developed aims to provide orders of magnitude for the contribution of each agricultural system to the local agri-food metabolism. These orders of magnitude are not self-evident. Without this objectification, it is easy to overestimate or underestimate agricultural flows, which can lead to an overvaluation of the operational impact of a specific leverage point. Moreover, understanding the immaterial dimension of flows and actors' decision-making logics serves as an additional safeguard against hasty conclusions regarding potential levers that may not be practically implementable. This approach also considers a territory as a stable entity and provides a snapshot at a given moment in time, which Gomiero (2018) refers to as a steady-state social metabolism approach. However, metabolism is constantly evolving. The results presented are valid for the study period (~ 2016) but have necessarily changed since then. For example, there have been conversions of farms to organic farming, as well as closures and the creation of new small-scale structures over this period.

8.4.2. *The design of territorial fictions and their appropriation by local actors*

In a second phase, we explored the potential of these analyses to stimulate reflection on the future of the territory and local planning strategies. The chosen approach involved formulating, within a collective of researchers gathered during the Saclay workshops[6], contrasting visions of the territory's possible futures, designed to stimulate debate among local actors (farmers, elected officials from local authorities, state representatives, association leaders, etc.). This deliberative process allowed participants to express points of convergence and divergence with the trajectory they envision for the territory[7].

The three territorial fictions serve as preliminary scenarios, each consisting of a narrative and a representation of the associated agri-food metabolism (see Figure 8.3). They are to be compared with the current state of the metabolism (see Figure 8.2), as analyzed in the previous section. The first fiction amplifies localist trends by increasing local food production, without aiming for complete self-sufficiency, intensifying material exchanges between different farming systems and using human excreta for agricultural purposes. The second fiction promotes agricultural systems oriented towards export, with increased use of technology to control production processes, further disconnecting production from consumption. The third fiction

6. See: https://www.youtube.com/watch?v=sUZv-89qmSI&t=8s.

7. The entire process, as well as the stakeholders' viewpoints, are reported in Petit et al. (2019) and can be accessed online at https://www.federation-fire.fr/sites/default/files/actu_fire/plaquette_saclay_2019_BDEF.pdf.

envisions the expanding influence of urban areas on the agricultural sphere, with the sector adapting to meet urban leisure demands while also repurposing part of the excreta recovered from recreational sites for agronomic use.

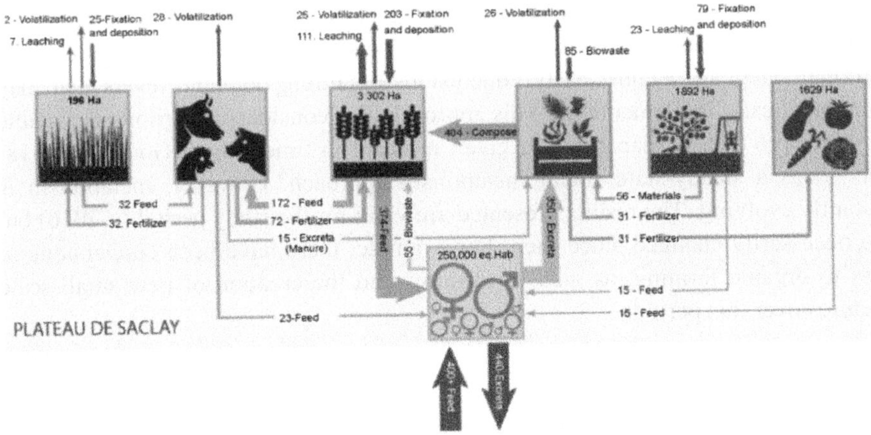

FICTION 1: "Autonomy and rural-urban synergies"

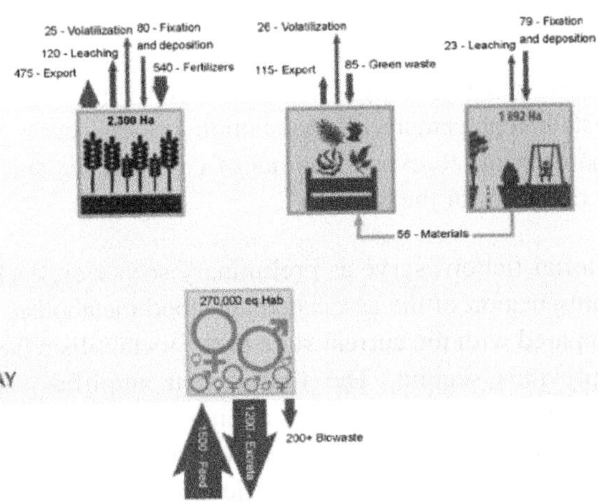

FICTION 2: "Spatial specialization and technological innovation"

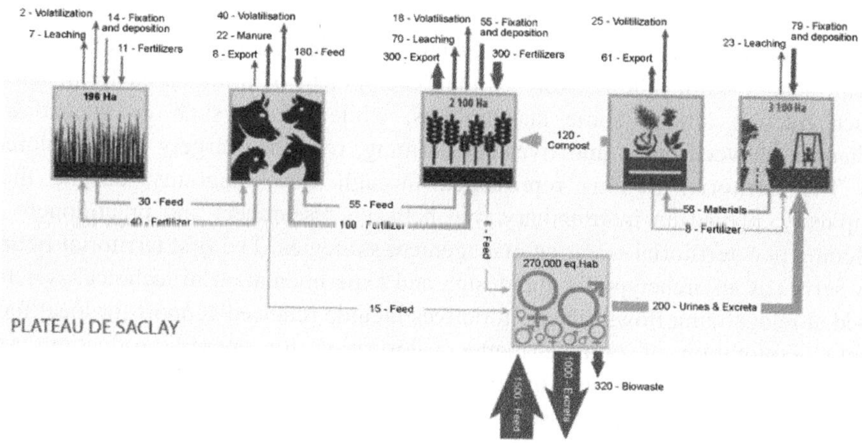

FICTION 3: "Scientific and recreational park"

Figure 8.3. *Three territorial fictions to support reflection on the future trajectory of the Saclay Plateau. Material flows are expressed in kilotons of nitrogen. Illustrations by Emilie Bonnet ©*

The Saclay workshops took place in a context where dialogue between researchers and local stakeholders had already been established for several years. They contributed to strengthening these interactions, which subsequently evolved into a living lab-type mechanism[8]. The territory thus provides a favorable environment for experimenting with the formalism of agri-food metabolism as an intermediary object in a participatory design process (Madelrieux et al. 2017; Gabriel et al. 2020). Despite initial skepticism from some stakeholders during the debate on metabolic scenarios, a gradual (individual or collective) appropriation of certain technical proposals presented in the scenarios, such as composting, anaerobic digestion, urine treatment and land application, was observed. The key lessons drawn from this process are threefold. (i) The organization diagrams of agri-food metabolism established a shared knowledge base of the territory, objectifying material cycles and providing orders of magnitude for the flows involved (e.g. the significance of imported agricultural fertilizers and wastewater treatment flows; the marginal nature of flows related to local short supply chains). (ii) They confirmed the potential impact of different interventions aimed at reconfiguring the metabolism, although some pathways, such as strengthening food self-sufficiency

8. A "living lab" is a participatory innovation process that includes the user. Its aim is to respond to complex, multidisciplinary problems on a territorial scale (source: agroecology dictionary: https://dicoagroecologie.fr/dictionnaire/living-lab/).

for specific food categories, were found to have limited quantitative potential. In terms of potential material synergies, certain strategies, such as the valorization of organic matter within the territory or the development of urine-based fertilization, gained traction among some stakeholders, while others, such as intensifying exchanges between crops and livestock farming, remained largely underexplored. (iii) The territorial fictions represented through flow diagrams, despite their complexity, played an intermediary role between researchers and practitioners in designing new territorial resource management strategies. The first territorial fiction now serves as an archetype for the design and experimentation of technical systems aimed at relocalizing flows. These initiatives include renewed support for local food supply, exploration of organic matter valorization for energy production, and experimentation with the use of urine as an agricultural fertilization following separate collection and safety assessment.

8.5. Discussion: avenues for further exploration and prospects for cross-disciplinary work between sociometabolic approaches and agronomy

In this final section, we discuss how the agri-food metabolism approach can open new research avenues in agronomy in relation to sociometabolic approaches.

8.5.1. *Accounting for the diversity of actors and their logics within agri-food metabolism*

The characterization of material and energy stock and flow dynamics related to agriculture in large-scale sociometabolic approaches necessarily relies on the use of public statistics (such as those provided by Agreste), which provide information on agricultural areas and yields. Working at the local scale presents an opportunity for agronomists to re-engage with and deepen their understanding of farming systems involved in territorial metabolism, focusing on the two traditional dimensions of agronomy: diagnosis and design. Global-scale approaches tend to prioritize the organization of production systems within a metabolism over the precise characterization of each production system. This perspective allows for a reevaluation and requalification of agricultural system diversity in light of their role in agri-food metabolism. Such characterization could contribute to the renewal of typological approaches in agronomy. It should also integrate emerging or understudied agricultural systems, such as intra-urban farming systems or domestic vegetable production. Exploring their potential to meet local food demand and/or contribute to autonomous and circular organic matter and fertility management (e.g. Bahers and Giacchè 2019; Edmondson et al. 2020) represents a new field for agronomy. This exploration should focus on the properties of these systems and their interactions to enhance agronomic and environmental performance at a territorial scale.

The identification of material flows and the actors governing them enables a better understanding of the biophysical structures of society, as defined by Haberl et al. (2004). The actors and structures involved in agri-food metabolism are highly diverse, and their identification is crucial in prospective approaches to pinpoint barriers to change and to better integrate the material and socio-economic dimensions of metabolism. A key challenge is determining the data needed to characterize these structures and their spatial distribution. Depending on the scale of analysis, data acquisition methods may vary (databases, field surveys, etc.). Several research studies have made significant advances in identifying and structuring relevant databases. For example, the SI-BOAT information system allows for a detailed characterization of a territory's agri-food system, including primary processing; Grillot et al. 2021). Final food consumption remains an area requiring further investigation due to its role in metabolism and its possible reconfiguration, both in terms of changing food demand and in waste emissions and management (Redlingshöfer 2022; Redlingshöfer et al. 2022).

8.5.2. *Territorial symbiosis: distinguishing agriculture from industry*

Project-based territories, such as Leader territories or those covered by Territorial Food Project (PAT) and Climate–Air–Energy Territorial Plans (PCAET), are areas where a network of stakeholders implements a collective project. These territories offer relevant scales for developing sociometabolic approaches that go beyond simple diagnosis and move towards the steering and even redesign of the agri-food system. To achieve this, research efforts must focus on diagnosing material interactions between actors, designing prospective scenarios and determining the conditions necessary to establish territorial symbioses.

As previously discussed, symbiotic strategies between actors have primarily developed in the industrial sector, aiming to create internal recycling loops within industrial sites for efficiency and optimization. While these strategies can sometimes include agricultural activities compatible with these industrial logics, such as the Bazancourt–Pomacle biorefinery in the Marne region (Thénot et al. 2018), they cannot be applied to the entire agricultural sector. Agriculture differs from industry at a fundamental level, even though current territorial organization focuses on concentrating and massifying biomass flows for national or global supply chains. Agricultural sectors are still highly dependent on pedoclimatic conditions and spread across vast production areas, unlike industries, which are more spatially concentrated. Moreover, agricultural materials are heterogeneous and not always interchangeable. Agriculture co-products generate diverse, diffuse and variable flows, both in quantity and quality. Unlike the industrial sector, the materials circulating within the agricultural system, such as manure, compost and urban residual products, are heterogeneous. While they can be standardized when commercialized (e.g. NFU 44-051 standard for compost), this is not always the case. Agricultural by-products face competition from multiple uses, such as soil fertility management, energy production, animal and human nutrition, construction, ecosystem services and biodiversity conservation. Finally, urbanization increases human population density, creating logistical constraints and intensifying material flows. Agricultural environments are also increasingly degraded, often due to farming practices themselves, impacting soils, biodiversity and water quality, which in turn puts ecological processes under strain.

The local scale is also relevant for assessing the quality and temporality of material exchanges between the different entities within the system. However, achieving effective circularity requires accounting for these variabilities, as well as the organizational and logistical constraints of the actors involved. It necessitates managing complexity and heterogeneity while balancing trade-offs between different functions and uses of agricultural materials. It is important to recognize that not all waste can be converted into resources at all times, as biophysical, economic and organizational constraints impose limitations.

8.5.3. *Considering input reduction and relocalization before circularity*

Building on industrial symbiosis experiences, current design efforts aim to partially replace mineral fertilizers with organic residual materials, particularly of urban origin, such as wastewater, green waste and household biowaste, often processed through composting or methanization. Developing appropriate cropping systems and agricultural practices requires further knowledge, along with the establishment of suitable organization frameworks at both farm and territorial levels. This approach aligns with a stock-flow logic, as described in the section 8.2.

However, sociometabolic approaches underline the urgency of a radical transformation in the relationship between human societies and the biosphere. They call for a "global systemic consideration", as emphasized by M. Fischer-Kowalski in 1998. The challenge of reducing the intensity of material and energy flows, i.e. reducing the absolute quantities of materials and energy used, appears as equally strategic as achieving circularity in these flows, if not more so.

How can this be concretely translated into the agricultural sector? It appears to revive debates on substitution and redesign strategies, as formalized by Hill and Mac Rae (1996) within the Efficiency–Substitution–Redesign (ESR) framework[9]. Indeed, the transformation of agricultural systems must integrate two issues that are currently studied rather separately: the substitution of synthetic fertilizers, on the one hand, and the reduction, or even elimination, of pesticides, on the other hand. Research on alternative fertilization strategies rarely addresses crop protection challenges, whereas studies on pesticide reduction implicitly emphasize the importance of balanced fertilization – since excessive fertilization can increase susceptibility to pests (Jacquet et al. 2022). Furthermore, a consensus is emerging in the scientific community: achieving ecological goals requires ambitious interventions and a deep redesign of farming techniques. A sector-based approach aligns with what Arnsperger and Bourg (2016) describe when discussing the implementation of circularity principles at the scale of a company or sector, which may be beneficial at the micro level, but could also generate negative macroeconomic effects, particularly through "rebound effects". Similarly, implementing circularity at the scale of a small territory may have limited impact on a broader scale, if it does not sufficiently challenge the system as a whole. While a practice may be beneficial at the farm level, it can lead to unintended negative consequences at the agricultural system level on a wider scale (as seen with the expansion of biogas production). Circularity alone does not guarantee sustainability, and its implementation at local scales may result in unforeseen adverse effects.

Before aiming for circularity, it is essential to prioritize autonomy and the reduction of inputs, particularly nitrogen fertilizers. This requires reconsidering the supremacy of agricultural yield and specialization. The transformation pathways explored in sociometabolic research align with the widespread adoption of agroecology (Billen et al. 2021), and practices that enhance material synergies (such as crop–livestock integration), and biological regulations (such as agroforestry) to preserve funds within the metabolism. However, as mentioned above, these approaches must be adapted to the local scale. To drive an ecological transition, we must be able to assess the consequences of changes across different scales, considering direct and indirect effects, as well as local, distant and global effects.

---

9. According to this framework, efficiency aims to reduce inputs and their negative impacts by drawing on available technological developments and available practices. Substitution seeks to replace inputs with alternative methods and more environmentally friendly products. Redesign tackles the causes of problems by integrating agronomic practices that combine partial effects. It aims to operate an agroecosystem based on a new set of ecological processes.

8.6. Conclusion

The complexity of globalized agri-food systems, which affect all regions of the world to varying degrees, reflects the industrial sociometabolic regime in which societies have evolved over the past two centuries. This regime is characterized by an intense use of resources and a linear flow of material and energy, meaning that resource extraction and waste disposal strategies prevail over strategies for reducing extraction, reuse and material recirculation. This results in significant environment, health and socio-economic impacts. The crucial question is how the agricultural and food supply sectors can and should adapt to these challenges.

It is essential to rethink agriculture in its relationship with the territory, to relocalize production as well as productive resources such as agricultural inputs, livestock feed, labor, water, etc. This also involves reconnecting agricultural production with food consumption at the territorial area. Reasoning in terms of agri-food metabolism makes it possible to situate the material flows involved in stakeholders' practices, from agricultural production to waste treatment, within a broader territorial system. It helps to quantify the relative importance of different material contributions, highlight the interconnections between subsystems and operators and, in other words, put them into perspective. This integrated and cross-cutting approach requires agronomists to develop diagnostics of material flows circulating within the territorial system, engage with a wide range of territorial actors to identify materials in circulation, and adopt prospective approaches to explore shifts in agri-food metabolism within territories.

According to the societal metabolism approach, agri-food systems are considered sustainable when they meet food needs while maintaining ecosystem functions, thereby preserving funds that provide services at a given rate (Tello and González de Molina 2018). Sociometabolic approaches offer a diverse range of methods for analyzing agri-food systems in this regard, and the resulting knowledge can serve as a lever for transforming these systems and guiding them towards strong sustainability.

8.7. References

Allenby, B.R. (1992). Achieving sustainable development through industrial ecology. *International Environmental Affairs*, 4(1), 56–68.

Arnsperger, C. and Bourg, D. (2016). Vers une économie authentiquement circulaire : réflexions sur les fondements d'un indicateur de circularité. *Revue de l'OFCE*, 1, 91–125.

Ayres, R.U. and Kneese, A.V., (1969). Production, consumption, and externalities. *The American Economic Review*, 59(3), 282–297.

Bahers, J.B. and Giacchè, G. (2019). Towards a metabolic rift analysis: The case of urban agriculture and organic waste management in Rennes (France). *Geoforum*, 98, 97–107.

Barles, S. (2007). Feeding the city: Food consumption and flow of nitrogen, Paris, 1801–1914. *Science of the Total Environment*, 375(1–3), 48–58.

Barles, S. (2010). Society, energy and materials: The contribution of urban metabolism studies to sustainable urban development issues. *Journal of Environmental Planning and Management*, 53(4), 439–455.

Barles, S. (2015). The main characteristics of urban socio-ecological trajectories: Paris (France) from the 18th to the 20th century. *Ecological Economics*, 118, 177–185.

Barles, S. (2017). Écologie territoriale et métabolisme urbain : quelques enjeux de la transition socioécologique. *Économie Régionale et Urbaine*, 5, 819–836.

Benoit, M., Chatzimpiros, P., Thieu, V. (2011). Les territoires d'alimentation des villes : empreinte alimentaire et bassin d'approvisionnement, deux concepts de l'agronomie des territoires. *Agronomie Environnement Société*, 1(2).

Billen, G., Garnier, J., Barles, S. (2012). History of the urban environmental imprint: Introduction to a multidisciplinary approach to the long-term relationships between Western cities and their hinterland. *Regional Environmental Change*, 12, 249–253.

Billen, G., Aguilera, E., Einarsson, R., Garnier, J., Gingrich, S., Grizzetti, B., Lassaletta, L., Le Noë, J., Sanz-Cobena, A. (2021). Reshaping the European agro-food system and closing its nitrogen cycle: The potential of combining dietary change, agroecology, and circularity. *One Earth*, 4(6), 839–850.

Bognon, S. (2014). Les transformations de l'approvisionnement alimentaire dans la métropole parisienne. Trajectoire socio-écologique et construction de proximités. Doctoral Thesis, Paris 1 Panthéon-Sorbonne University.

Bonaudo, T., Piraux, M., Gameiro, A. (2021). Analysing intensification, autonomy and efficiencies of livestock production through nitrogen flows: A case study of an emblematic Amazonian territory. *Agricultural Systems*, 190, 103072. doi: 10.1016/j.agsy.2021.103072.

Bonnefoy, S. and Brand, C. (2014). Régulation politique et territorialisation du fait alimentaire: de l'agriculture à l'agri-alimentaire. *Géocarrefour*, 89(89/1–2), 95–103.

Brullot, S., Junqua, G., Zuindeau, B. (2017). Écologie industrielle et territoriale à l'heure de la transition écologique et sociale de l'économie. *Revue d'Economie Regionale Urbaine*, 5, 771–796.

Buclet, N. (2011). *Écologie industrielle et territoriale : stratégies locales pour un développement durable*. Presses Universitaires du Septentrion, Villeneuve-d'Ascq.

Buclet, N. (ed.) (2015). *Essai d'écologie territoriale. L'exemple d'Aussois en Savoie*. CNRS Éditions, Paris. doi: 10.4000/books.editionscnrs.28318.

Cochet, H. (2011). Comment construire des typologies d'exploitations agricoles? In *L'agriculture comparée*, Cochet, H. (ed.). Éditions Quæ, Versailles.

Crippa, M., Solazzo, E., Guizzardi, D., Monforti-Ferrario, F., Tubiello, F.N., Leip, A. (2021). Food systems are responsible for a third of global anthropogenic GHG emissions. *Nature Food*, 2(3), 198–209.

Domingues, J.P., Gameiro, A., Bonaudo, T., Gabrielle, B., Tichit, M. (2019). How current social, environmental and cultural services provided by livestock are linked to past intensification trajectories? *Animal*, 14(3), 598–608. doi: 10.1017/S1751731119001952.

Edmondson, J.L., Childs, D.Z., Dobson, M.C., Gaston, K.J., Warren, P.H., Leake, J.R. (2020). Feeding a city – Leicester as a case study of the importance of allotments for horticultural production in the UK. *Science of the Total Environment*, 705, 135930.

Esculier, F., Le Noë, J., Barles, S., Billen, G., Créno, B., Garnier, J., Lesavre, J., Petit, L., Tabuchi, J.P. (2019). The biogeochemical imprint of human metabolism in Paris Megacity: A regionalized analysis of a water-agro-food system. *Journal of Hydrology*, 573, 1028–1045.

FAO (2019). *The State of Food and Agriculture 2019: Moving Forward on Food Loss and Waste Reduction*. United Nations, Rome.

Fernandez-Mena, H., Nesme, T., Pellerin, S. (2016). Towards an agro-industrial ecology: A review of nutrient flow modelling and assessment tools in agro-food systems at the local scale. *Science of the Total Environment*, 543, 467–479.

Fernandez-Mena, H., Gaudou, B., Pellerin, S., MacDonald, G.K., Nesme, T. (2020). Flows in agro-food networks (FAN): An agent-based model to simulate local agricultural material flows. *Agricultural Systems*, 180, 102718.

Fischer-Kowalski, M. (1998). Society's metabolism: The intellectual history of materials flow analysis, Part I, 1860–1970. *Journal of Industrial Ecology*, 2(1), 61–78.

Fischer-Kowalski, M. and Haberl, H. (eds) (2007a). *Socioecological Transitions and Global Change: Trajectories of Social Metabolism and Land Use*. Edward Elgar, Cheltenham.

Fischer-Kowalski, M. and Haberl, H. (2007b). Conceptualizing, observing and comparing socioecological transitions. In *Socioecological Transitions and Global Change: Trajectories of Social Metabolism and Land Use*, Fischer-Kowalski, M. and Haberl, H. (eds). Edward Elgar, Cheltenham.

Fraňková, E., Haas, W., Singh, S.J. (eds) (2018). Introduction: Key concepts, debates and approaches in analysing the sustainability of agri-food systems. In *Socio-Metabolic Perspectives on the Sustainability of Local Food Systems: Insights for Science, Policy and Practice*. Springer, Cham.

Gabriel, A.W., Madelrieux, S., Lescoat, P. (2020). A review of socio-economic metabolism representations and their links to action: Cases in agri-food studies. *Ecological Economics*, 178, 106765.

Gameiro, A.H., Bonaudo, T., Tichit, M. (2019). Nitrogen, phosphorus and potassium accounts in the Brazilian livestock agro-industrial system. *Regional Environmental Change*, 19(3), 893–905.

Georgescu-Roegen, N. (1971). *The Entropy Law and the Economic Process*. Harvard University Press, Cambridge, MA.

Giampietro, M., Mayumi, K., Ramos-Martin, J. (2009). Multi-scale integrated analysis of societal and ecosystem metabolism (MuSIASEM): Theoretical concepts and basic rationale. *Energy*, 34(3), 313–322.

Gomiero, T. (2018). Biophysical analysis of Agri-food systems: Scales, energy efficiency, power and metabolism of society. In *Socio-Metabolic Perspectives on the Sustainability of Local Food Systems: Insights for Science, Policy and Practice*, Fraňková, E., Haas, W., Singh, S.J. (eds). Springer, Cham.

Grillot, M., Ruault, J.F., Torre, A., Bray, F., Madelrieux, S. (2021). Le proto-métabolisme : approche du fonctionnement bioéconomique d'un territoire agricole. Économie rurale. *Agricultures, alimentations, territoires*, (376), 55–75.

Güldner, D. and Krausmann, F. (2017). Nutrient recycling and soil fertility management in the course of the industrial transition of traditional, organic agriculture: The case of Bruck estate, 1787–1906. *Agriculture, Ecosystems & Environment*, 249, 80–90.

Haberl, H., Fischer-Kowalski, M., Krausmann, F., Weisz, H., Winiwarter, V. (2004). Progress towards sustainability? What the conceptual framework of material and energy flow accounting (MEFA) can offer. *Land Use Policy*, 21(3), 199–213.

Haberl, H., Wiedenhofer, D., Pauliuk, S., Krausmann, F., Müller, D.B., Fischer-Kowalski, M. (2019). Contributions of sociometabolic research to sustainability science. *Nature Sustainability*, 2(3), 173–184.

Harchaoui, S. and Chatzimpiros, P. (2018). Energy, nitrogen, and farm surplus transitions in agriculture from historical data modeling. France, 1882–2013. *Journal of Industrial Ecology*. doi: 10.1111/jiec.12760.

Haribey, J.M. (2012). La portée écologiste de l'oeuvre de Marx. *Actuel Marx*, 2, 121–129.

Hill, S.B. and MacRae, R.J. (1996). Conceptual framework for the transition from conventional to sustainable agriculture. *Journal of Sustainable Agriculture*, 7, 81–87.

Jacobsen, N.B. (2006). Industrial symbiosis in Kalundborg, Denmark: A quantitative assessment of economic and environmental aspects. *Journal of Industrial Ecology*, 10(1–2), 239–255.

Jacquet, F., Jeuffroy, M.H., Jouan, J., Le Cadre, E., Malausa, T., Reboud, X., Huyghe, C. (2022). *Zéro pesticide : un nouveau paradigme de recherche pour une agriculture durable*. Éditions Quæ, Versailles.

Jambou, M., Herbelin, A., Chebbi, A. (2018). Écologie industrielle, écologie territoriale : les jeunes chercheurs s'emparent des questions épistémologiques et d'interdisciplinarité. *Natural Sciences Society*, 26(4), 454–462.

Jarrige, F., Le Roux, T., Le Lay, S. (2016). Le rôle des déchets dans l'histoire. *Mouvements: des idées et des luttes*, 87, 59–68.

Krausmann, F., Fischer-Kowalski, M., Schandl, H., Eisenmenger, N. (2008). The global sociometabolic transition: Past and present metabolic profiles and their future trajectories. *Journal of Industrial Ecology*, 12(5–6), 637–656.

Lassaletta, L., Billen, G., Grizzetti, B., Anglade, J., Garnier, J. (2014). 50 year trends in nitrogen use efficiency of world cropping systems: The relationship between yield and nitrogen input to cropland. *Environmental Research Letters*, 9(10), 105011.

Le Noë, J., Billen, G., Garnier, J. (2017). How the structure of agro-food systems shapes nitrogen, phosphorus, and carbon fluxes: the generalized representation of agro-food system applied at the regional scale in France. *Science of the Total Environment*, 586, 42–55.

Liu, Y., Pan, X., Li, J. (2015). A 1961–2010 record of fertilizer use, pesticide application and cereal yields: A review. *Agronomy for Sustainable Development*, 35, 83–93.

Madelrieux, S. and Redlingshöfer, B. (2023). Métabolisme associé aux systèmes agri-alimentaires : enjeux et diversité d'approches dans la communauté de recherche française. *Cahiers Agricultures*, 32, 8.

Madelrieux, S., Buclet, N., Lescoat, P., Moraine, M. (2017). Écologie et économie des interactions entre filières agricoles et territoire: quels concepts et cadre d'analyse ? *Cahiers Agricultures*, 26(2), 24001.

Marty, P., Dermine-Brullot, S., Madelrieux, S., Fleuet, J., Lescoat, P. (2022). Transformation of socioeconomic metabolism due to development of the bioeconomy: The case of northern Aube (France). *European Planning Studies*, 30(7), 1212–1229.

Meadows, D., Meadows, D., Randers, J., Behrens, W. (1972). *The Limits to Growth*. Universe Books, New York.

Missemer, A. (2015). *Nicholas Georgescu-Roegen, pour une révolution bioéconomique. Suivi de De la science économique à la bioéconomie par Nicholas Georgescu-Roegen*. ENS Éditions, Lyon.

Moinard, V., Levavasseur, F., Houot, S. (2021). Current and potential recycling of exogenous organic matter as fertilizers and amendments in a French peri-urban territory. *Resources, Conservation and Recycling*, 169, 105523.

Moraine, M., Duru, M., Therond, O. (2017). A social-ecological framework for analyzing and designing integrated crop–livestock systems from farm to territory levels. *Renewable Agriculture and Food Systems*, 32(1), 43–56.

Nesme, T. and Withers, P.J.A. (2016). Sustainable strategies towards a phosphorus circular economy. *Nutrient Cycling in Agroecosystems*, 104, 259–264.

Niang, A., Bourdin, S., Torre, A. (2020). L'économie circulaire, quels enjeux de développement pour les territoires ? Développement durable et territoires. *Économie, géographie, politique, droit, sociologie*, 11(1). doi: 10.4000/developpementdurable.16902.

Odum, E.P. (1989). *Ecology and Our Endangered Life-Support Systems*. Sinauer Associates Inc., Sunderland, MA.

Pauliuk, S. and Hertwich, E.G., (2015). Socioeconomic metabolism as paradigm for studying the biophysical basis of human societies. *Ecological Economics*, 119, 83–93.

Petit, C. (2021). Le métabolisme agri-alimentaire pour une contribution de l'agronomie aux approches socio-métaboliques. *Agronomie, Environnement & Sociétés*, 11(2), 53–68.

Petit, C., Levavasseur, F., Verger, Y. (2018). Reconnecter villes et campagnes : vers des symbioses agricoles et agri-urbaines ? Une exploration sur le Plateau de Saclay. *VertigO: la revue électronique en sciences de l'environnement*, 31.

Petit, C., Gimat, C., Vettoretti, A. (2019). Les Ateliers de Saclay, Réflexion en écologie territoriale sur le cas du plateau de Saclay : Synthèse.

Rastoin, J.L. and Ghersi, G. (2010). *Le système alimentaire mondial : concepts et méthodes, analyses et dynamiques*. Éditions Quæ, Versailles.

Rastoin, J.L. and Meynard, J.M. (2020). L'urgence de systèmes alimentaires territorialisés. *The Conversation* [Online]. Available at: https://theconversation.com/lurgence-de-systemes-alimentaires-territorialises-136445.

Redlingshöfer, B. (2022). Food waste in cities: An urban metabolism approach applied to Paris and Île-de-France. Thesis, Humboldt University of Berlin..

Redlingshöfer, B., Petit, C., Pinet, F., André, G., Deschamps, J., Delcroix, F. (2022). The eating population: Addressing food consumption in urban metabolism studies. In *Transforming Socio-Economic Metabolism in Times of Multiple Crises: Conference of the Socio-Economic Section of the International Society for Industrial Ecology*. International Society for Industrial Ecology, Vienna.

Rockström, J., Steffen, W., Noone, K., Persson, Å., Chapin, F.S., Lambin, E., Lenton, T.M., Scheffer, M., Folke, C., Schellnhuber, H.J. et al. (2009). Planetary boundaries: Exploring the safe operating space for humanity. *Ecology and Society*, 14(2).

Tanguy, A., Bahers, J.B., Athanassiadis, A. (2020). Outsourcing of urban metabolisms and its consequences: A multiscale energy flow analysis of a French port-city. *Resources, Conservation and Recycling*, 161, 104951.

Tedesco, C., Petit, C., Billen, G., Garnier, J., Personne, E. (2017). Potential for recoupling production and consumption in peri-urban territories: The case-study of the Plateau de Saclay near Paris, France. *Food Policy*, 69, 35–45.

Tello, E. and González de Molina, M. (2018). Methodological challenges and general criteria for assessing and designing local sustainable agri-food systems: A socio-ecological approach at landscape level. In *Socio-Metabolic Perspectives on the Sustainability of Local Food Systems: Insights for Science, Policy and Practice*, Fraňková, E., Haas, W., Singh, S.J. (eds). Springer, Cham.

Thénot, M., Bouteiller, C., Lescieux-Katir, H. (2018). Des coopératives agricoles agents de symbiose industrielle : étude de la bioraffinerie de Bazancourt-Pomacle (Marne, France). *RECMA*, 1, 31–47.

United Nations Environment Programme (2021). Food Waste Index Report 2021. Nairobi.

Verger, Y., Petit, C., Barles, S., Billen, G., Garnier, J., Esculier, F., Maugis, P. (2018). A N, P, C, and water flows metabolism study in a peri-urban territory in France: The case-study of the Plateau de Saclay. *Resources, Conservation and Recycling*, 137, 200–213.

Wohlfahrt, J., Ferchaud, F., Gabrielle, B., Godard, C., Kurek, B., Loyce, C., Therond, O. (2019). Characteristics of bioeconomy systems and sustainability issues at the territorial scale. A review. *Journal of Cleaner Production*, 232, 898–909.

Wolman, A. (1965). The metabolism of cities. *Scientific American*, 213(3), 178–193.

9

Farms in the Transition to Sustainable Food in Urban Areas: The Role of Collective Catering

Christine AUBRY and Emmanuel RAYNAUD
Paris Saclay University, AgroParisTech, INRAE, UMR SADAPT, Palaiseau, France

9.1. Introduction

The United Nations' second sustainable development goal is to improve access to healthy and ecologically viable food for the entire world population. With this agroecological and food transition in mind, action strategies aimed at transforming food supply (production and processing of agricultural products) and demand (towards a healthier diet) must therefore be developed and proposed. Public intervention is expected to provide a regulatory framework and public policy instruments conducive to a transition in food systems. At the national level, the recent EGAlim law (section 9.2.2) imposes food supply objectives for all public catering. At the municipal level, numerous initiatives are also being taken to develop more sustainable local food systems. The question of food supply for school canteens is often an important element of proposed food projects. The same is true of initiatives taken at departmental and regional levels. The PAT[1] tool is also

1. Projets Alimentaires Territorialisés: Territorialized Food Projects.

Farms and Territory,
coordinated by Philippe MARTIN and Solène PISSONNIER.

sufficiently flexible to enable food strategies to be implemented at different territorial levels.

In this chapter, we examine the role of catering establishments as vehicles for disseminating practices that improve the sustainability of food systems, on both the supply and demand sides. To meet ambitions in terms of food sustainability and the constraints of the EGAlim law on public catering, how can school catering stakeholders and suppliers at different stages of the supply chain change their practices, organize and coordinate to facilitate the re-territorialization of supply? How can public food procurement be organized to promote organic and local sourcing, while respecting the constraints of the public procurement code? How can school supply chains be reorganized to optimize logistics from upstream to downstream, and ensure "fair" remuneration for those involved?

9.2. Collective catering: a vector for sustainable food practices

In this section, we focus on the foodservice sector. We identify the various segments of the public catering sector, as well as the broad economic trends (section 9.2.1). We then briefly present the regulatory framework of this activity (section 9.2.2), as well as the relevant elements of the public procurement code (section 9.2.3). Finally, based on existing literature and surveys, we present the main obstacles identified to the supply of so-called sustainable products to collective catering, in particular school catering (section 9.2.4).

9.2.1. *Defining collective catering in France and its main challenges*

Collective catering is part of the "*hors foyer*" or "*hors domicile*" (out-of-home) catering sector. It aims to produce meals for the guests of a private or public community. It accounts for approximately 3.7 billion meals a year (GiraFoodService 2019) and generated sales of 25 billion euros in 2019 (INSEE 2020). It is distinguished from other forms of out-of-home catering (e.g. commercial catering) by its social nature, which invites the offer of a meal at a relatively moderate price. Collective catering covers four sectors. We indicate their respective shares of the catering market: schools, from crèches to universities (35%), medical-social services (47%), company and administrative catering (11%) and "other" types of collective catering (vacation centers, armies). The collective catering business may be directly managed by the local authority or outsourced to a collective catering company[2]. Due to the number of meals it provides and the associated need for products from the

2. In France, these are known as "*société de restauration collective*" (SRCs).

agri-food sector, collective catering can potentially have a major impact on both supply chain trends (on the supply side) and eating habits (on the demand side). It is also a sector at the crossroads of many social issues.

In terms of school catering, the challenge of food education is linked to a public health issue, against a backdrop of increasing diet-related illnesses such as obesity. In France, school catering accounts for approximately 1.1 billion meals a year, and over 70% of pupils in public secondary schools attend the canteen. A school-age child's diet is important for growth, psychomotor development and learning capacity. The school meal is also sometimes the only balanced meal children have during the day. Moreover, eating habits seem to be characterized by a high degree of resilience, which argues in favor of food education from the earliest age. School catering has been identified as a prime venue for transmitting educational and civic values (CNA 2017) such as the fight against waste (estimated at 25–30% in school catering) or product discovery. It is a mechanism for raising awareness among children, as much by contributing to ecological awareness as by promoting healthy eating.

Food and beverage services also represent an environmental challenge, as they are responsible for a quarter of greenhouse gas emissions (ADEME 2021), mainly as a result of transport and production methods. On this point, a Senate report highlights the ecological impact of freight transport, identifying it as one of the major causes of environmental degradation (Senate 2021). The quest for sustainability within the catering industry involves reconsidering supply strategies: rethinking these channels helps to reduce the carbon footprint of a meal.

Finally, catering is also at the heart of societal and economic issues. As mentioned above, the price of catering is stable and relatively moderate. However, the canteen manager must guarantee access to all families who request it, which can be a major challenge for some local authorities, since the user's contribution to financing the service (approximately 23%) is lower than its actual cost (Cour des Comptes 2020). The service is thus financed more by the taxpayer than by the guest. This dimension of social justice can be found in all branches of the catering industry. From an economic point of view, collective catering represents a major outlet for foodstuffs from the agri-food sector. Public food procurement contracts are often seen as a promising mechanism for local economic development, in particular to enable SMEs to expand (see section 9.2.3.). They can also be mobilized to support a particular sector, as is directly the case for organic farming (though not exclusively at the national level).

9.2.2. *Regulatory framework for collective catering at the national level*

The collective catering sector is highly regulated: health and nutrition regulations, economic regulations with the obligations of the public procurement code, etc. Other types of documents do not have the status of regulations or laws but are nonetheless important for guiding the sector. Together, they form what we call the regulatory framework of the sector. This framework refers to different dimensions of sustainability. The aim here is not to provide an exhaustive list of the various components of the regulatory framework, but to highlight a few of the key ones.

In addition to the Common Agricultural Policy (CAP), its objectives and tools, France has made a number of commitments in the climate and environmental fields, which were reflected in the Grenelle Environment Round Table in 2007, leading to the Grenelle I law in 2009. In terms of agriculture, the law set a target of 20% of utilized arable area (UAA) to be converted to organic farming by 2020. In 2014, the future law for agriculture, food and forestry aimed to promote agroecological practices.

In terms of public health, the focus is mainly, but not exclusively, on nutrition. In France, the National Nutrition and Health Program (PNNS[3]), was initially set up in 2001, with various updates. It aimed to "improve the state of health of the entire population by acting on one of the major determinants which is nutrition" (Ministère des Solidarités et de la Santé 2019, authors' translation) by promulgating several dietary recommendations, the most famous of which is that relating to the daily consumption of fruit and vegetables.

In addition, since 2009, the GEMRCN[4] has been producing a practical guide to the nutritional quality of meals served in catering establishments (schools, crèches, adult and elderly catering establishments). The aim is to help catering professionals make nutritional choices according to the type of guest (children, adults, etc.) and the type of establishment (crèches, schools, EPHAD, hospitals, etc.), by controlling the frequency with which dishes are served and the recommended grammage according to age and type of establishment. These recommendations provide a framework for the preparation of menus by local authorities.

3. Programme National Nutrition et Santé.

4. Groupement d'Etude des Marchés en Restauration Collective et Nutrition: Collective Catering and Nutrition Market Study Group.

Promulgated in 2018, the EGAlim law for "balanced trade relations in the agricultural and food sector and healthy, sustainable food accessible to all" is the result of the consultation launched in 2017 around the General State on Food (États Généraux de l'Alimentation). Although its scope is much broader than collective catering, it nevertheless contains a section specifically devoted to this sector[5]. In relation to food procurement, the EGAlim law set a fairly ambitious target of at least 50% sustainable or quality products, including 20% AB products from January 2022[6]. These requirements have been reinforced by the 2021 Climate and Resilience Act, which stipulates new requirements from January 2024: 60% of meat and fish supplies must be sustainable and quality products; fair trade products and those acquired primarily on the basis of their environmental and direct sourcing performance are included in the sustainable and quality products category; perpetuation of the weekly vegetarian menu in school catering.

9.2.3. *Public food procurement and the public procurement code*

Public procurement covers all contracts entered into for consideration by a public purchaser or a granting authority with a public service mission (contracting authorities or entities), to meet its needs in terms of works, supplies or services, with one or more economic operators. Contracts can take the form of public procurement contracts (the entire service is financed by the public authority) or concessions (delegation of a public service activity, where the manager is partly remunerated by the service activity). In the case of concessions, the contracting authority delegates the performance of the activity to another entity, often a catering company. The contract between the two parties sets out the contracting authority's requirements. However, the public procurement code no longer applies to this arrangement as the organization of food supplies is carried out by the service provider (although the delegation contract may contain clauses concerning this supply). The French public procurement code does, however, apply when the public authority itself handles procurement. The aim of this code is to ensure the best use of public funds. The fundamental principles of this code are freedom of access to public procurement, equal treatment of candidates and transparency of procedures.

The code at the national level is largely conditioned by rules at the community level. These EU rules have recently undergone changes to better integrate

5. Initially applicable only to public catering, the EGAlim law has recently been extended to all collective catering.

6. For a comprehensive list of products considered sustainable and of the following quality: https://ma-cantine.agriculture.gouv.fr/mesures-phares/qualite-des-produits.

sustainability issues. The EU has promoted the introduction of Green Public Procurement (GPP). GPP is defined as "a process by which public authorities seek to purchase goods, services and works that have a reduced environmental impact throughout their life cycle, compared with goods, services and works having the same principal function that would otherwise be obtained" (authors' translation). For different markets or products, the commission has established criteria that set out the requirements for green public procurement. In the case of food procurement, for example, organic farming is considered an important criterion[7]. The EU has also modified its procurement regulations. EU Directive 2021/14 introduced the possibility of using award criteria other than price. Criteria including environmental, social and quality aspects are now permitted. However, the code places a limit on the choice of criteria. It prohibits public purchasers from using any geographical criteria. Indeed, any mention of local preference would be contrary to the very principles of the code, namely equal treatment of candidates and freedom of access to public procurement[8].

At the national level, two important changes have also been made to the French public procurement code. Firstly, the allotment of public procurement contracts has become the general rule, although exemptions are possible. Allotment consists of dividing the contract into smaller sub-contracts, which are more accessible to SMEs that might be unable to respond to a contract that is too large. The lots of a contract are organized in such a way that each lot is associated with a homogeneous product category. The most common types of organization include allotment by product family (egg products, vegetables, meats, etc.), by packaging (frozen, fresh, etc.) and by production method (organic, label, etc.), but there are even wider possibilities (depending on the number of intermediaries, etc.). A lot can be more or less "broad" (e.g. vegetable lot) or "fine" (e.g. organic fresh vegetable lot or equivalent), depending on the diversity of products it contains.

Secondly, the procurement procedure depends on the size of the contract. The larger the contract, the more onerous the procedures. Table 9.1 summarizes the different types of procedure according to the value of the contract in question (according to the Conseil National de la Restauration Collective[9]).

---

7. For more detailed information, see: https://green-business.ec.europa.eu/green-public-procurement/gpp-criteria-and-requirements_en.
8. However, some local authorities use some of these criteria to de facto favor local sourcing, for example by including criteria for product freshness (in the case of fruit and vegetables).
9. National Council for Collective Catering.

Procedures	"Simplified" procedure (or mutual agreement)	Adapted procedure (MAPA)		Formalized procedure
HT thresholds	Up to €39,999	From €40,000 to €89,999	From €90,000 to €214,999	From €215,000
Advertising and competitive bidding	Optional advertising or negotiated contract without advertising or competitive bidding Obligation to respect the three fundamental principles of public procurement	Advertising appropriate to the amount and purpose of the contract	Advertisement in BOAMP or JAL	Advertising in BOAMP and JOUE
Contractualization requirements	Payment on invoice without prior order letter and not in writing below €25,000 excluding VAT	The need for appropriate written content		Technical specifications
Award procedure		Dematerialized		Dematerialized

Table 9.1. *Diversity of public procurement procedures*

The public purchaser enjoys discretionary power over the selection of suppliers, which decreases with the value of the contract. Contracts worth less than €40,000 enjoy the greatest flexibility, while complying with minimum competitive bidding requirements.

Recent developments in public procurement therefore seem to be more favorable to the introduction of products that are no longer assessed solely on the basis of price. This greater flexibility enables public purchasers to integrate sustainability aspects more fully into the design of public food procurement.

9.2.4. *Barriers to sourcing "sustainable" products*[10]

As suggested earlier, the purchasing power of the foodservice industry is often seen as a powerful mechanism for changing behavior on both the supply and demand sides. However, transforming this potential into concrete action is hampered by a number of factors. The literature on this subject has mainly focused on the supply of public-school meals. In terms of product types, it focuses on the difficulty of sourcing AB and/or local products, even though the local nature of a product is not taken into account by the EGAlim law and contravenes the French public procurement code. Supply difficulties seem to be particularly acute when introducing organic and local products. Local authorities committed to sustainability tend to associate organic and local products, which is why we deemed it relevant to refer to them.

One of the difficulties reported in the vast majority of literature relates to the *extra cost* involved in purchasing AB products (Mikkelsen and Sylvest 2012; FNH survey and Réseau Restau'Co 2019). The price of organic products is in fact estimated to be between 10% and 25% higher than the price of conventional products (Filippini et al. 2018; FNH 2019). The AMF[11] survey of 2020 indicates that 82% of local authorities in the sample reported difficulties, particularly with regard to the additional cost of these products. However, this finding needs to be qualified. Indeed, the additional cost was cited as the number one obstacle by local authorities not yet committed to organic procurement, rather than those already doing so. This confirms the finding of Leresche (2009) that "price is more of an obstacle to the initiation of a project to introduce organic produce into the catering sector than it is to the sustainability of the project" (authors' translation). Some local authorities have managed to adapt to this proven cost constraint.

Organic food comes at a price, and the cost it represents for canteens is either deemed to be amortizable by canteens serving a large number of meals per day or must be offset by financial assistance (DRAAF Nouvelle-Aquitaine 2021). Larger canteens in fact have greater resources at their disposal to implement actions to offset the extra cost. The implementation of these measures also depends on the political commitment of the local authority's elected representatives, and the *lack of political support* can thus be identified as another obstacle to the introduction of organic food (ADEME 2021). Indeed, school catering as a public service depends on a public actor who must be sensitive to a sustainability approach.

10. This section largely represents the work of Anna Ferreira (2022).
11. Association des Maires de France: Association of French Mayors.

Furthermore, to meet the 20% organic requirement enshrined in the EGAlim law, public purchasers are often tempted to combine organic and local origin products, although there is no consensus on the definition of local (Le Velly et al. 2011). On this point, one of the obstacles highlighted in the literature is the *unavailability of local organic supply* (in terms of volume and/or diversity). Unavailability refers to the fact that organic UAA in France will represent just 10.4% of total UAA in 2023 (source: Agence Bio). The lack of local supply[12] may explain the low uptake of organic produce in some local authorities. However, this argument is mainly put forward by communes with little information on local organic availability (Esnault 2016). This lack, real or assumed, can lead elected officials to think that organic provision represents an insurmountable challenge. Therefore, many writings point to the inadequacy and lack of awareness of the organic offer, as well as difficulty in supplying collective catering in terms of volumes (Terra Nova report 2020), or point the finger at the lack of structure in the sector (Esnault 2016; DRAAF Nouvelle-Aquitaine 2021). In particular, producers sometimes find themselves unable to meet the demand from mass caterers, given the volumes that are required, especially by large local authorities. The diversity of establishments, whether in terms of size or geographical location, makes logistics complicated. A producer, especially if they are on their own, will not necessarily have the time to deliver their products from canteen to canteen. They will be more inclined to work in short distribution channels, an outlet that gives greater economic value to their production, as these consumers demand quality and are therefore more inclined to pay a higher price (Le Velly et al. 2010).

The literature also studies the *mismatch between school catering requirements and agricultural supply* as a factor in producers' disinterest in working with school catering. Indeed, supply periods can be a constraint for producers, given school vacations, when fruit and vegetable production is high in summer. Furthermore, school canteens are demanding in their requests. In the case of pork, for example, they will only choose certain specific cuts, and the producer will have to find another outlet to sell the whole carcass (ADEME 2021). They also sometimes demand fourth and fifth range products, which require pre-preparation and therefore processing facilities. Their sanitary requirements are also more stringent: respect for the cold chain, labeling, all these are additional constraints for producers. The school catering sector is not very attractive to producers and is therefore competing with other existing marketing channels. As a result, the main suppliers continue to be wholesalers who source their products on a regional scale and beyond, which hinders local organic sourcing (Le Velly et al. 2010).

12. Local is to be understood here in the sense of regional (or smaller) scale.

As seen, sourcing local products is prohibited by the French public procurement code. Changes to this code, for example in the selection criteria for bids received, make it possible to circumvent this prohibition to some extent. Despite the existence of guides (e.g. Ministère de l'Agriculture 2015), this requires a certain amount of expertise in the field, which brings us to another difficulty: *the complexity of public procurement procedures and the lack of training among buyers* (Bouwer et al. 2006). Difficulties relating to public procurement procedures cover a number of different aspects: difficulties in accessing contracts, poor transmission of information to local producers, lack of knowledge to respond, the need to know the local production fabric, etc. Lack of knowledge of procedures prevents buyers from using mechanisms to facilitate the introduction of organic products in their orders (allotment, criteria to be mobilized, etc.). Contract procedures are sometimes seen as a source of administrative complications compared with direct agreement (DRAAF Nouvelle-Aquitaine 2021). Moreover, this lack of knowledge is partly responsible for the quantity of imported and standardized products in canteens, since buyers are unaware of the mechanisms that can be used to incorporate greater sustainability into their purchases. This constraint is even greater for municipalities with fewer than 10,000 inhabitants (AMF survey 2020), which have fewer resources and expertise than larger municipalities. Difficulties relating to public procurement are also found among producers. Indeed, as mentioned, public procurement represents a considerable workload that producers are sometimes unable to take on, notably due to lack of time.

Additional costs
Lack of political support
Lack of organic supply in the region
Lack of awareness of local organic supply
Difficulty for farmers to meet volume demand
Competition from other, more profitable outlets
Unstable demand from school caterers (vacations)
School catering demands mono-products
Complex public procurement procedures and lack of training for buyers

Table 9.2. *Barriers to organic and/or local food sourcing*

Finally, the size of local authorities reveals inequalities between them. The largest (30,000 inhabitants or more) have more resources to carry out their sourcing and develop a detailed knowledge of the local agricultural and food supply (Terra

Nova report 2019), but encounter difficulties in obtaining supplies, particularly in terms of volumes, a difficulty that would take precedence over that of additional cost (AMF survey 2020). Smaller municipalities (with fewer than 10,000 inhabitants) face more difficulties in terms of sourcing and public procurement constraints (AMF survey 2020), should their order volumes not fall below regulatory thresholds. Table 9.2 summarizes, but is not exhaustive of, the various obstacles mentioned in the literature on organic and/or local produce sourcing.

9.3. Adapting agricultural production to meet growing demand for sustainable food: the example of Île-de-France

Although some metropolises, and even some French cities such as Marseille[13], have significant agricultural areas within their boundaries, urban areas are more often than not characterized by a concentration of inhabitants, and therefore of food demand, combined with a low or even very low presence of in-house farms. The growing demand for "sustainable" food among urban dwellers, particularly in the collective catering sector, often includes, but is not limited to, a demand for "local" production. We should keep in mind that, in the not-too-distant past, cities produced the majority of their food consumption, particularly fresh produce (De Biasi and Roy 2020), both within the city and in their immediate "hinterland", and participated directly in agricultural production via their organic inputs (Barles 2005; Esculier et al. 2019), following Von Thünen's (1826) model of spatial distribution.

9.3.1. *The challenges of defining "local production" in urban areas*

What is produced in the city? Around the city? How far can we go to find farms that contribute to this "local" production? Can the city, particularly through its demand for food services, support or even guide the development of farms in what it considers to be its "hinterland"?

Urban agriculture, in its most widely accepted definition, includes forms of agriculture developing in intra-urban environments, but also certain forms of agriculture on the urban periphery, with a functional link (notably food) to the city by sharing resources with it (Moustier and M'Baye 1999; Nahmias and Le Caro 2012). Intra-urban agriculture (Aubry et al. 2022) is (re)developing at an often-sustained pace in all French cities, with a diversity of productions, both food

---

13. See: https://www.marseille.fr/environnement/actualites/protection-des-terres-agricoles-dans-marseille-l-exemple-de-sainte-marthe.

(vegetables, fruit, berries, herbs, mushrooms, eggs) and non-food (cut flowers, plants, etc.). However, due to space constraints, the most common forms of gardening (widely acclaimed community gardens, participatory microfarms) and its multiple functions (social ties, education, living environment, environmental and cultural services, etc.), which make it essential to many urban expectations (Grard et al. 2021), also make it a minor contributor, except in special cases, to the quantitative food supply of urban dwellers (Pourias et al. 2015, 2016). Indoor systems (vertical farms), which were once heralded by the media as a major solution to the local supply of fresh vegetables to cities, are highly productive per unit area, but they consume a lot of energy and are therefore not very environmentally efficient (Dorr 2022). In France at least, and with the exception of a few products (mushrooms, endives, microgreens) in niche markets, they have not found their economic model, despite massive public and private support. In fact, those that remain are reorienting themselves towards aromatic or even non-food products with very high added value (cosmetics, medicinal products). As a result, they are not major candidates for local food production, especially not for collective catering, and all the more so at a time when we are seeking to reduce energy consumption.

Peri-urban agriculture, often greatly reduced by urban sprawl in the late 20th and early 21st centuries (Poulot 2014), is not always geared to supplying the nearby city. The example of the Paris region is telling in this respect: although agriculture occupies almost half of the Île-de-France territory, agri-food systems are very open, including in peri-urban areas where agriculture is strongly present (Tedesco et al. 2017 on the Plateau de Saclay). Farms in these areas, which are mainly large scale, are largely geared towards the national market and exports, rather than local consumers. For example, bread wheat production in the Île-de-France region is more than four times greater than the needs of its 12 million inhabitants, but is largely exported as grain or flour, while the region imports approximately half its flour requirements (Toullalan 2012; IAU 2015). This situation, which raises questions today, is the result of the interplay of markets in which local supply was not at all favored, with a few exceptions (see below)[14]. Another example is lettuce, for which Île-de-France is virtually self-sufficient thanks to locally well-established and productive cereal-salad production systems (plaine de Montesson, plaine de Chailly en Bière, etc.): but these Île-de-France lettuces end up, for the most part, mixed with lettuces from other origins in the national interest market at Rungis, and are then redistributed through supermarket channels. The "local" character of the lettuce is therefore not highlighted for consumers, except in the context of the Carreau des Producteurs (a specific area of Markets of National Interest (MINs) intended for commercial catering, relatively confidential) or in short, direct farm sales circuits:

14. During the first confinement of 2020, flours stamped "Ile de France" appeared in supermarkets in the Paris region, but they have become rarer today (author's observation).

fairground markets with producer-sellers (very rarefied since the beginning of the 21st century), AMAP[15]-type fresh produce deliveries (Aubry and Kebir 2013), or forms of distribution that are increasing but which only encompass a small minority of Île-de-France farms and consumers (IAU 2015).

As a result, the "definition" of a local product for cities, and in particular for the supply of foodservices, involves a variety of geographical extensions: from "what can be produced as nearby as possible" (Aubry and Kebir 2013; Darly and Aubry 2014, authors' translation) to the delimitation of a geographical area designed to encompass several rural agricultural production basins, such as the "250 kms around Paris" predefined as part of the Agri Paris Seine initiative (source: EGAD Ville de Paris 2022)[16]. In some cases, the city can directly support rural farms, or even create them: most recently, communes in the Greater Paris metropolitan area have sought to acquire peri-urban land, either to install a private farmer under contract or to farm it directly themselves, in particular for part of their catering needs[17]. The public water company Eau de Paris has largely financed the transition to organic farming and the diversification of farms in the Vanne basin in Yonne, which contributes almost 20% of its water supply (Eau de Paris 2019). To achieve this, it has used the secure purchase of products from farmers via the public-school catering services of certain Paris arrondissements. In a very similar vein, the city of Rennes, via the Régie Eau du bassin rennais, acquires farmland, and accompanies and supports conversion to organic farming via the introduction of the Terres de Sources® label (Eau du bassin rennais 2019)[18].

Nevertheless, for urban areas, a major challenge for sustainable local supply, particularly for the volumes demanded by mass catering, is the controlled development of peri-urban agriculture, which needs:

– On the one hand, to protect agricultural land from urban sprawl: today, certain public policies aim to "protect" this agricultural land in urban planning documents (Loudiyi et al. 2022), and the recent introduction of the ZAN[19] scheme, even if it is causing a stir, particularly in peri-urban areas, should go in the same direction. It

15. The Association pour le Maintien de l'Agriculture Paysanne is a form of Community Supported Agriculture (CSA) that connect consumers directly with local farmers, ensuring access to fresh, seasonal produce while supporting sustainable farming practices.

16. These appropriately include part of the Val de Loire Maraicher, livestock farming areas (Centre, Normandy) and sea fishing areas.

17. This initiative also aims to supply fresh produce to the most food insecure populations (e.g. the city of Villejuif).

18. In both cases, the fact that water supply and catering, or part of them, are managed by local authorities makes it easier for towns to translate their strategy into support for local agricultural development.

19. Zéro Artificialisation Nette: Zero Net Artificialization.

should be noted that this consumption of peri-urban agricultural land for urban expansion is not a phenomenon of the past, as demonstrated by the ongoing example of the Triangle de Gonesse in the Val d'Oise (Macé le Ficher 2024).

– On the other hand, to support farms in developing their production and marketing channels, to "re-territorialize" them, to increase their "degree of urbanity" (Donadieu 1998; Donadieu and Fleury 2003), i.e. to make them more relevant to urban food needs. Collective catering, as we will see, can be a powerful mechanism for achieving this. We shall take the Île-de-France region as an example, and the need to increase the diversification potential of local production through new market opportunities.

9.3.2. *How could farms adapt to the growing demand for collective catering in local areas?*

A number of recent initiatives in the Île-de-France region have attempted to forge more or less fruitful links between the region's agriculture and its catering industry, notably through the development of organic farming channels, which were targeted as early as the Grenelle Environment Forum in 2007.

Echoing the high level of bread wheat production in the Île-de-France region, one of the first initiatives to bring the region closer together was launched in 2004, through the Pain Bio d'Ile-de-France[20] network, supported by the region, starting with Île-de-France high school canteens and now extending to other sectors, from municipal catering to private bakeries. We have already mentioned the initiative taken by the Eau de Paris[21] water authority, but it has not yet been implemented on its catchments in the Paris region, due to the many obstacles to converting to organic farming in these areas. Other initiatives have been launched over the past 10 years in the Paris region: for example, in 2010, the Agence d'Espaces Verts de la Région Ile-de-France[22] set up three organic farmers (one cereal grower, two market gardeners) in Les Mureaux on 117 ha, followed in 2014 by the creation of a vegetable factory via a catering company. This farm and its vegetable factory have been linked up with the school catering services of several local authorities in the Paris region[23]. More sustainable was the creation in 2014 of Coop Bio Ile-de-France[24], a cooperative of more than 60 producers, which produces and processes fruit, vegetables, pulses,

20. Ile-de-France Organic Bread.
21. Paris Water.
22. Green Spaces Agency for the Île-de-France Region.
23. The project had a checkered history, with the vegetable factory passing through the hands of the SOS group before being taken over by Coop Bio Île de France.
24. Ile-de-France Organic Cooperative.

field vegetables and flour in its new *légumerie* (2020), mainly for public catering, with several local authorities and school boards as members of the SCIC[25] formed for this cooperative. Today, Coop Bio Ile-de-France aims to be a driving force behind the conversion to organic farming and diversification of field crop farms in Île-de-France, in conjunction with the Groupement des Agriculteurs Biologiques d'Ile-de-France[26].

It is likely that, in addition to the overall incentive provided by the two successive regional Plans Bio[27] campaigns (2009–2013 then 2014–2020), these initiatives, strongly driven by demand from the Île-de-France catering sector, have contributed to the sharp increase in the surface area and number of organic farms in the region. Starting from a very low base in 2007 (84 farms and barely 1% of the UAA), by 2022 organic farming in the region represented 617 farms (out of 4,425) with just over 6% of the UAA (GAB IdF2022; Agreste 2022).

The region, like others, is also the scene of a resurgence in diversified market gardening, most often in organic farming or other non-conventional systems and on small acreages[28]. Despite their strong ties to the territory (often assisted by local authorities for access to land), these diversified market gardeners are not potential suppliers to the regional catering industry, due to the limited volumes produced and the low prices offered by this outlet. Their outlets are much more in response to very strong local demand, basket systems (such as AMAP), farm gate sales, markets and/or other forms of short circuits aimed at private individuals (with substantial purchasing power). Only certain forms of direct, strong involvement by local authorities (e.g. Mouans-Sartoux in the Alpes Maritimes, Grande Synthe in the Pas de Calais – Aubry and Paturel 2022), possibly in the form of a farm management company, can establish a link between these diversified market-garden farms and public catering, but in cases where demand is limited by the size of the commune (Moissy Cramayel – Seine et Marne) and/or the restrictions on the types of catering involved (Gennevilliers – Hauts de Seine – as part of a PAT and Villejuif – Val de Marne).

In Île-de-France, under the impetus of growing demand from the foodservice sector and thanks to the downstream instruments already mentioned (including the Coop Bio d'Ile-de-France), we are gradually seeing the diversification of field crop

25. *Société coopérative d'intérêt collectif:* Collective Social Interest Cooperative.
26. Ile-de-France Organic Farmers Group.
27. Organic Plans.
28. Out of an average of 110 start-ups per year, the majority of which (70) are family take-overs in arable farming, in 2020 and 2021, there will be 35 to 40 start-ups outside the family framework, half of them in market gardening, beef farming and mixed farming (https://driaaf.ile-de-france.agriculture.gouv.fr/IMG/pdf/RA2020_zoom_installations_cle012a66.pdf).

farms into legumes, field vegetables and even fruit crops, all of which are in great demand by the foodservice sector. Field vegetables, in particular, are relatively "easy" to integrate into the crop rotations and successions of large-scale cereal farms (lettuces, carrots, potatoes, beet), as they are easy to mechanize and do not require major restructuring of the workforce, except in specific cases. There are signs of this diversification, as in the 2020 agricultural census, the proportion of "other crops" such as field vegetables increased between 2010 and 2020 for field crop farms (Agreste 2022). However, this diversification is not self-evident: specific requirements for equipment (potential use of agricultural contractors, CUMAs[29]) or irrigation may emerge. Work is underway to study the adaptation of this vegetable diversification (vegetable and market gardening systems) to climate change, highlighting their dependence on irrigation (Climaleg 2021–2022 and Climaleg-Eau 2022–2023 projects, supported by DRIAAF, the Terre et Cité association and Paris Saclay University (Touili et al. 2024)).

The introduction of legumes in crop rotation may prove to be a resilience factor for field crop farms, at a time when the price of nitrogen fertilizers is rising considerably – a situation that existed before, but was reinforced in 2022 by the war in Ukraine and its impact on energy prices. Further consideration is required to link urban areas to changes in farming practices, particularly in field crops, in order to limit their dependence on nitrogen (and phosphate) fertilizers.

Since the Ministry of Agriculture's post-Covid-19 Recovery Plan of 2021, the number of territorial food projects has been multiplying in France, with approximately one-third of them coming from urban territories (RnPat 2022). Île-de-France is no exception, with 12 PATs approved by 2023. All of them include a major component on the supply of food for collective catering, and all of them focus on the desired "local" sources of this supply, while noting the major difficulties currently encountered in finding them. But today, there is (still?) no shared reflection on how to act on, and above all with, Île-de-France agriculture, as well as on mechanisms and financial, human and logistical means to encourage the desired changes. Recent history shows that initiatives such as the Coop Bio d'Ile de France, particularly when they have been set up between farmers and urban demand for collective catering, have been an interesting mechanism for territorializing agriculture that has long been "disconnected" from its territory. The deployment of these many "urban" PATs in France over the coming years will provide a unique laboratory for analyzing the phenomena of (re)territorialization.

29. Coopérative d'Utilisation de Matériel Agricole: Cooperative for the Use of Agricultural Equipment.

9.4. Conclusion

There is an abundance of initiatives to improve the sustainability of food systems at different territorial scales. Some are older than others, more ambitious than others, focusing on different dimensions of food sustainability. Public catering, whether for schools or other establishments, often plays an important role in these initiatives. However, developing a sustainable and local food supply for public catering is not self-evident. Despite the significant potential in terms of outlets for agricultural products, the requirements and constraints of the public catering sector do not come together spontaneously with agricultural supply and suffer from what economists refer to as transaction costs. This reflection is part of a more general movement to (re)localize agricultural supply in order to promote territorialized food systems. This chapter has highlighted some of the obstacles to the development of a more sustainable and local food supply in public catering. It has also presented the adaptations to agricultural supply and production that are needed to facilitate the above-mentioned encounter, in a context of essentially urban territories.

We certainly lack the hindsight to characterize in detail the diversity of the initiatives observed and draw generalizable conclusions, or at the very least, to identify all the parameters that affect the performance of these initiatives. A preliminary stage of survey and comparison is necessary. The identification of "best practices", which may be contingent on certain local parameters such as community size, is nevertheless important if we are to be in a position to support public decision-making, whether at the local or national level, to improve the sustainability of food systems.

As the obstacles and mechanisms to developing a more sustainable and local food supply are of different natures, the analysis can only be made through multidisciplinary work. This chapter, for example, was written by an agronomist and an economist. Furthermore, as the notion of a sustainable food system suggests, this system is made up of different elements. This suggests that the panoply of public policy instruments that can be mobilized is relatively broad (Lascaux Program 2016). These range from the local authority's land policy (through urban planning documents) to its agricultural policy (e.g. creation of an agricultural board) to its purchasing policy (carrying out appropriate sourcing, allotting its public procurement contract). This raises the question of how to match public policy instruments to food ambitions, and the overall coherence of the various public interventions.

9.5. References

ADEME, A(R)GILE, Bien communs, Framhein, Effet2Levier, Maiom (2021). Freins et leviers pour une restauration scolaire plus durable [Online]. Available at: https://librairie.ademe.fr/consommer-autrement/4556-freins-et-leviers-pour-une-restauration-collective-scolaire-plus-durable.html.

Agence bio (2021). Le bio, acteur incontournable de la souveraineté alimentaire. *Les chiffres 2020 du secteur bio* [Online]. Available at: https://www.agencebio.org/vos-outils/les-chiffres-cles/.

Agreste (2022). Premiers résultats du recensement agricole en Ile de France : la synthèse en 15 points clés [Online]. Available at: https://driaaf.ile-de-france.agriculture.gouv.fr/IMG/pdf/220222_Essentiel_cle4f41ff.pdf.

Association des Maires de France (2020). Panorama de la restauration scolaire après Egalim. Enquête 2020 [Online]. Available at: https://medias.amf.asso.fr/docs/DOCUMENTS/a484303fc444b994cdbc6bdc8086e0c5.pdf.

Aubry, C. and Kebir, L. (2013). Shortening food supply chains: A way for maintening agriculture close to urban areas? The case of the French metropolitan areas of Paris. *Food Policy*, 41, 85–93. doi: 10.1016/j.foodpol.2013.04.006.

Aubry, C., Giacché, G., Maxime, F., Soulard, C. (2022a). Une politique d'agriculture urbaine, entre innovation sociale et démocratie alimentaire, le cas de Grande-Synthe. In *Les agricultures urbaines en France : comprendre les dynamiques, accompagner les acteurs*, Aubry, C., Giacchè, G., Maxime, F., Soulard, C.-T. (eds). Éditions Quæ, Versailles.

Aubry, C., Giacché, G., Maxime, F., Soulard, C. (2022b). La diversité des formes d'AU et de leurs modèles économiques. In *Les agricultures urbaines en France : comprendre les dynamiques, accompagner les acteurs*, Aubry, C., Giacchè, G., Maxime, F., Soulard, C.-T. (eds). Éditions Quæ, Versailles.

Barles, S. (2005). *L'invention des déchets urbains : France 1790–1970*. Champ Vallon, Seyssel.

Bouwer, M., Jonk, M., Berman, T., Bersani, R., Lusser, H., Nappa, V., Nissinen, A., Parikka, K., Szuppinger, P., Viganò, C. (2006). *Green Public Procurement in Europe 2006 – Conclusions and Recommendations*. Virage Milieu & Management bv, Haarlem.

Conseil National de l'Alimentation (2017). Les enjeux de la restauration collective en milieu scolaire. Avis 77 [Online]. Available at: https://www.cna-alimentation.fr/v2/wp-content/uploads/2017/09/CNA_Avis77.pdf.

Cour des Comptes (2020). Rapport public annuel. Tome I : finances, politiques, et gestions publiques [Online]. Available at: https://www.ccomptes.fr/system/files/2020-02/20200225-RPA-2020-tome-I_0.pdf.

Darly, S. and Aubry, C. (2014). La demande en produits locaux de la restauration collective : quels liens avec l'offre de proximité dans une région d'agriculture industrielle ? Le cas de l'Ile-de-France. *Géocarrefour*, 145–157.

De Biasi, L. and Roy, J.-M. (2020). La Grande Hitsoire des légumes et de leurs terroirs en Ile-de-France. Note no. 868, Institut Paris Region.

Donadieu, P. (1998). *Campagnes urbaines*. Actes Sud, Arles.

Donadieu, P. and Fleury, A. (2003). La construction contemporaine de la ville-campagne. *Revue de Géographie Alpine*, 91(4), 19–30.

Dorr, E. (2022). Environmental performance of urban agriculture: How to apply life cycle assessment, and the knowledge and questions generated. PhD Thesis, Paris Saclay University.

DRAAF Nouvelle-Aquitaine (2021). Projets alimentaires territoriaux : restauration collective et loi EGAlim. Panorama des actions menées en Nouvelle-Aquitaine.

Eau de Paris (2019). AAC des sources de la Vallée de la Vanne [Online]. Available at: https://territoiresbio.fr/decouvrez-leurs-experiences/eau-de-paris-aac-des-sources-de-la-vallee-de-la-vanne/.

Eau du bassin Rennais (2019). Terres de sources [Online]. Available at: https://www.eaudubassinrennais-collectivite.fr/la-solidarite-internationale/labellocal-terres-de-sources/ [Accessed 1 October 2022].

Esculier, F., Le Noe, J., Barles, S., Billen, G., Créno, B., Garnier, J., Lesavre, J., Petit, L., Tabuchi, J.-P. (2019). The biogeochemical imprint of human metabolism in Paris Megacity: A regionalized analysis of a water-agro-food system. *Journal of Hydrology*, 573, 1028–1045. doi: 10.1016/j.jhydrol.2018.02.043.

Esnault, M. (2016). La restauration scolaire dans le premier degré : quelle place pour un approvisionnement bio de proximité ? Étude centrée sur six communes du Calvados.

Filippini, R., De Noni, I., Corsi, S., Spigarolo, R., Bocchi, S. et al. (2018). Sustainable school food procurement: What factors do affect the introduction and the increase of organic food? *Food Policy*, 76, 109–119.

FNH and Réseau Restau'Co (2019). Quels besoins d'investissement en restauration collective pour engager la transition agricole et alimentaire dans les territoires ? ADEME, INTERFEL.

GAB Ile de France (2022). Observatoire régional de l'Agriculture Biologique en Ile de France. Edition 2022 – Chiffres-clés au 31 décembre 2021 [Online]. Available at: https://www.bioiledefrance.fr/documents/2022-observatoire-final.pdf [Accessed 1 October 2022].

GiraFoodService (2019). Analyse du marché de la restauration. Report.

Grard, B.J-P., Joimel, S., Vieublé Gonod, L., Giacche, G., Aubry, C., Consales, J.-N., Séré, G., Manouchehri, N., Haudin, C-S., Auclerc, A. et al. (2021). Evaluation des services écosystémiques fournis par les sols de micro-fermes urbaines : méthodologie et retours d'expériences. *Etude et Gestion des Sols*, 28, 31–47.

IAU (2015). Les filières courtes de proximité dans le système alimentaire francilien. Note de synthèse et 8 fascicules [Online]. Available at: https://www.institutparisregion.fr/nos-travaux/publications/les-filieres-courtes-de-proximite-au-sein-du-systeme-alimentaire-francilien/.

Le Velly, R. (2011). *Comment se fait la rencontre entre offres et demandes locales pour la restauration collective publique ? Premiers enseignements*. Éducagri éditions, Dijon.

Le Velly, R. (2012). La relocalisation des approvisionnements de la restauration collective et le code des marchés publics. *Pour*, 215–216(3–4), 269–274.

Loudiyi, S., Duvernoy, I., Banzo, M. (2022). Les systèmes alimentaires territorialisés : de nouveaux cadres d'intégration des agricultures urbaines. In *Les agricultures urbaines en France : comprendre les dynamiques, accompagner les acteurs*, Aubry, C., Giacchè, G., Maxime, F., Soulard, C.-T. (eds). Éditions Quæ, Versailles.

Macé le Ficher, P. (2024). Militer pour l'agriculture périurbaine : le projet CARMA sur le Triangle de Gonesse. *Territoire en mouvemential. Revue de géographie et aménagement* [Online]. Available at: http://journals.openedition.org/tem/11237 [Accessed 2 May 2024].

Mikkelsen, B. and Sylvest, J. (2012). Organic foods on the public plate: Technical challenge or organizational change? *Journal of Food Service Business Research*, 15, 64–83.

Ministère de l'Agriculture (2015). Favoriser l'approvisionnement local et de qualité en restauration collective : guide pratique.

Ministère de l'agriculture, de l'agroalimentaire et de la forêt (2016). Typologie des politiques publiques de restauration collective et durable.

Ministère des Solidarités et de la Santé. (2019). Programme national nutrition santé [Online]. Available at: https://sante.gouv.fr/IMG/pdf/pnns4_2019-2023.pdf.

Moustier, P. and M'Baye, A. (1999). Introduction. In *Agriculture périurbaine en Afrique subsaharienne : actes de l'atelier international du 20 au 24 avril 1998*, Moustier, P., Mbaye, A., De Bon, H., Guérin, H., Pagès, J. (eds). CIRAD, Montpellier.

Nahmías, P. and Le Caro, Y. (2012). Pour une définition de l'agriculture urbaine : réciprocité fonctionnelle et diversité des formes spatiales. *Environnement urbain/Urban Environment*, 6, 1–16.

Poulot, M. (2014). Agriculture et ville : des relations spatiales et fonctionnelles en réaménagement. Une approche diachronique. *Pour*, 4(224), 51–66.

Pourias, J., Aubry, C., Duchemin, E. (2015). Is food a motivation for urban gardeners? Multifunctionality and the relative importance of the food function in urban collective gardens of Paris and Montreal. *Agriculture and Human Values*. doi: 10.1007/s10460-015-9606-y.

Pourias, J., Duchemin, E., Aubry, C. (2016). Products from urban collective gardens: Food for thought or for consumption? Insights from Paris and Montreal. *Journal of Agriculture, Food Systems, and Community Development*, 5(2), 1–25.

Programme Lascaux (2016). Typologie des politiques publiques de restauration collective locale et durable [Online]. Available at: www.observatoire-restoco-durable.org.

Réseau National des PAT (2022). L'observatoire national des PAT [Online]. Available at: https://rnpat.fr/projets-alimentaires-territoriaux-pat/ [Accessed 1 October 2022].

Sénat (2021). Transport de marchandises : se donner les moyens d'une transition nécessaire. Information Report 604, 19 May [Online]. Available at: http://www.senat.fr/rap/r20-604/r20-6041.pdf.

Tedesco, C., Petit, C., Billen, G., Garnier, J., Personne, E. (2017). Potential for recoupling production and consumption in peri-urban territories: The case-study of the Saclay plateau near Paris, France. *Food Policy*, 69, 35–45.

Terra Nova (2020). Les enfants à table : accélérer la transition alimentaire dans les cantines scolaires.

von Thünen, J.H. (1826). *Der Isolierte Staat ln Beziehung auf Landwirtschaft und Nationalöconomie*. Perthes, Hamburg.

Touili, N., Morel, K., Aubry, C., de Noblet Ducoudré, N. (2024). What do vegetable farmers expect from climate services to adapt to climate change by 2060? A case study from the Parisian region. *Climate Services*, 34, 100474.

Toullalan, M. (2012). Les enjeux de la production et de l'approvisionnement alimentaire en Ile-de-France. Report to CESER, March [Online]. Available at: https://www.lagazettedes communes.com/telechargements/rapport-TOULLALAN.pdf.

10

Farming and the Circular Economy in Territories: Recycling Urban Waste in Agriculture

Florent LEVAVASSEUR and Sabine HOUOT

Paris Saclay University, AgroParisTech, INRAE, UMR ECOSYS, Palaiseau, France

10.1. Introduction

Recycling organic waste in urban agriculture meets a number of challenges to which the agroecological transition is seeking to respond. It can contribute to the closure of biogeochemical cycles by recycling nutrients (nitrogen, phosphorus, etc.). In conventional systems, it can replace mineral fertilizers, which either come from mines with finite reserves unevenly distributed around the world (phosphorus, potash), or are synthesized using large quantities of fossil energy (nitrogen). In cases where mineral fertilizers account for a significant proportion of farmers' expenses (11% in field crops, for example; Lecuyer et al. 2013), recycling can make economic sense. By definition, recycling organic waste increases the return of organic matter to the soil, thereby helping to maintain or even increase its carbon content. Maintaining or improving this stock is essential for maintaining or restoring the various components of fertility: chemical (supply of nutrients), physical (stability of structure, porosity) and biological (substrate for soil organisms), which can lead to greater productivity (Oldfield et al. 2020). Finally, recycling means storing carbon in soils, which can also help mitigate climate change (Amelung et al. 2020).

Farms and Territory,
coordinated by Philippe MARTIN and Solène PISSONNIER. © ISTE Ltd 2025.

However, recycling urban waste in agriculture can also present environmental and health risks: the main ones being soil, water and food chain contamination (Fijalkowski et al. 2017), nitrogen losses to water or air and greenhouse gas emissions (Nicholson et al. 2017).

From the city's point of view, recycling urban organic waste in agriculture is an interesting treatment and recovery solution. For health reasons, this waste must be evacuated from cities to avoid water pollution or the spread of disease, as the possibilities for recovery in the city are generally insufficient.

Historically, recycling urban organic waste into agriculture was a common practice, whether food waste or human excreta (Esculier 2018). With the advent of mineral fertilizers, the development of other waste treatment techniques (wastewater treatment plants (WWTP), incineration, etc.) and the contamination of some of this organic waste, recycling was partly lost. However, alternative methods of managing organic urban waste do have environmental impacts: particulate matter and greenhouse gases emitted by burning or incineration, methane emitted by landfilling, greenhouse gas emissions from WWTP and residual discharges into rivers, poor biogeochemical cycle looping, etc. Because of these impacts and the challenges of the agroecological transition mentioned above, the recycling of urban waste in agriculture is enjoying renewed interest. For example, in France, the anti-waste law for a circular economy (known as the AGEC law) of February 10, 2020, provides for the generalization of source sorting of biowaste (food waste and green waste) by 2024 and its recovery by composting (centralized or not) or anerobic digestion (through a biogas plant). According to the waste hierarchy (Directive 2008/98/EC of the European Parliament and of the Council of November 19, 2008, on waste), the priority remains prevention (limiting waste production), then reuse, before recycling.

The general aim of this chapter is to present, from an agronomic and environmental point of view, the principles of recycling urban organic waste in agriculture. More specifically, the aim is to present (i) the diversity of urban organic waste, the possible treatments and the characteristics of the resulting organic residual products (ORPs), (ii) some information on the main regulations governing the agricultural recycling of urban ORPs, (iii) the contribution of this recycling to the needs of agriculture and (iv) the associated constraints, (v) some information on the environmental impact of recycling urban organic waste in agriculture, compared with other treatment methods, (vi) an overview of the various actors involved in these recovery methods and (vii) more specifically, some information on farmers' perceptions of this type of recycling.

10.2. From urban waste to organic fertilizer

10.2.1. *Main (peri-)urban organic waste and their sources*

Urban activities generate a wide range of organic waste. There are several main sources (see Figure 10.1):

– food waste from private households, catering establishments and food shops;

– green waste from households, gardens and parks, etc.;

– human feces and urine;

– equine manure associated with urban leisure activities, depending on the acceptance of the notion of urban;

– feces and urine from domestic animals (cats, dogs).

Other organic wastes present in peri-urban areas but not specific to urban areas are not considered here (livestock or agro-industrial effluents in particular). The quantities of each type of organic waste produced vary from one author to another. Figure 10.1 is intended only to give an order of magnitude.

Figure 10.1. *Main sources of municipal organic waste in France (data source: Menigaux (2004), Doligez (2015), FranceAgrimer (2015), Esculier (2018), Espelia – SOLAGRO (2021), Ministère de l'agriculture (2021))*

The vast majority of food waste is collected, but only a small proportion selectively at source. Approximately 16×10^6 t/year of residual household waste (RHW) are produced each year by the French people. Of this, putrescible waste accounts for 5×10^6 t/year, or 83 kg/capita/year (Espelia – SOLAGRO 2021). It is mainly made up of meal leftovers, peelings and uneaten food (approximately 80%, or 70 kg/capita/year), as well as garden waste (20%) (Redlingshöfer 2022). The priority of public policy is to reduce food waste, even if better recovery of this waste is also an objective. Sorted waste can be recycled in a number of ways: individual or collective composting, collection and then centralized treatment by composting or anerobic digestion.

Collected green waste (voluntary drop-off, door-to-door collection) represents a national deposit of 4.10^6 t/year of raw material (FranceAgrimer 2015), or approximately 60 kg/capita/year. The theoretical production of green waste given the surface area of French green spaces is approximately 60×10^6 t/year, highlighting the fact that most green waste is recovered in situ or not extracted (FranceAgrimer 2015). Additional collection potential could therefore exist. What is more, 20% of putrescible household waste is still green waste, as mentioned above. Regional particularities also exist, with production increasing with population density. This production is quite variable, both in terms of production timing and quality (leaves, clippings, branches, etc., more or less rich in water, woody matter, etc.).

Each human excretes approximately 0.5 m^3/year of urine and 50 kg/year of feces (Esculier 2018), representing a theoretical national deposit of approximately 30×10^6 m^3 of urine and 3×10^6 t of feces. These materials are almost entirely collected with wastewater, via collective (80%) or non-collective (20%) sanitation systems. Dry toilets or source-separated urine collection with separate toilets or urinals are still rare.

There are approximately 1.2 million horses in France, most of them involved in urban activities such as leisure, sport and racing (French Ministry of Agriculture 2021). At an average of 12 t/year/horse of manure, the total equine manure production in France is estimated at several million tons per year (Doligez 2015), or approximately 200 kg/capita/year (for a French population of 65 million). This manure is made up of dung and the bedding used (straw, wood shavings, etc.).

With approximately 7 million dogs in France in 2004, and 300 g of feces and 0.5 l of urine per animal per day (Menigaux 2004), the estimated annual quantity is 1.1×10^6 t/year of feces and 1.8×10^6 t/year of urine, i.e. approximately 20 kg/inhabitants/year of canine feces and 30 kg/inhabitants/year of canine urine. Flows linked to feline feces are considered to be lower (Menigaux 2004). Some authors have also shown that these flows are not negligible compared to other organic waste flows on a European scale (van Dijk et al. 2016) and should therefore be considered. The proportion of dog and cat feces collected selectively is a priori almost nil, although little information on the subject has been found. Selective collections are being experimented with within shelters (Maisonneuve 2017). These droppings are otherwise either "lost" in parks and gardens or collected with household waste.

We can therefore see that among the various organic wastes, the proportion collected and the selectivity of their collection are variable, which conditions their subsequent treatment and agronomic valorization.

10.2.2. *Treatment of organic urban waste collected for recycling in agriculture*

The organic waste collected does not necessarily lend itself to agronomic recovery for various reasons:

– insufficient sanitary quality, with the presence of various pathogens;

– too low concentration of fertilizing elements or organic matter;

– non-stabilized organic matter that can give rise to odor problems or nitrogen immobilization during decomposition.

Various treatments aim to modify the state of this organic waste to improve its agronomic value. The aim of these treatments can also be to produce energy (biogas plant). The products obtained from these treatments and intended for agricultural use are known as organic waste products (OWPs).

Wastewater is a special case. In this case, the aim of treatment is to produce purified water for discharge into the environment. Sewage sludge is simply the residue of this treatment. These sludges are generally obtained after biological treatment to partially eliminate carbon and nitrogen (with gaseous discharge into the atmosphere: CO_2, N_2, N_2O, etc.); these elements are thus only slightly preserved in the sludge. Phosphorus may be treated by precipitation and end up in the sludge. Once the sludge has been produced, various additional treatments can follow: dewatering, liming (hygienization and adding a liming value to the sludge), composting, anaerobic digestion, etc.

Composting involves the aerobic fermentation (in the presence of oxygen) of organic matter. During composting, organic matter is broken down and transformed into more stable organic molecules (more resistant to degradation by micro-organisms). An initial thermophilic phase (50–70°C) hygienizes the compost, while a second, lower-temperature phase matures it. During composting, a significant proportion of carbon is lost in the form of CO_2 (and methane), while some nitrogen is also lost in gaseous form (notably ammonia). There are several types of composting:

– rustic composting, individual or collective;

– industrial composting on a composting platform;

– semi-industrial composting (e.g. electro-mechanical).

Various types of organic urban waste can be composted: green waste, food waste, sewage sludge, horse manure and dry toilet contents. For sludge and food waste, the addition of a structuring agent (woody green waste, for example) is necessary to ensure aerobic conditions for composting.

Methanization is an anaerobic fermentation process (in the absence of oxygen), resulting in the production of biogas (mainly a mixture of methane and carbon dioxide) and digestate. The biogas can be valorized to produce energy (electricity and heat in cogeneration, injection of methane into the gas network after purification, etc.). Digestate contains partially stabilized residual organic matter and all nutrients. Anaerobic digestion can be carried out either dry or wet (the latter being the most common method). In the wet process, post-processing by phase separation can take place, resulting in a solid and a liquid digestate. Most organic materials can be anaerobically digested. However, the economic interest of biogas depends on their methanogenic power, which is high for oils and low for wood waste, for example.

The principle of pyrolysis is the thermal degradation of organic matter at high temperatures under anaerobic conditions. Pyrolysis produces gases, oils and a residual solid, biochar. Biochar contains a highly stabilized form of carbon. Numerous studies envisage the use of biochar for long-term soil carbon storage to offset anthropogenic greenhouse gas emissions. However, the large-scale use of biochar has not yet been developed in France (or anywhere else in the world).

At present, Table 10.1 lists the main municipal waste treatments used or planned in France, and the resulting OWPs used in agriculture.

The vast majority of human urine and feces are collected mixed with wastewater and treated in WWTPs, either directly via collective sanitation or after emptying of non-collective sanitation (NCS) systems. Sewage sludge represents a significant source, approximately $1{\times}10^6$ t DM/year. A significant proportion (approximately 70%) is recycled in agriculture, around half of which after composting (Houot et al. 2014).

As most household food waste is still not sorted at source, most of it ends up being incinerated and is not recycled for agricultural use. A few mechanical–biological treatment (MBT) facilities do exist, enabling fermentable waste to be sorted after collection, and then composted or methanized. However, this little-developed treatment method could eventually disappear as a result of regulatory changes requiring sorting at source (AGEC law). Finally, where source separation already exists, various recovery solutions are available. In the absence of centralized collection and processing, biowaste can be recycled after individual or collective composting in gardens and parks, with no direct link to professional agriculture. On the other hand, centralized collection and processing means that biowaste compost or digestate (often mixed with other waste in both cases) can be returned to the soil for agricultural use. These volumes are set to increase in the near future as source separation becomes more widespread.

The vast majority of selectively collected green waste is composted, either on its own or as a mixture, and then recycled for agricultural use. A non-negligible proportion, however, is recycled on site (as mulch or individual composting), as we have already seen.

Lastly, due to the virtual absence of selective collection, dog and cat waste is not currently recycled in agriculture, although composting experiments associated with shelters are underway (Maisonneuve 2017).

Municipal organic waste	Collection and treatment	Corresponding OWP	Estimated deposition (Houot et al. 2014)
Human urine and feces	Mixing with wastewater and biological treatment in wastewater treatment plants, followed by dehydration, digestion, liming, etc.	Liquid WWTP sludge, dewatered, limed, composted, etc.	$0.3.10^6$ t DM/yr
	Mixing with wastewater and biological treatment in a wastewater treatment plant, followed by dehydration, digestion, liming, etc., then composting with woody waste.	Sewage sludge compost	$0.3.10^6$ t DM/yr
	Mixing with wastewater, NCS, then emptying and treatment in a wastewater treatment plant	Sludge, sludge compost, etc.	Included in the above tonnages
	Selective urine collection	Stored urine	Negligible
	Dry toilet, then composting	Dry toilet compost	Negligible
Food waste	Mixed with household waste, then incinerated	None	-
	Mixed with household waste, then MBT of the fermentable fraction followed by composting or biogas	Compost or digestate MBT	Negligible?

Municipal organic waste	Collection and treatment	Corresponding OWP	Estimated deposition (Houot et al. 2014)
	Sorting at source, then individual or collective composting	Biowaste compost	Downward trend forecast by regulations
	Selective collection followed by semi-industrial composting	Biowaste compost	?
	Selective collection followed by industrial composting with green waste	Biowaste compost	Negligible?
	Selective collection followed by biogas with other waste or not	Biogas	?
Green waste	Mixed with household waste, then incinerated	No	-
	Selective collection, then individual or collective composting	Green waste compost	?
	Separate collection, then industrial composting	Green waste compost	2.10^6 t RM/yr
	Selective collection, then industrial composting (with other waste or not)	Compost from green waste (and other waste)	?
	Selective collection, then biogas with other waste	Biogas	Negligible?
	Selective collection, then pyrolysis	Biochar	Negligible
Horse manure	Simple storage	Horse manure	10.10^6 t RM/yr
	Composting	Horse manure compost	
	Biogas with other waste	Biogas	?
Dog and cat droppings	Mixed with household waste, then incinerated	None	-
	Selective collection and composting	Compost from dog and cat waste	Negligible

Table 10.1. *Main current or possible treatments by the type of urban organic waste and corresponding organic waste products (OWPs). NCS: non-collective sanitation, RM: raw material, DM: dry matter, WWTP: wastewater treatment plant, MBT: mechanical–biological treatment*

10.2.3. *A variety of organic fertilizers and enhancers*

The diversity of urban organic waste sources and the treatments applied to them result in the production of OWPs with highly variable characteristics. These products can be characterized according to the variability of their fertilizing (nutrient supply) and amending (soil organic matter maintenance) values, which can be approximated by their dry matter, C, total N, ammoniacal N, P and K contents (Table 10.2). Dry matter contents vary from a few percent for liquid OWPs (liquid urban sludge, raw digestate, urine) to over 50% for most composts. C content correlates fairly well with dry matter content. Conversely, total nitrogen content is fairly similar between all OWPs, ranging from 5 to 10 g/kg of raw material The proportion of nitrogen in mineral form is low, except for urine, digestate and, to a lesser extent, liquid sludge. Phosphorus is more prevalent in solid OWPs and those derived from wastewater treatment.

Nutrient availability varies according to the type of OWP. For nitrogen, for example, it depends on the fraction in mineral form and the mineralization of organic nitrogen. Availability will also depend on the application methods used: timing of application in relation to crop needs, weather conditions that favor ammonia volatilization and/or organic nitrogen mineralization to a greater or lesser extent, equipment or burial methods that limit volatilization, etc. Mineral fertilizer equivalent (MFE) reflects the variability of nitrogen availability from OWPs (COMIFER 2013). Overall, composting treatments will tend to stabilize organic matter and result in composts with low nitrogen availability (MFE < 15%). Conversely, biogas, while also stabilizing organic matter, leads to mineralization of the nitrogen in organic waste and therefore to greater nitrogen availability. The MFE of urine is also high (80%), due to a very high fraction of nitrogen in mineral form. The MFE of sludge is intermediate (35–45%), notably due to significant mineralization of its organic nitrogen.

Phosphorus availability varies between different urban OWPs. Although OWPs are organic products, phosphorus in OWPs is predominantly in mineral form, and its availability depends on its solubility, which in turn depends on calcium, aluminum and iron contents (to which P binds) and pH (Brod et al. 2022). The P availability of green waste and/or biowaste composts is approximately 50%, while that of sludge is over 75% (Morel 2017).

The stability of OWP organic matter, i.e. the fraction that will contribute to soil organic matter after land application, can be approximated using the indicator of potential residual organic carbon (IROC) (Lashermes et al. 2009). As a result of the stabilization of organic matter during composting, the IROC of composts is high (>75%) (with the exception of RHW composts, which contain less woody matter). Anerobic digestion of biowaste also stabilizes organic matter and results in a relatively high IROC (65%), while that of sludge or horse manure is lower (approximately 50%). In all cases, the stability of organic matter in urban biowaste is superior to that of crop residues (Levavasseur et al. 2020).

OWP	DM	C	Nt	N-NH_4	P	K	MFE	IROC
	% RM			g/kg RM			%N	%C
Green waste compost	56.6	12.0	8.3	0.5	1.9	4.2	0–10	82
Green waste and biowaste compost (industrial)	62.3	14.7	10.1	0.0	3.2	9.3	0–10	77
Residual household waste compost	65.7	11.1	4.4	0.5	1.6	3.5	5–10	53
Raw biowaste digestate (only)	4	2	8	5	0.5	1.3	50–90	65
Liquid WWTP sludge	5.2	9.0	1.6	0.5	0.7	0.14	35–45	46
Dewatered WWTP sludge	16.5	5.0	9	0.1	3.7	0.7	35–45	46
Limed dewatered WWTP sludge	25	5.5	9.3	0.1	5.2	0.8	35–45	46
Green waste and sewage sludge compost	49.2	12.4	10.8	1.3	4.9	2.7	15	79
Human urine	1	2	7	6	0.3	2	80	?
Dry toilet compost	33.7	14.9	5.4	0	2.4	2.9	?	?
Horse manure	54	18.3	7.5	1.1	3.4	14.6	10–20	49
Horse manure compost	41	7	5.2	0	3.7	7.9	5–20	80

Table 10.2. *Main urban organic waste products (OWPs) and their fertilizing and amending characteristics (COMIFER 2013; Martin-Rosset et al. 2013; Brun et al. 2014; Houot et al. 2014; Levavasseur et al. 2020; Martin 2020). The values given are only orders of magnitude, as these characteristics are variable for a given OWP. As MFE and IROC values are not provided in the references given for all OWPs, some values have been estimated by reconciliation with other OWPs. IROC: organic matter stability indicator; MFE: fertilizer equivalence coefficient; RM: raw matter; DM: dry matter*

OWPs are also highly variable in terms of metal and organic contamination (PAHs[1] or PCBs[2], drug residues, etc.). Given the diversity of contaminants and OWPs, it is difficult to summarize the results of the literature. However, OWPs from non-selectively collected organic waste (urban WWTP sludge, RHW compost) generally present higher levels of contamination than those from selective collection (green waste and/or biowaste composts), whether for metals (Michaud et al. 2019) or pharmaceutical residues (Bourdat-Deschamps et al. 2017); for example, OWPs from selective collection are not, however, free of contaminants, as illustrated by the problem of microplastics in green waste composts (Braun et al. 2021). Finally, it should be noted that, in order to be spread in agriculture, these OWPs must comply with thresholds for certain contaminants, depending on the regulations to which they are subject (see section 10.3). Elements on the main contaminants in these OWPs can also be found in Houot et al. (2014). Their potential environmental impacts are discussed in section 10.5.4.

10.3. Regulations governing the return of urban OWPs to the soil in agriculture

Only the main regulations specifically governing the return of municipal solid waste back to the soil are recalled here, and not all the regulations relating to the management of municipal organic waste (mandatory sorting of biowaste at source, for example). The regulations governing the return of municipal organic waste back into the soil are in a state of flux. At the beginning of 2022, the main regulations governing the return of municipal solid waste back into the soil are still those shown in Table 10.3.

The decree of January 8, 1998, sets out the technical requirements applicable to the spreading of sludge on agricultural soils. It defines the content of the pre-spreading study, which in particular defines the characteristics and quantities of sludge to be spread, and the plots suitable for spreading according to environmental constraints (soil, natural areas, distance from dwellings or water points, etc.). The decree also defines maximum levels and input flows for trace metals (TMs) and trace organic compounds (TrOCs) (PAHs and PCBs) in sludge and in the soil receiving the sludge. Lastly, it sets out the procedures for storage, spreading and monitoring (frequency of analysis of sludge and soil on spread plots, etc.). Together, these provisions make up the mandatory sludge spreading plan.

Green waste and food waste composts are covered by a mandatory standard, NF U44-051 (2006), while urban sewage sludge composts are covered by NF U44-095

1. Polycyclic aromatic hydrocarbons.
2. Polychlorinated biphenyls.

(2008). Both standards set minimum or maximum thresholds for their agronomic value (e.g. N + P2O5 + K2O < 7% RM), as well as for certain contaminants (TM, TrOC, undesirable inert and pathogenic elements). Once standardized, the farmer is responsible for the proper use of the product, and its spreading does not require a spreading plan as is the case for sludge (no study of suitability for spreading, traceability of spread plots, etc.).

Digestates incorporating food waste are not subject to the same regulations as digestates from agricultural and agri-food waste. The French decree of November 10, 2009, setting the technical rules to be met by anaerobic digestion facilities subject to authorization under Title I of Book V of the Environment Code indicates that the spreading of digestates must comply with the provisions of the French decree of February 2, 1998, on water withdrawals and consumption, as well as emissions of all kinds from facilities classified for environmental protection and subject to authorization. The provisions are quite similar to those of the decree of January 8, 1998, on sewage sludge. As is the case for sludge, land application is subject to a mandatory land application plan.

Horse manure is governed by departmental health regulations, specific to each department, which set out the conditions for storage and spreading. These regulations include provisions concerning areas where spreading is prohibited, distances to be respected from water sources or dwellings, and whether or not to bury the OWP after spreading. A spreading plan is not mandatory.

Finally, human urine is not currently covered by any specific regulations, despite the fact that it is already collected and spread in non-negligible quantities from time to time (e.g. at festivals). In the absence of a specific text, the decree of January 8, 1998, may apply, following exchanges with certain competent authorities.

Urban OWP	Key legislation concerning OWPs and their spreading	Mandatory spreading plan
WWTP sludge	French decree of January 8, 1998, laying down the technical requirements applicable to the spreading of sludge on agricultural land	Yes
Compost from green waste and sewage sludge	NF U44-095: Organic amendments – composts containing materials of agronomic interest from water treatment	No
Compost from green waste and/or food waste	NF U44-051: Organic soil improvers – denominations, specifications and marking	No

Biowaste digestate (alone or mixed)	French decree of February 2, 1998, on water withdrawals and consumption, as well as emissions of all kinds from facilities classified for environmental protection and subject to authorization.	Yes
Horse manure	Departmental sanitary regulations	No
Human urine	No specific text (applicable sludge regulations?)	?

Table 10.3. *Key legislations by the type of urban organic waste products (OWPs)*

In addition to regulations specific to each type of OWP, land spreading is subject to more general regulations, such as the European "nitrates" directive and its associated action programs. For each type of OWP, certain spreading periods are prohibited. The supply of nitrogen by OWPs must be taken into account in farmers' fertilization plans. Regulations governing the organic farming label also prohibit the use of certain urban OWPs in agriculture (in particular WWTP sludge, sludge compost and urine). Certain production contracts signed between agro-industries and farmers also restrict the use of certain urban OWPs (SEDE – VEOLIA and SEVESC 2019).

The regulatory framework as a whole is likely to evolve in the very near future, with the common base decree for fertilizing materials and growing media. This decree will define a priori different classes of fertilizing materials that will have to comply with rules in terms of agronomic composition, safety, traceability, etc. This decree follows on from European Regulation 2019/1009 of the European Parliament and of the Council of June 5, 2019, laying down rules for making fertilizers available on the market.

10.4. How to meet the needs of agriculture

10.4.1. *Contributing to fertilization*

As mentioned above, one of the major benefits of recycling urban OWPs in agriculture is to contribute to the closure of nutrient cycles and thus reduce the need for mineral fertilizers in agriculture, which are non-renewable resources. At the plot level, the contribution of urban OWPs to fertilizer requirements will depend on the nutrient content of the OWP, the availability of these nutrients, the application rates and the application periods (see section 10.2.3).

Table 10.4 shows a number of classic or potential urban OWP application practices on soft winter wheat. It can be seen that the contribution of OWP inputs to the crop's nitrogen requirements is highly variable, ranging from almost nil for a green waste compost input before sowing, to almost half for a urine input at the end of winter. It should be noted, however, that repeated applications of amending OWPs (such as compost) ultimately increase the soil's supply of nitrogen, which also helps to meet crop needs.

For example, on the long-term QualiAgro experimental site in the Yvelines region, the mineralization surplus of urban compost treatments increases steadily over time, compared with an unfertilized and unamended control, following repeated OWP applications (see Figure 10.2).

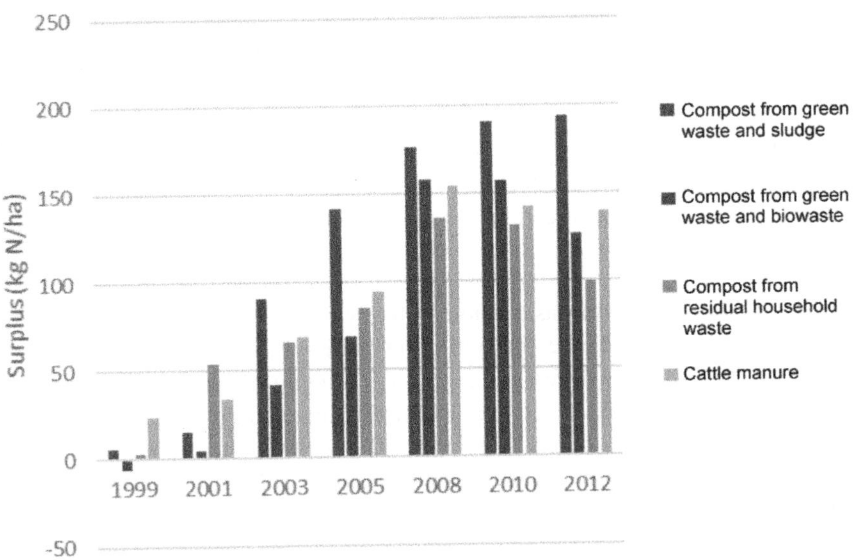

Figure 10.2. *Surplus in apparent nitrogen mineralization on the QualiAgro site between treatments with organic amendment (4 t C/ha every two years) and the unfertilized control. For a color version of this figure, see www.iste.co.uk/martin/farmsandterritory.zip*

As far as phosphorus is concerned, all inputs provide significant quantities in relation to wheat requirements (Table 10.4). Some OWPs even provide quantities in excess of requirements, with surpluses contributing to subsequent crops (phosphorus is partly managed on a crop rotation basis).

OWP	Input period	Dose (t RM/ha)	N					P₂O₅				
			N Content (kg/t RM)	MFE (%)[1]	Input (kg /ha)	Available input (kg/ha)	Needs covered (%)	Content (kg/t RM)	MFE (%)[2]	Input (kg/ha)	Available input (kg/ha)	Needs covered (%)
Limed dehydrated urban sludge	Before sowing	10	9.3	10	93	9.3	4	11.9	80	119	95	182
Green waste compost	Before sowing	20	8.3	5	166	8.3	3	4.4	50	88	44	84
Biowaste digestate	Out of winter	20	8.0	65	160	104	43	1.1	90	22	20	38
Horse manure	Before sowing	20	7.5	10	150	15	6	7.8	80	156	125	240
Human urine	Out of winter	20	7.0	80	140	112	47	0.7	100	14	14	27

Table 10.4. *Nutrient inputs for different urban organic waste product (OWP) applications compared with the needs of a wheat crop with a yield of 8 t/ha (240 kg N/ha and 52 kg P2O5/ha).* [1]*Comifer (2013) and Martin (2020);* [2]*Agricultural digestate values were used for biowaste digestate and cattle manure values for horse manure (Jordan-Meille et al. 2017)*

On the scale of a peri-urban area that concentrates numerous urban waste treatment facilities, the recycling of urban OWPs can represent a significant proportion of agriculture's phosphate and potassium fertilizer requirements (Moinard et al. 2021) (see Figure 10.3). However, these urban residues account for only a small proportion of the nitrogen required by crops, as the nitrogen in these residues is mainly in organic form and slow to mineralize (especially composts). Recycling additional resources such as human urine and food biowaste in the form of digestate could significantly increase the proportion of needs covered (see Figure 10.3). On a global scale, Martin (2020), based on Trimmer et al. (2019), also estimates that almost 20% of nitrogen needs could be covered by recycling human urine in agriculture. However, the contribution of urban OWPs to agricultural needs is highly variable depending on the territory, with nutrient flows from urban OWPs being low in highly agricultural regions relative to needs (Senthilkumar et al. 2012). The contribution of urban OWPs to agricultural needs could also be more significant in the context of changes in cropping systems. For example, in the context of a complete conversion to organic farming, Moinard et al. (2021) show that on the plain of Versailles, the recovery of the area's food biowaste could almost be sufficient to cover N requirements. On a global scale, Barbieri et al. (2021) estimate that urine recycling would cover the N needs of an entirely organic world agriculture.

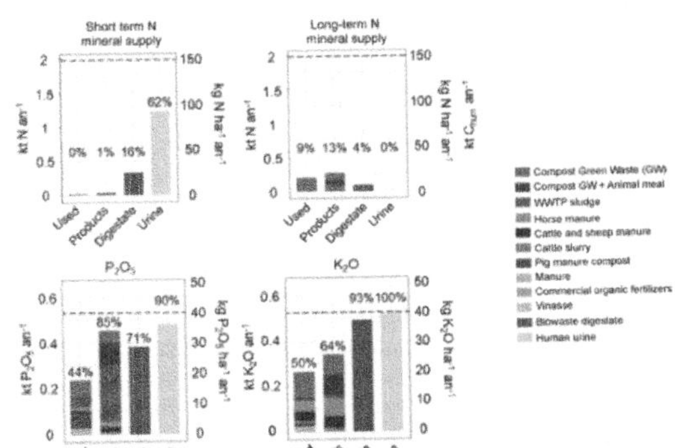

Figure 10.3. *Nutrients available through recycling of OWPs on the plain of Versailles, according to whether OWPs are currently used (used = produced - exports + imports) or produced on the plain, as well as for two prospective scenarios: anerobic digestion of biowaste currently incinerated on the plain (digestates) and recycling of human urine currently treated in the plain's wastewater treatment plants. The short-term supply of mineral N represents the supply of N in the year of OWP input for a single input, while the long-term supply represents the surplus N input by the soil following repeated OWP inputs. The numbers indicate the proportion of demand (N, P, K) met by the OWP. Adapted from Moinard et al. (2021). For a color version of this figure, see www.iste.co.uk/martin/farmsandterritory.zip*

10.4.2. *Contribution to soil carbon storage*

Recycling urban OWPs in agriculture contributes to soil carbon storage. Indeed, increasing carbon inputs is considered the main lever for increasing these stocks (Chenu et al. 2019). Increasing these stocks is beneficial both from a soil fertility point of view and potentially for climate change mitigation. Beyond C inputs, the C quality of OWPs conditions C storage in soils, as seen above (Table 10.2). On the scale of an agricultural plot, OWP inputs can represent the main humified C inputs in certain specific cases (frequent manure inputs in a succession of corn silage and wheat with exported straw). However, at conventional agronomic rates, urban OWP inputs generally represent lower C inputs than above- and below-ground crop residues, even in the case of regular inputs (Chen et al. 2022).

On a regional scale, the contribution of urban OWPs to soil C storage varies widely depending on the type of area. It can be substantial in peri-urban areas, where urban waste treatment facilities are concentrated, and can account for almost a quarter of the carbon storage provided by crop residues (Moinard et al. 2021). However, it remains generally limited in other areas (see Figure 10.4), also in relation to livestock effluents. It should be noted that as most urban waste is already treated and then spread in agriculture (green waste compost and sludge in particular), this carbon storage does not constitute additional storage that could contribute to the 4p1000[3] initiative, but maintaining these existing stocks is nevertheless important.

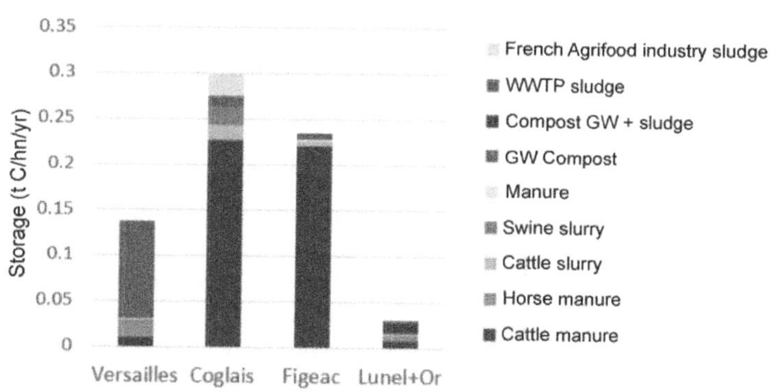

Figure 10.4. *Average annual carbon storage over 30 years enabled by organic waste products (OWPs) in four contrasting territories, peri-urban (plaine de Versailles and Pays de l'Or et de Lunel) or rural with livestock (Coglais, Figeac). Adapted from Berger et al. (2022). For a color version of this figure, see www.iste.co.uk/martin/farmsandterritory.zip*

3. The 4p1000 initiative aims to encourage soil carbon storage to contribute to food security and climate change mitigation (see: https://4p1000.org/decouvrir/).

10.5. What are the challenges associated with the agricultural recycling of municipal waste?

10.5.1. *Partial treatment efficiency, nutrient losses and associated impacts*

As mentioned above, most urban OWPs require one or more treatments before they can be used in agriculture. The effectiveness of these treatments, whether in terms of pathogen abatement or organic matter stabilization, is not necessarily complete. As a result, many contaminants and pathogens may be present in urban OWP (see section 10.5.4). The organic matter in OWPs is not always stabilized and can in some cases lead to nitrogen starvation, for example, with immature green waste composts with high C/N (Amlinger et al. 2003).

In terms of nutrient recycling, these treatments enable cycles to come full circle to a greater or lesser extent. Composting loses around a third of total substrate nitrogen (Zhao et al. 2020), while other major nutrients are more or less conserved. Anerobic digestion is fairly conservative for all major nutrients (excluding the loss of ammoniacal nitrogen from uncovered digestate storage) (Bareha et al. 2021). With regard to wastewater treatment in WWTP, Esculier (2018) estimates that only 4% of the nitrogen contained in wastewater is recycled via the agricultural recycling of sewage sludge in Île-de-France, the majority being lost to the air (denitrification in WWTP) or in the treated water discharged. However, the same author estimates that 41% of phosphorus is recycled. With regard to carbon, losses are more or less significant during treatment and depending on the waste in question. The variability of these losses during treatment is partly offset by the decomposition of OWP carbon after land application, so that the quantity of carbon from a given substrate that will ultimately be stored in the soil is considered to be not very dependent on the type of treatment undergone prior to land application (Thomsen et al. 2013).

Furthermore, the treatment of municipal waste to produce OWPs can also have impacts (Avadí 2020): greenhouse gas emissions linked to waste transport, greenhouse gas emissions during the OWP process or storage (e.g. nitrous oxide during composting or methane leakage from a methanizer), and the energy used for the process (electricity to run a methanizer or dewater sludge). Accidental pollution can also occur (e.g. river pollution following a storage rupture). Despite the impact of treatment, anaerobic digestion has the advantage of producing biogas, which enables energy substitution and avoids the impacts associated with energy production.

Nutrient losses and treatment-related impacts must be taken into account in the environmental assessment of recycling processes (see section 10.6).

10.5.2. *Difficulties in fertilization control*

As we have seen, the availability of OWP nutrients can be quite variable, depending on the type of OWP, the way it is incorporated into cropping systems, soil and climatic conditions, etc. For the farmer using OWPs, this means difficulty in controlling fertilization, compared with mineral fertilizers whose availability is less variable. This can lead them to underestimate the mineral fertilizer savings accessible with OWP to secure yield (Aubry et al. 2014).

Another difficulty in fertilizing with urban OWPs is their relatively low nutrient concentration, which calls for the use of adapted equipment: liquid OWP spreaders with or without a tank (with trailing hoses, injector) (see Figure 10.5), manure spreaders (with or without spreading table), etc. This equipment is generally heavy and cannot enter plots in conditions of limited bearing capacity. This can be a limiting factor when fertilizing winter cereals or rapeseed, for example. The repeated passage of heavy machinery can also lead to soil compaction problems with an impact on agricultural production (Lantz and Börjesson 2014). Soil compaction can be limited to some extent with umbilical spreading. Because of the specific nature of machinery and the often-limited periods of use during the year, many farmers are not themselves equipped with spreading solutions but use agricultural contractors. In the case of spreading plans (for urban sewage sludge, for example), spreading is generally handled by the sludge producer, with a partner agricultural contractor.

Figure 10.5. *Spreading digestate from anaerobic digestion with a slurry tanker fitted with trailing hoses*

Finally, for the majority of farmers who do not themselves control the supply of urban OWP, the question of the variable availability of deposits can be an obstacle to their use and to good control of the fertilization plan. In the case of sewage treatment plans, however, the assurance of having materials to spread regularly is reinforced (sewage sludge is often spread every three to four years, for example, on the same plot).

10.5.3. *Environmental impacts of nutrient losses in the field*

As with any other nitrogen fertilizer, spreading urban OWPs can lead to nitrogen losses (see Figure 10.6).

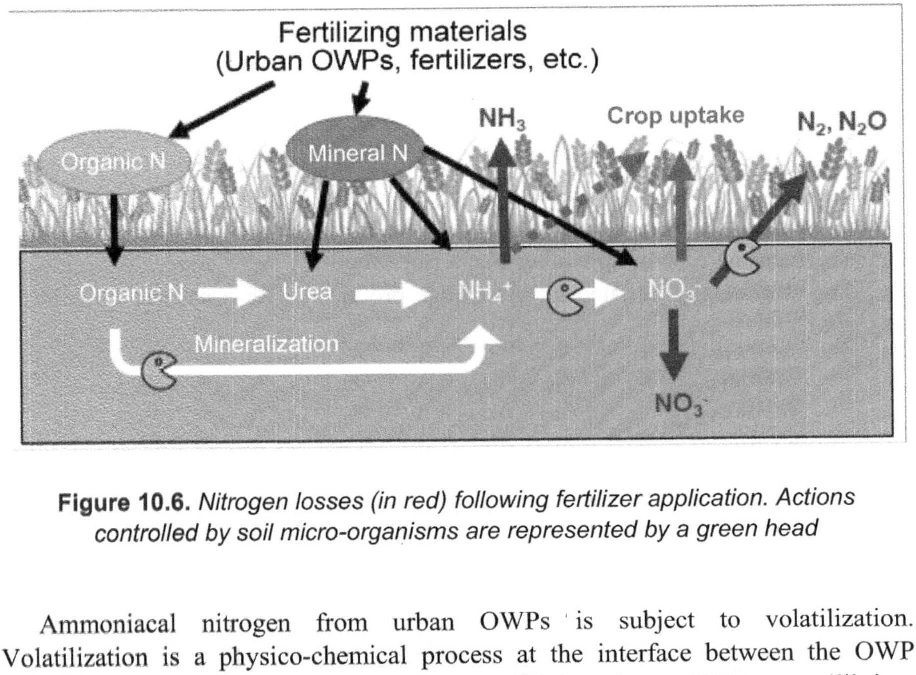

Figure 10.6. *Nitrogen losses (in red) following fertilizer application. Actions controlled by soil micro-organisms are represented by a green head*

Ammoniacal nitrogen from urban OWPs is subject to volatilization. Volatilization is a physico-chemical process at the interface between the OWP and the air, which depends on two equilibria: the acid–base equilibrium (ammonium/ammonia) and the solubility of ammonia in the aqueous phase. The more ammoniacal nitrogen a OWP contains, the more sensitive it will be to volatilization. Furthermore, for a given quantity of ammoniacal nitrogen, volatilization will increase with the viscosity of the OWP (a liquid OWP infiltrates the soil and is less sensitive to volatilization), its pH, temperature and wind in particular (Sommer et al. 2003). Under unfavorable conditions, ammoniacal volatilization can account for almost all the ammoniacal nitrogen in OWPs within a few hours (Parnaudeau and Génermont 2014). These losses represent both a loss of OWP fertilizer value and a deterioration in air quality, as ammonia is a precursor of fine particles. Among urban OWPs, digestates (Moinard 2021) and human urine (Martin 2020) are particularly sensitive to this volatilization due to their high proportion of nitrogen in ammonia form and their high pH. They should therefore be spread in conditions that limit volatilization (cool, windless weather, just before rain) and/or buried in the soil, which greatly reduces volatilization. Conversely, green waste composts, which are virtually free of ammoniacal nitrogen, do not lead to significant ammonia emissions.

Nitrification of ammoniacal nitrogen under aerobic conditions, and denitrification of nitric nitrogen under anaerobic conditions by soil micro-organisms, are sources of N_2O, a greenhouse gas with a high global warming potential (100-year GWP[4] equal to 265). These emissions are highly variable, depending on soil and climatic conditions, application periods and rates, type of OWP, etc., so that the N_2O emission differential between OWP and mineral fertilizers is still poorly understood (Walling and Vaneeckhaute 2020). These emissions are nevertheless important to consider in the GHG balance of OWP recycling (urban or otherwise) (Obriot 2016), especially as the long-term increase in soil organic matter following repeated OWP inputs could also lead to an increase in N_2O emissions (Lv et al. 2020).

Once in nitric form in soils, nitrogen from urban OWPs is subject to leaching and can lead to groundwater pollution, just like nitrogen from other fertilizers. In the short term, if OWP inputs are reasoned, both in terms of doses and input periods according to crop needs, leaching risks are generally not increased compared to mineral fertilization (Morvan et al. 2016). However, over the long term, the increase in soil organic matter following repeated applications of urban amending OWPs can lead to greater mineralization of organic nitrogen in soils and therefore to an increased risk of leaching if the plant cover does not allow this nitrogen to be valorized (Asada et al. 2015; Noirot-Cosson et al. 2016).

In addition to nitrogen losses, as urban OWPs contain several nutrients, determining their application rate based on nitrogen can lead to significant surpluses of another nutrient. For example, determining the application rate of urban OWPs based on nitrogen alone can lead to phosphorus surpluses, as is the case for livestock effluents (Rosemarin et al. 2021). These phosphorus surpluses can lead to greater losses of phosphorus (through runoff and soil erosion in particular), creating a risk of eutrophication.

---

4. Global warming potential.

10.5.4. *End-of-life for contaminants*

Bünemann et al. (2023) recently published a review of the contaminants present in OWPs and the associated risks (Table 10.5). The risks vary widely depending on the contaminants considered and include aspects of toxicity to soil life, crops, livestock and humans via accumulation in the food chain. Although the authors stress the potential risks associated with emerging contaminants (PFAS[5], antibiotic resistance genes, microplastics), they consider that soils are often more resilient than expected. It should also be noted that OWPs are not the only contaminants found in agricultural soils. Phosphate fertilizers are the main source of cadmium, while pesticides are a major source of copper. However, limiting contamination at the source of OWPs (selective sorting, reducing the use of incriminating molecules) and continuing ecotoxicological assessment work are necessary in order to establish sustainable recycling channels.

Group of contaminants	Main OWPs concerned	Chief risks
Trace metals	All	Accumulation in soils. Deleterious effects on soil life (Cu, Zn), risk of transfer into the food chain (Cd)
Organic trace compounds	(Cu and Zn mainly for livestock OWPs)	Compounds subject to high volatilization or degradation in soils: limited risks
Pathogens and antibiotic resistance genes	All, but more contaminated non-selectively collected urban residues	Most persistent compounds: accumulation in soils but no significant transfer to plants (e.g. PCBs). Risk of transfer to animals (grazing)
Microplastics	Mainly OWP from human and animal dejecta	Persistent and mobile compounds (e.g. PFAs): risk of transfer to water and the food chain

Table 10.5. *Main contaminants in urban OWPs and associated risks (according to Bünemann et al. (2023))*

10.6. Environmental balance sheet

Recycling municipal waste in agriculture provides a number of environmental benefits (carbon storage, substitution for mineral fertilizers, etc.), as well as generating various impacts (greenhouse gas emissions, air pollution, etc.). Various methods are available to assess these benefits and impacts, compared with those of

5. Per- and polyfluorinated alkyl substances.

other urban waste treatment methods. Life cycle assessment (LCA) is the main reference method for environmental evaluation. It is a standardized method (ISO 14040 and 14044). It records all the material and energy flows associated with a product or activity, from "cradle to grave". LCA quantifies the impacts associated with these flows using various indicators: contribution to climate change, stratospheric ozone depletion, freshwater eutrophication, fine particle formation, ecotoxicological impact, toxicological impact on humans, etc. For the recycling of urban OWPs in agriculture, different activities need to be considered, depending on the boundaries of the system under consideration: waste collection and transport, waste treatment, OWP storage, transport to the plot and spreading. Emissions associated with these activities and with the manufacture of the necessary machinery (transport truck, spreader, etc.) are considered, as are those in the field following spreading (nitrous oxide emissions, nitrate leaching, etc.).

Numerous LCAs have compared different urban waste treatment processes, or different agricultural production systems, with or without fertilization with urban OWPs. De Sadeleer et al. (2020) show, for example, that the greenhouse gas balance for the anerobic digestion of food waste is better than that of incineration with energy recovery, even disregarding fertilizer savings linked to digestate spreading. This result is notably due to the type of energy substituted (gas from biogas vs. electricity with incineration). The authors show, however, that reducing food waste would be of even greater benefit. Le Pera et al. (2022) indicate that the anerobic digestion of green waste followed by composting of the digestate limits environmental impacts (contribution to climate change, formation of fine particles) compared with direct composting, which itself limits most impacts compared with open-air combustion. However, from the point of view of agricultural production, Avadí et al. (2021) show that, compared with mineral fertilization, the use of green waste and sewage sludge composts can degrade several LCA indicators (including the greenhouse gas balance), due to the emissions associated with processing these OWPs. If emissions from the alternative treatment of sludge and green waste (incineration and landfill) are included in the OWP-free scenario, the agricultural production balances nevertheless become similar with or without OWP. Yoshida et al. (2017) indicate a limited benefit of sludge land application compared with sludge incineration, regardless of sludge treatment, and even a deterioration in toxicity-related indicators (for humans or the ecosystem). Martin (2020) shows that the use of human urine in agriculture improves several LCA impact indicators, in particular thanks to the impacts avoided as a result of the lower volume to be treated in WWTP.

Despite the standardized nature of LCAs, the boundaries of the system considered may vary from one study to another. Taking into account the multi-functionality of processes (waste treatment, energy and fertilizer production) also leads to different methodological choices being made by different authors, which can have a significant impact on the results. The alternative channel with which agricultural recycling is compared also varies between studies (incineration with or without energy recovery, landfill). Finally, some impacts are more or less well considered, depending on the impact assessment methods used. For example, the fate and impacts of contaminants are often imperfectly taken into account in LCA, as are fertilizer substitution and/or soil carbon storage. These reasons partly explain the variability of results from LCA studies of urban OWP recycling. Laurent et al. (2014) consider that it is difficult to generalize LCA results for solid waste treatment (not only organic) due to the specificities of each study.

10.7. Value chain organization

Many different actors are involved in the recycling of municipal organic waste. Upstream, the user, the local authority or the company are the waste producers (urine, green waste, food waste, etc.) and are responsible for sorting and, in part, collecting it. Municipalities and their groupings organize the collection and treatment of waste, either directly, or via a concession or public service delegation to private companies. Farmers can be involved right from the processing stage, for example, if they operate a composting or biogas plant. Farmers using OWPs may then be responsible for transporting and spreading the OWPs, or an agricultural contractor may take charge of this stage on behalf of the OWP producer. The entire sector is also subject to regulatory oversight by the public authorities, who issue the relevant regulations (section 10.3) and monitor their application. Independent bodies (chambers of agriculture, trade unions, etc.) are responsible for controlling the spreading plans for OWPs subject to these regulations. Other actors are also involved on a case-by-case basis, such as consulting firms for the preparation of spreading plans (agricultural prospecting, study of soil suitability for spreading) or analysis laboratories for the characterization of urban OWPs and soils.

10.8. Farmers' perception of agricultural recycling of urban OWPs and social acceptability

Many parameters will influence farmers' perception of urban OWPs (Aubry et al. 2014): type of OWP (disgust/trust it inspires), sensitivity to soil quality, farmer owning or renting land, availability of spreadable surfaces (competition with livestock OWPs and/or suitability of land), involvement in marketing channels that limit the use of urban OWPs, parceling of land, neighborhood problems, etc. For example, a cereal farmer who owns their own land, is keen to increase the organic matter in their soils and has no access to livestock OWPs, might a priori be inclined to use green waste composts. Conversely, over and above regulatory restrictions, market gardeners have shown little inclination to use urban OWPs that inspire disgust (biowaste digestate, human urine), notably for fear of consumer reaction (Boros et al. 2022). Farmers' perceptions of the agricultural recycling of urban OWPs can also be influenced locally by past experiences that have "ended badly", as in the case of the excessive spreading of wastewater in Paris, which led to a ban on growing vegetables in the spreading plain following soil contamination, particularly by trace metal elements. Finally, beyond seeking to understand the factors influencing farmers in their choice of whether or not to use urban OWPs, it seems important not to try to convince them at all costs of the merits of the practice. Instead, we should seek to engage them in a constructive dialogue to understand their apprehension about returning OWPs to the soil and develop new research questions, with a view to environmental democracy (Barbier and Larrue 2011).

As far as (peri-)urban populations are concerned, the social acceptability of agricultural recycling of urban OWPs also depends on many factors, even if the notion itself seems difficult to define (Aubry et al. 2014). This acceptability applies both to OWP production and storage facilities, and to land application itself. Proximity to facilities and/or land application may or may not play a role, as may the level of knowledge of these processes. Overall, urban dwellers' knowledge of the agricultural recycling of urban OWPs seems low. The level of media coverage may therefore influence this acceptability. For people living in the immediate vicinity of land-spreading operations, the issue of odors seems to be paramount (Lowman et al. 2013). Lastly, a number of studies suggest that populations would be rather in favor of agricultural recycling of "new" urban waste such as source-separated food waste (Borrello et al. 2017) or even human urine (Segrè Cohen et al. 2020).

10.9. Conclusion

Agricultural recycling of urban waste helps to close cycles and maintain soil fertility. It can help address the scarcity of fossil resources on which the majority of agricultural systems rely today. Although agricultural recycling of urban waste is now commonplace, there is considerable potential for increasing the loop of cycles: increasing the recovery of waste already recovered (urban sludge, green waste), generalizing the recovery of food waste and, much more prospectively, recovering human urine.

The urban OWPs resulting from the collection and treatment of organic urban waste have a wide range of characteristics, both in terms of fertilizing and amending value, and in terms of contamination. Their inclusion in cropping systems needs to be carefully considered in order to limit the associated impacts (nitrogen losses, soil compaction, accumulation of contaminants in the food chain). The impacts associated with treatment must also be taken into account in the environmental assessment of the return of urban OWPs to the soil, in comparison with other methods. Cropping systems that maximize the benefits of urban OWP recycling and limit the associated impacts still need to be designed with farmers, in order to contribute to the agroecological transition, in conjunction with other changes in agricultural and food systems.

10.10. Reference

Amelung, W., Bossio, D., de Vries, W., Kögel-Knabner, I., Lehmann, J., Amundson, R., Bol, R., Collins, C., Lal, R., Leifeld, J. et al. (2020). Towards a global-scale soil climate mitigation strategy. *Nature Communications*, 11(1). doi: 10.1038/s41467-020-18887-7.

Amlinger, F., Götz, B., Dreher, P., Geszti, J., Weissteiner, C. (2003). Nitrogen in biowaste and yard waste compost: Dynamics of mobilisation and availability—A review. *European Journal of Soil Biology*, 39(3), 107–116. doi: 10.1016/S1164-5563(03)00026-8.

Asada, K., Eguchi, S., Tsunekawa, A., Tsuji, M., Itahashi, S., Katou, H. (2015). Predicting nitrogen leaching with the modified LEACHM model: Validation in soils receiving long-term application of animal manure composts. *Nutrient Cycling in Agroecosystems*, 102(2), 209–225. doi: 10.1007/s10705-015-9690-9.

Aubry, C., Barbier, R., Lupton, S., Pradel, M. (2014). Dimensions économique et sociotechnique de l'épandage des Mafor. In *Valorisation des matières fertilisantes d'origine résiduaire sur les sols à usage agricole ou forestier. Impacts agronomiques, environnementaux, socio-économiques. Rapport final de l'expertise scientifique collective*. INRA, Paris.

Avadí, A. (2020). Screening LCA of French organic amendments and fertilisers. *The International Journal of Life Cycle Assessment*. doi: 10.1007/s11367-020-01732-w.

Avadí, A., Galland, V., Parnaudeau, V., Colomb, V., Paillat, J.-M. (2021). Agricultural eco-design scenarios based on AGRIBALYSE® residual organic fertiliser inventories. *Journal of Cleaner Production*, 318, 128506. doi: 10.1016/j.jclepro.2021.128506.

Barbier, R. and Larrue, C. (2011). Démocratie environnementale et territoires : un bilan d'étape, Abstract. *Participations*, 1, 67–104. doi: 10.3917/parti.001.0067.

Barbieri, P., Pellerin, S., Seufert, V., Smith, L., Ramankutty, N., Nesme, T. (2021). Global option space for organic agriculture is delimited by nitrogen availability. *Nature Food*, 1–10. doi: 10.1038/s43016-021-00276-y.

Bareha, Y., Affes, R., Moinard, V., Buffet, J., Girault, R. (2021). A simple mass balance tool to predict carbon and nitrogen fluxes in anaerobic digestion systems. *Waste Management*, 135, 47–59. doi: 10.1016/j.wasman.2021.08.020.

Berger, S., Sicard, L., Levavasseur, F., Houot, S. (2022). *Contribution des Produits Résiduaires Organiques à la séquestration du carbone dans les sols*. RECORD, Saint-Denis [Online]. Available at: https://librairie.ademe.fr/changement-climatique-et-energie/6152-contribution-des-produits-residuaires-organiques-a-la-sequestration-du-carbone-dans-les-sols.html.

Boros, L., Levavasseur, F., Kebalo, L.F., Bijon, N., Morel, K. (2022). Recycling local organic waste in peri-urban horticulture: A case-study in the Parisian region. *Acta Horticulturae*, 1356. doi: 10.17660/ActaHortic.2022.1356.37.

Borrello, M., Caracciolo, F., Lombardi, A., Pascucci, S., Cembalo, L. (2017). Consumers' perspective on circular economy strategy for reducing food waste. *Sustainability*, 9(1). doi: 10.3390/su9010141.

Bourdat-Deschamps, M., Ferhi, S., Bernet, N., Feder, F., Crouzet, O., Patureau, D., Montenach, D., Moussard, G.D., Mercier, V., Benoit, P. et al. (2017). Fate and impacts of pharmaceuticals and personal care products after repeated applications of organic waste products in long-term field experiments. *Science of the Total Environment*, 607–608, 271–280. doi: 10.1016/j.scitotenv.2017.06.240.

Braun, M., Mail, M., Heyse, R., Amelung, W. (2021). Plastic in compost: Prevalence and potential input into agricultural and horticultural soils. *Science of the Total Environment*, 760, 143335. doi: 10.1016/j.scitotenv.2020.143335.

Brod, E., Ogaard, A.F., Mueller-Stover, D.S., Rubaek, G.H. (2022). Considering inorganic P binding in bio-based products improves prediction of their P fertiliser value. *Science of the Total Environment*, 836, 155590. doi: 10.1016/j.scitotenv.2022.155590.

Brun, F., Leguen, S., Oxarango, L., Delmaire, A., Déportes, I., Gran-Aymerich, L., Bour, L., Jousse, S., Mazaud, D., Agenet, P. et al. (2014). Gestion des sous-produits de toilettes seches familiales : etude sur le traitement des matieres par compostage. *Toilettes Du Monde* [Online]. Available at: https://halarchives-ouvertes.fr/hal-01803047.

Bünemann, E.K., Reimer, M., Smolders, E., Smith, S.R., Bigalke, M., Palmqvist, A., Brandt, K.K., Möller, K., Harder, R., Hermann, L. et al. (2023). Do contaminants compromise the use of recycled nutrients in organic agriculture? A review and synthesis of current knowledge on contaminant concentrations, fate in the environment and risk assessment. *Science of the Total Environment*, 168901. doi: 10.1016/j.scitotenv.2023.168901.

Chen, H., Levavasseur, F., Montenach, D., Lollier, M., Morel, C., Houot, S. (2022). An 18-year field experiment to assess how various types of organic waste used at European regulatory rates sustain crop yields and C, N, P, and K dynamics in a French calcareous soil. *Soil and Tillage Research*, 221, 105415. doi: 10.1016/j.still.2022.105415.

Chenu, C., Angers, D.A., Barré, P., Derrien, D., Arrouays, D., Balesdent, J. (2019). Increasing organic stocks in agricultural soils: Knowledge gaps and potential innovations. *Soil and Tillage Research*, 188, 41–52. doi: 10.1016/j.still.2018.04.011.

COMIFER (2013). Calcul de la fertilisation azotée. Guide méthodologique pour l'établissement des prescriptions locales. Cultures annuelles et prairies.

van Dijk, K.C., Lesschen, J.P., Oenema, O. (2016). Phosphorus flows and balances of the European Union Member States. *Science of the Total Environment*, 542, 1078–1093. doi: 10.1016/j.scitotenv.2015.08.048.

Doligez, P. (2015). Production et caractéristiques du fumier de cheval [Online]. Available at: https://equipedia.ifce.fr/infrastructure-et-equipement/installation-et-environnement/effluents-delevage/production-et-caracteristiques-du-fumier-de-cheval.

Esculier, F. (2018). Le système alimentation/excrétion des territoires urbains : régimes et transitions socio-écologiques. PhD Thesis, Paris-East Créteil University [Online]. Available at: https://pastel.archives-ouvertes.fr/tel-01976550.

Espelia – SOLAGRO (2021). Biodéchets : du tri à la source jusqu'à la méthanisation. Guide à destination des collectivités pour réussir le tri à la source à des biodéchets dès 2024.

Fijalkowski, K., Rorat, A., Grobelak, A., Kacprzak, M.J. (2017). The presence of contaminations in sewage sludge – The current situation. *Journal of Environmental Management*, 203(Part 3), 1126–1136. doi: 10.1016/j.jenvman.2017.05.068.

FranceAgrimer (2015). Valorisation des déchets verts en France. Les synthèses de FranceAgriMer.

Houot, S., Pons, M.N., Pradel, M. (2014). Valorisation des matières fertilisantes d'origine résiduaire sur les sols à usage agricole ou forestier. Impacts agronomiques, environnementaux, socio-économiques. Rapport final de l'expertise scientifique collective. Report, INRA, Paris.

Jordan-Meille, L., Morel, C., Salducci, X., Michaud, J. (2017). Valeur agronomique (C, N, P) de digestats de méthanisation d'origine agricole et agro-alimentaire de Dordogne. In *13èmes Rencontres de la fertilisation raisonnée et de l'analyse Comifer Gemas*, 8–9 November, Nantes.

Lantz, M. and Börjesson, P. (2014). Greenhouse gas and energyassessment of the biogas from co-digestion injected into the natural gas grid: A Swedish case-study including effects on soil properties. *Renewable Energy*, 71, 387–395. doi: 10.1016/j.renene.2014.05.048.

Lashermes, G., Nicolardot, B., Parnaudeau, V., Thuriès, L., Chaussod, R., Guillotin, M.L., Linères, M., Mary, B., Metzger, L., Morvan, T. et al. (2009). Indicator of potential residual carbon in soils after exogenous organic matter application. *European Journal of Soil Science*, 60(2), 297–310. doi: 10.1111/j.1365-2389.2008.01110.x.

Laurent, A., Bakas, I., Clavreul, J., Bernstad, A., Niero, M., Gentil, E., Hauschild, M.Z., Christensen, T.H. (2014). Review of LCA studies of solid waste management systems – Part I: Lessons learned and perspectives. *Waste Management*, 34(3), 573–588. doi: 10.1016/j.wasman.2013.10.045.

Le Pera, A., Sellaro, M., Bencivenni, E. (2022). Composting food waste or digestate? Characteristics, statistical and life cycle assessment study based on an Italian composting plant. *Journal of Cleaner Production*, 350, 131552. doi: 10.1016/j.jclepro.2022.131552.

Lecuyer, B., Chatellier, V., Daniel, K. (2013). Les engrais minéraux dans les exploitations agricoles françaises et européennes. *Économie rurale. Agricultures, alimentations, territoires*, 333, 151–161. doi: 10.4000/economierurale.3845.

Levavasseur, F., Mary, B., Christensen, B.T., Duparque, A., Ferchaud, F., Kätterer, T., Lagrange, H., Montenach, D., Resseguier, C., Houot, S. (2020). The simple AMG model accurately simulates organic carbon storage in soils after repeated application of exogenous organic matter. *Nutrient Cycling in Agroecosystems*, 117, 215–229. doi: 10.1007/s10705-020-10065-x.

Lowman, A., McDonald, M.A., Wing, S., Muhammad, N. (2013). Land application of treated sewage sludge: Community health and environmental justice. *Environmental Health Perspectives*, 121(5), 537–542. doi: 10.1289/ehp.1205470.

Lv, F., Song, J., Giltrap, D., Feng, Y., Yang, X., Zhang, S. (2020). Crop yield and N_2O emission affected by long-term organic manure substitution fertilizer under winter wheat-summer maize cropping system. *Science of the Total Environment*, 732, 139321. doi: 10.1016/j.scitotenv.2020.139321.

Maisonneuve, M. (2017). Politiques urbaines et réponses aux problèmes des déjections canines en ville. Retour d'expérience du lombicomposage sur 2 refuges canins. Thesis, Claude Bernard University Lyon 1.

Martin, T. (2020). L'urine humaine en agriculture : des filières variées pour contribuer à une fertilisation azotée durable. Thesis, Paris Saclay University [Online]. Available at: https://halarchives-ouvertes.fr/tel-03189185.

Martin-Rosset, W., Vermorel, M., Fleurance, G., Doligez, P. (2013). Evaluation et prévision de différentes sources de pollution issues de l'élevage et de l'utilisation du cheval. In *39ème Journée de la Recherche Equine*, Paris.

Menigaux, H. (2004). Les vétérinaires et l'environnement. Enquête nationale auprès de vétérinaires salariés et libéraux. Thesis, Ecole Nationale Vétérinaire d'Alfort.

Michaud, A.M., Cambier, P., Sappin-Didier, V., Deltreil, V., Mercier, V., Rampon, J.-N., Houot, S. (2019). Mass balance and long-term soil accumulation of trace elements in arable crop systems amended with urban composts or cattle manure during 17 years. *Environmental Science and Pollution Research*. doi: 10.1007/s11356-019-07166-8.

Ministère de l'agriculture (2021). Le bien-être et la protection des chevaux. Ministère de l'Agriculture et de la Souveraineté alimentaire [Online]. Available at: https://agriculture.gouv.fr/le-bien-etre-et-la-protection-des-chevaux.

Moinard, V. (2021). Conséquences de l'introduction de la méthanisation dans une exploitation de polyculture-élevage sur les cycles du carbone et de l'azote. Combinaison de l'expérimentation et de la modélisation à l'échelle de la ferme. PhD Thesis, Paris Saclay University [Online]. Available at: https://pastel.archives-ouvertes.fr/tel-03485490.

Moinard, V., Levavasseur, F., Houot, S. (2021). Current and potential recycling of exogenous organic matter as fertilizers and amendments in a French peri-urban territory. *Resources, Conservation and Recycling*, 169, 105523. doi: 10.1016/j.resconrec.2021.105523.

Morel, C. (2017). Qualité et valeur agronomique : disponibilité du P, forme, contaminants. *Journée "Le phosphore recyclé en agriculture"*. AgroParisTech, Paris.

Morvan, T., Houot, S., Levavasseur, F., Parnaudeau, V., Trochard, R., Hanocq, D. (2016). Comment des systèmes avec apports organiques restent aujourd'hui source de pollution par le nitrate, pistes pour demain. In *Séminaire RMT SdCi. Changer la gestion de l'azote dans les champs cultivés : comment accompagner les agriculteurs dans leurs choix stratégiques* ?, Paris.

Nicholson, F., Bhogal, A., Cardenas, L., Chadwick, D., Misselbrook, T., Rollett, A., Taylor, M., Thorman, R., Williams, J. (2017). Nitrogen losses to the environment following food-based digestate and compost applications to agricultural land. *Environmental Pollution*, 228, 504–516. doi: 10.1016/j.envpol.2017.05.023.

Noirot-Cosson, P.E., Vaudour, E., Gilliot, J.M., Gabrielle, B., Houot, S. (2016). Modelling the long-term effect of urban waste compost applications on carbon and nitrogen dynamics in temperate cropland. *Soil Biology and Biochemistry*, 94, 138–153. doi: 10.1016/j.soilbio.2015.11.014.

Obriot, F. (2016). Epandage de produits residuaires organiques et fonctionnement biologique des sols : de la quantification des impacts sur les cycles carbone et azote a l'evaluation multicritere de la pratique a l'echelle de la parcelle. PhD Thesis, AgroParisTech [Online]. Available at: http://prodinra.inra.fr/?locale=en#!ConsultNotice:356978.

Oldfield, E.E., Wood, S.A., Bradford, M.A. (2020). Direct evidence using a controlled greenhouse study for threshold effects of soil organic matter on crop growth. *Ecological Applications*, 30(4), e02073. doi: 10.1002/eap.2073.

Parnaudeau, V. and Génermont, S. (2014). Valeur fertilisante azotée et pertes azotées influençant la valeur fertilisante. In *Valorisation des matières fertilisantes d'origine résiduaire sur les sols à usage agricole ou forestier. Impacts agronomiques, environnementaux, socio-économiques. Rapport final de l'expertise scientifique collective*. INRA, Paris.

Redlingshöfer, B. (2022). Food waste in cities: An urban metabolism approach applied to Paris and Île-de-France. Deutsche Digitale Bibliothek, University of Berlin [Online]. Available at: http://www.deutsche-digitale-bibliothek.de/item/I7FQWCDMYFMQAYPU6YSQ3YJ5H5YJHD7F.

Rosemarin, A., Ekane, N., Andersson, K. (2021). Phosphorus flows, surpluses, and N/P agronomic balancing when using manure from pig and poultry farms. *Agronomy-Basel*, 11(11), 2228. doi: 10.3390/agronomy11112228.

de Sadeleer, I., Brattebø, H., Callewaert, P. (2020). Waste prevention, energy recovery or recycling—Directions for household food waste management in light of circular economy policy. *Resources, Conservation and Recycling*, 160, 104908. doi: 10.1016/j.resconrec.2020.104908.

SEDE – VEOLIA & SEVESC. (2019). Mémoire en réponse. Enquête publique du 24 janvier au 25 février 2019. Demande d'Autorisation environnementale. Epandage des boues d'épuration de la station Carré de Réunion [Online]. Available at: https://www.yvelines.gouv.fr/contenu/telechargement/18681/114664/file/20190319_Me%CC%81moire+en+re%CC%81ponse.pdf.

Segrè Cohen, A., Love, N.G., Nace, K.K., Árvai, J. (2020). Consumers' acceptance of agricultural fertilizers derived from diverted and recycled human urine. *Environmental Science & Technology*, 54(8), 5297–5305. doi: 10.1021/acs.est.0c00576.

Senthilkumar, K., Nesme, T., Mollier, A., Pellerin, S. (2012). Regional-scale phosphorus flows and budgets within France: The importance of agricultural production systems. *Nutrient Cycling in Agroecosystems*, 92(2), 145–159. doi: 10.1007/s10705-011-9478-5.

Sommer, S.G., Génermont, S., Cellier, P., Hutchings, N.J., Olesen, J.E., Morvan, T. (2003). Processes controlling ammonia emission from livestock slurry in the field. *European Journal of Agronomy*, 19(4), 465–486. doi: 10.1016/S1161-0301(03)00037-6.

Thomsen, I.K., Olesen, J.E., Møller, H.B., Sørensen, P., Christensen, B.T. (2013). Carbon dynamics and retention in soil after anaerobic digestion of dairy cattle feed and faeces. *Soil Biology and Biochemistry*, 58, 82–87. doi: 10.1016/j.soilbio.2012.11.006.

Trimmer, J.T., Margenot, A.J., Cusick, R.D., Guest, J.S. (2019). Aligning product chemistry and soil context for agronomic reuse of human-derived resources. *Environmental Science & Technology*, 53(11), 6501–6510. doi: 10.1021/acs.est.9b00504.

Walling, E. and Vaneeckhaute, C. (2020). Greenhouse gas emissions from inorganic and organic fertilizer production and use: A review of emission factors and their variability. *Journal of Environmental Management*, 276, 111211. doi: 10.1016/j.jenvman.2020.111211.

Yoshida, H., ten Hoeve, M., Christensen, T.H., Bruun, S., Jensen, L.S., Scheutz, C. (2017). Life cycle assessment of sewage sludge management options including long-term impacts after land application. *Journal of Cleaner Production*. doi: 10.1016/j.jclepro.2017.10.175.

Zhao, S., Schmidt, S., Qin, W., Li, J., Li, G., Zhang, W. (2020). Towards the circular nitrogen economy – A global meta-analysis of composting technologies reveals much potential for mitigating nitrogen losses. *Science of the Total Environment*, 704, 135401. doi: 10.1016/j.scitotenv.2019.135401.

11

Biogas and Biomass Valorization Circuits at the Farm and Territory Levels

Jeanne CADIOU[1] and Jean-Marc MEYNARD[2]

[1] *AgroParisTech Innovation, IDEAS, UMR SADAPT, Palaiseau, France*

[2] *Paris Saclay University, AgroParisTech, INRAE, UMR SADAPT, Palaiseau, France*

11.1. Introduction

Agricultural biogas is developing rapidly worldwide, and particularly in Europe. It is at the crossroads of multiple issues (energy, agriculture, waste, transport), which are reflected in a variety of public policies. In France, following selective experimentation in the 20th century, the development of biogas in the early 2000s was driven by a desire to recycle waste. The regulatory framework was strengthened in 2006 and then 2011, as part of a drive to develop renewable energies in agriculture. Cogeneration of heat and power (CHP) biogas (valorization of biogas into electricity and heat) then dominated. Between 2010 and 2020, biogas upgrading (valorization by injecting biomethane into the gas network) also developed, with strong momentum. This development took place as part of a strong relationship with the agricultural sector, and projections made by actors in the sector identified a future deposit that is predominantly agricultural (ADEME et al. 2018, 2019; GRDF et al. 2018). The public policies that set these development targets are primarily in the energy and waste fields, aiming to develop renewable energy production and valorize various organic wastes. However, the 2013 Bio-Methane Support Policy (Energie Méthanisation Autonomie Azote (EMAA)) plan outlines a French model

Farms and Territory,
coordinated by Philippe MARTIN and Solène PISSONNIER. © ISTE Ltd 2025.

for agricultural biogas, in synergy with more sustainable livestock farming. This French model should avoid reproducing the agri-environmental excesses (corn monoculture, competition for land use) of the development of biogas in Germany.

Figure 11.1. *Photo of an agricultural biogas digester. The materials to be digested are supplied by the farmer. It may come from the farmer's own farm (livestock manure, crops) or from the local area (agro-industrial byproducts, biowaste or biomass from neighboring farms). The digestate is then spread on the land of the project owner or other farmers in the area*

However, the objectives and implementation of France's biogas policy are controversial. Beyond the issues of local nuisance and NIMBY[1]-type protests, the place of biogas in agricultural and energy transitions is being debated by various actors in these sectors. Here, we discuss some of the new challenges (benefits and risks) of biomass conversion to anaerobic digestion, and its implications for France's conversion to sustainable agriculture.

In this chapter, we focus on agricultural biogas. From a legal point of view, in France, biogas is defined as agricultural when it is derived from the digestion of at least 50% by volume of inputs from agriculture, and the capital of the biogas unit is at least 50% owned by actors carrying out an agricultural activity (Article D311-18 – French Rural and Maritime Fishing Code 2017). Our analysis focuses in particular on the development of agricultural biogas in the Grand Est region, based on case studies and interviews in Bas-Rhin and Vosges.

1. Not In My BackYard. This acronym refers to the phenomenon of opposition to projects if they directly affect a person's environment. In principle, a person is not opposed to a project/action if it takes place far from their home.

11.2. New biomass flows find new uses in agriculture

The arrival of anaerobic digestion on an agricultural territory modifies the relationships between biogas farmer and other farmers in the area. New biomass flows (methanogenic inputs and digestates) develop, which can be exchanged for other biomass, or monetized purchase/sale.

11.2.1. *Biomass used for agricultural biogas production*

The decision by a farmer, or more generally, a developer of an anaerobic digestion project, to use biomass for anaerobic digestion is based on a number of criteria. Anaerobic digestion is the first criterion: a product has a high potential if it is highly composed of organic molecules whose labile carbon is biodegradable by bacteria in an anaerobic environment. This biomethane potential characterizes the maximum biogas production potential per unit of material digested in the anaerobic digestion unit.

Conventional agricultural biogas feedstocks can be divided into five main types: energy crops and "energy cover crops"; livestock manure; agricultural crop byproducts (crop residues); agri-food byproducts; biowaste. These inputs have a varying biomethane potential: an energy crop will have approximately two to three times more biomethane potential than cattle manure (in CH_4 produced per ton of raw feedstock); an agri-food byproduct can have a highly variable biomethane potential (see Table 11.1).

	Biomethane potential Nm^3 CH_4/tRM
Corn silage	90
Energy crops and energy cover crops: triticale-single-oats	60.6
Fresh cattle manure	40.5
Fresh pig manure	8
Agri-food byproducts: whey	23
Agricultural byproducts: silo waste	318
Beet pulp	30
Biowaste: edible oil	784
Biowaste: WWTP sludge	10–250

Table 11.1. *Table of anaerobic digestion power of some inputs. The anaerobic digestive power of an input is expressed in cubic normo-meters of methane produced per ton of raw material (tRM) of digested input. These values are indicative and may vary over a wider range depending on the source. These values have been estimated using data from the following sources (Morin 2015; Esnouf et al. 2021; FranceAgriMer 2022)*

Agricultural biogas digesters use these feedstocks according to their needs in anaerobic digestate material, and according to the set of socio-technical constraints in which they operate (the cost of feedstock, the work associated with the production of feedstock, physical availability):

– Energy crops: chosen for their biomethane potential, within the French regulatory limit of 15% of total supply tonnage. The most common in eastern France are corn, triticale and barley as main crops, and rye and triticale as energy cover crops, sometimes mixed with other species. They are often preserved by silage. Producing crops for biogas production requires the farmer to devote time to it, which not all farmers are able or willing to do. These energy crops are generally produced by the biogas farmer, either as a substitute for food crops, or as a double crop. They can also be purchased from neighboring farmers. The production of energy cover crops has been growing rapidly in France since 2018–2019: these are winter or summer cover crops that theoretically require less treatment and fertilization than main crops, even if empirical studies show that they are cultivated in a variety of ways (see Carton and Levavasseur (2022) for Île-de-France). The use of energy crops in anaerobic digestion depends on agricultural market prices, production costs and biogas production subsidies. Before the energy crisis linked to the war in Ukraine, the price of corn silage for biogas was fairly stable in the Grand Est region, at approximately €31 per ton of raw material: at the biogas buy-back price in 2021, it was worthwhile producing it for biogas as an energy cover crop as well as a dedicated crop.

– Livestock manure: this includes all manure and slurry produced by livestock farms. These agricultural byproducts are legally considered as waste and have a fertilizing and amending value that is more or less well used in agriculture. In livestock farming areas, manure digestion is common, either on the farms of biogas farmers, or as part of manure-digestate exchanges with neighboring farms. Their biomethane potential is lower than that of a crop or certain byproducts, but manure, a livestock byproduct, is considered "free", as it is already present on the livestock farm. However, some livestock manure, such as pig slurry, have little biomethane potential. The proximity of manure production to the anaerobic digester is key to its valorization: farmers who build their individual anaerobic digesters directly next to their livestock buildings have a particular economic interest in valorizing these effluents because it means few transportation costs. Manure is rarely monetized, but farmers may sell it to a biogas farmer nearby (in one case surveyed in eastern France, the farmer bought €3/t of cattle manure from their neighbor).

– Agricultural crop byproducts: these are all crop residues from agricultural production (crop residues, corn stalks, as well as fodder of degraded quality), which have an interesting biomethane potential. They are available in varying quantities throughout the territory and sometimes require additional work (harvesting of chaff) or pre-treatment (shredding of corn stalks). Their cost depends on the conditions under which these byproducts are recovered. Green waste from local authorities also falls into this category but is not always clean enough to be digested.

– Agri-food byproducts: these are very diverse (from cereals, vegetable fats, whey, brewery grains, etc.). They are all the more interesting as they have biomethane potential, and their liquid character can also be sought after by biogas farmers to liquefy the biogas unit ration. They are particularly appreciated by farmers, as they are feedstocks delivered to the farm on demand and cost the farmer very little work. However, their price has risen sharply in recent years (see below), and their supply can be highly irregular. In fact, it is uncommon for farmers to sign contracts of more than one year with byproduct suppliers.

– Biowaste: this refers to all waste derived from organic biomass (agri-food biowaste, green waste), but which has the status of waste prior to anaerobic digestion. To use them, biogas farmers must have a hygienization unit to limit pathogenic risks. They are less subject to speculation than food byproducts but present the same risks of irregular supply.

11.2.2. *Digestate, a byproduct of biogas production: a new effluent with high nitrogen fertilizing power*

The composition of the digestate resulting from the anaerobic digestion process varies according to the feedstocks digested and the fermentation parameters. However, the main characteristics of biomass evolution in an anaerobic digestion unit are well known: the labile carbon in the inputs is strongly degraded during anaerobic digestion, so the digestate has proportionally less labile organic carbon[2] and more refractory carbon, with amending power (Möller 2015; Moinard 2021). Some of the organic nitrogen also passes into the form of mineral nitrogen, giving the digestate a nitrogen fertilizing value approaching that of synthetic mineral nitrogen. Farmers can carry out post-treatments on digestate, such as phase separation, to differentiate between solid and liquid digestate, with amending and fertilizing values respectively.

Digestate management is a new activity on the farm. The volumes of digestate produced are generally greater than the volumes of effluent managed before biogas on the farm. This is because new biomass is transferred to farms via anaerobic digestion units in order to meet gas production targets, and farmers frequently import inputs from outside their farms. Digestate use as fertilizer on the farm is

2. Labile organic carbon is the portion of soil organic carbon that can be readily decomposed by soil organisms.

governed by a spreading plan. In practice, digestate spreading techniques vary from one farm to another. Spreading equipment, spreading periods and doses depend, in particular, on the capacity of the biogas famers to invest and organize spreading: for example, spreading without a tanker[3] is chosen to be able to easily spread digestate in spring on winter cereals without compacting the soil. However, this requires considerable resources (manpower, financing) and is not always profitable for small plots. In the eastern part of France, spreading plans frequently involve plots of land belonging to neighboring farmers. In this case, the spreading service may or may not be monetized: it is often the subject of exchanges such as "contribution of effluent/crop residues to biogas" in exchange for "spreading of digestate".

Depending on their technical choices, objectives and constraints, farmers are able to substitute their mineral fertilization with digestate to varying degrees. The fertilizing properties of digestate are particularly attractive to organic farmers, for whom it is an effective organic fertilizer. However, for digestate to be used in organic farming, inputs for anaerobic digestion must meet strict specifications. "Organic" digestate must be derived from the anaerobic digestion of crop matter, effluents and/or milk processing byproducts. Effluents must not come from "industrial" livestock farming (Atee Club Biogaz 2018), but crops and crop residues can be produced in conventional agriculture.

The arrival of an anaerobic digester on a farm or in a given territory therefore induces new logics for the purchase or exchange of biomass between farmers and with other actors in the territory. We expand upon this in the following section.

11.3. By modifying the flows and uses of biomass, biogas production renews the links between farms and territory: a case study from Grand Est

11.3.1. *The evolution of cropping systems*

Surveys carried out among biogas farmers in the Bas-Rhin and Vosges regions of France reveal the different approaches to biomass conversion. Calculating the biogas profitability of an input is at the heart of the supply of high biomethane potential products. For most agricultural biogas, the benchmark in terms of price-to-biogas ratio is corn silage. They will buy external feedstocks (agro-industrial

3. Spreading without a tanker site is one that enables digestate (or slurry) to be spread on a field without having to tow a slurry tanker across the field. The slurry tank (or box) containing the liquid to be spread remains at the edge of the field and is connected by a hose to a tractor to which a spreading boom is attached. The farmer can therefore spread the digestate with minimum weight in the field (Cadiou 2023).

byproducts or biowaste) if gas production per ton of material is cheaper than for corn silage:

> Our aim is to make it no more expensive than corn silage, […] a material that quickly produces gas. We're trying to get some, because if we see that the biogas units are going down, we'll go back up, but we're trying to stay at 7–8 centimes per kilowatt. For the time being, it's going up slowly (Biogas farmer, Vosges, authors' translation).

Other farmers calculate a price not to be exceeded per kWh of gas. In this way, corn silage becomes a yardstick for determining the profitability of an input. This profitability threshold is highly dependent on subsidy policies for biogas and can vary from one biogas unit to another.

Crops used for biogas production can come from a variety of sources and form part of a variety of cropping systems. On some farms, without any change in cropping system, we can observe a reorientation of the outlet of certain existing crop productions to meet the needs of the biogas units. This change in biomass use can occur at the level of farms that have invested in biogas units (e.g. a farmer will produce a little less corn for their livestock), or at the level of territories (e.g. some farmers now sell corn to neighboring agricultural biogas digesters). On other farms, however, we see changes in land use, with some farmers adapting their cropping systems, to a greater or lesser extent, to produce more biomass for biogas. They may, for example, choose to introduce energy cover crops, which means replacing winter crops with spring crops preceded by a winter energy cover crop, or shifting the sowing dates of the food crop that follows. For example, some agricultural biogas digesters are replacing 5–10% of their sales crop area with corn followed by energy cover crops or temporary grassland for biogas. For others, if the potential of the land allows it, the farmer can keep their initial crops by intensifying soil use through the introduction of energy cover crops. Less frequent changes can also be observed on neighboring farms, which will introduce cover crops in return for a digestate spreading service. Farmers who choose to make fewer changes to their cropping systems will often, in return, source external high feedstocks for anaerobic digestion (territorial agricultural byproducts, agri-food byproducts and biowaste). On the other hand, those who produce more of their own crops seek to free themselves as much as possible from the market for external byproducts. Access to local sources of inputs for anaerobic digestion can thus influence the crop rotation of biogas farmers, as explained below. Depending on the farmer's strategy, which depends on their resources and objectives, feed-food-energy competition appears to vary considerably.

11.3.2. *Access to methanogenic agro-industrial byproducts*

Cadiou (2023) shows that some agricultural biogas digesters have found it more difficult in recent years to access byproducts for anaerobic digestion. Ten years ago, access to local byproducts was easy, with manufacturers sometimes paying fees for farmers to take charge of their waste. In recent years, the agricultural biogas digesters that are the oldest (2012–2016) have noted an increasing rise in the price of these byproducts, which they link to the strong development of biogas in the territory (see Table 11.2, Figure 11.1).

	Verbatim
Biogas farmer 1 – June 2021	"At one time, you could have as much product as you wanted, and you could negotiate the rates you wanted. Now, with the explosion in biogas, we can see that there's starting to be competition, in terms of waste in fact. Some of the feed we used to get is starting to come in less."
Biogas farmer 2 – June 2021	"We have to raise the price more and more, because you've got industrialists [...] who are buying more and more expensively, and then they issue invitations to tender every year, [...] that's the way it is, that's business."

Table 11.2. *Translated verbatim from biogas farmers (authors' translation)*

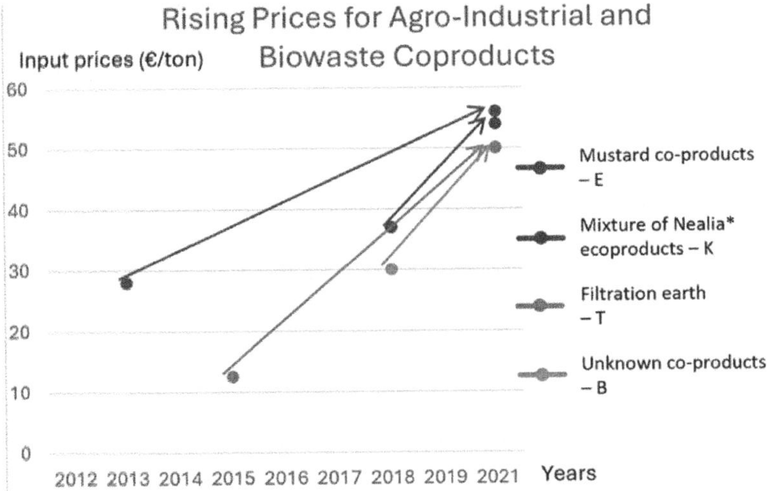

Figure 11.2. *A 10-year price trend for four agro-industrial byproducts and biowaste (source: semi-directed interviews with farmers). For a color version of this figure, see www.iste.co.uk/martin/farmsandterritory.zip*

The link between these agricultural biogas and suppliers of local byproducts is therefore rather fragile. The purchase of these byproducts is very rarely contractualized, or only for a short period of one or two years. The stability of a commercial relationship between biogas farmers and feedstock suppliers depends more on the farmers' working relationships with these industrialists: for example, if they have been working together for a long time, are located nearby, or offer a working relationship that facilitates the industrialist (ability to collect inputs at weekends, no refusal by the farmers of inputs offered by the industrialist).

Competition between farmers for these byproducts raises the question of the local depletion of certain resources, and therefore indirectly of the territorial perimeter within which a byproduct can be valorized. Data from the Grand Est region in 2020 show that byproducts and external waste are valorized within a perimeter of approximately 30 km for agricultural and territorial units, with maximum distances of over 200 km in territorial units (ADEME and Région Grand Est 2022). The cost of transporting byproducts limits their long-distance travel, but increased competition for these byproducts could encourage less local valorization.

A number of actors in the sector have raised the issue of changing uses: beet pulp, an agro-industrial byproduct historically used in animal feed, is now proving to be methanogenic. Some producers and resellers of byproducts, as well as some animal feed sellers, are now reselling them to biogas units. For the moment, the effects on farmers' access to these byproducts seem limited (IDELE, n. d.), but the strong momentum of project construction in the Grand Est region suggests that competition for use will increase. Changes in the use of certain byproducts have already led to the weakening of existing local industries. This is the case for the alfalfa industry in the Aube region, where the digestion of beet pulp has undermined the economic model of the dehydration plant that depended on the dehydration of this pulp. The alfalfa industry is thus undermined by the weakening of the dehydration plant's economy (Madelrieux et al. 2020).

11.3.3. *Exchanges of crop and animal agricultural byproducts between farms*

Agricultural byproducts are often the subject of local exchanges between farmers, for mutual benefit. Agricultural biogas farmers appreciate the digestion of livestock effluent from neighboring farmers, which is economically attractive if the effluent is harvested nearby and the associated digestate is spread nearby, as transport costs are then limited. Neighboring farmers may be interested in a "crop residue for digestate" exchange, to preserve the return of a certain amount of organic matter to the soil. Others prefer to keep their crop residues and effluents, for fear of depleting their soils of organic matter. The scientific literature on this subject shows that, in terms of soil carbon storage, biomass digestion does not alter the long-term balance (provided the digestate is returned to the plots where the biomass was harvested); however, it does alter the nature of the carbon returned to the soil in the short term. These latter effects on soil life have yet to be fully documented in the medium term.

Neighboring farmers can use their livestock effluent in an anaerobic digestion unit, in exchange for a digestate spreading service. For some farmers, having their manure handled by an anaerobic digestion unit is attractive, as it facilitates manure management (storage and spreading) and enables them to obtain a digestate richer in mineral nitrogen than the incoming material. Others, however, are wary of the effects of digestate on soils, but are often convinced by its fertilizing power. Digestate is a particularly interesting organic fertilizer for organic farming, as it can significantly increase crop yields, at least in the short term.

In these configurations, the smooth operation of biogas digesters depends directly on the links that the biogas farmer maintains with neighboring farmers and is less dependent on markets for agricultural products and waste. The exchange rules that emerge depend on the territory and reflect the mutual interests of farmers in participating in these exchanges.

11.3.4. *Biowaste: a source of supply diversification*

In all territories, biowaste is a source of feedstock that is less frequently valorized in agricultural biogas. Its legal status as "waste" makes it more difficult to recycle in agriculture, as most farmers have to hygienize[4] it (or buy it hygienized from treatment platforms) in order to spread the digestate on their fields.

Because of the costs involved, few units in the Grand Est 2020 region (18 out of 214 units) process biowaste, even if competition for byproducts appears to be encouraging some farmers to seek access to the biowaste market. Some biowaste that had previously been composted is now destined for anaerobic digestion. The 2020 anti-waste law generalizes source sorting of food biowaste from January 1, 2024, which will stimulate the mobilization of this source in the coming years. Some farmers are co-investing in biogas units with biowaste manufacturers, giving them easier access to these inputs. However, biowaste digestion means that the digestate can no longer be used for organic production, which holds back some farmers.

4. Thermal treatment of biomass to reduce pathogen load and associated health risks.

Farm biogas production is thus giving rise to new uses for biomass in territories, influencing the trajectories of farms, as well as the links between farms in a given area, and between local agricultural and non-agricultural actors. These reshuffles linked to the biomass needs of biogas units show the importance of thinking about the hierarchy of uses at the territorial level, in connection with the dynamics of territorial regulation that are being put in place. For example, the growing positioning of actors such as cooperatives in the biogas sector could alter certain territorial balances. Their ability to invest and gain access to agro-industrial byproducts or biowaste is potentially greater than that of farmers, since they are major economic actors in the agricultural sector. These actors are also helping to develop new anaerobic digestion models in which farmers are essentially feedstock suppliers, implying new changes in agricultural work and technical systems.

11.4. New sustainability challenges emerge on farms and in territories

These new forms of biomass/biowaste recovery are bringing about socio-technical and socio-economic transformations at the farm and territorial levels.

At the farm level, biogas production is a new type of workshop that can generate an interesting income, one which is better, according to many biogas farmers surveyed in the Grand Est region, than a livestock workshop for the same amount of work. However, some biogas farmers point out that this workshop can also generate significant economic losses if it does not work well. When it is set up, and throughout the life of the project, it requires the allocation of production factors (on the order of several million euros of investment at installation). This new workshop therefore plays an important role in the farmer's management strategy and resource allocation choices. It can be attached to a variety of systems, and can reinforce different trajectories, from intensification of production to transition to organic farming. These transformations at the farm level depend in particular on the links between the farms and the territories, for example, on the availability of byproducts in the territories. Cadiou's work (2023) shows that a diversity of technical choices can be documented among biogas farmers in the Vosges and Bas-Rhin regions. Implementing optimal practices for the environment is not always feasible or desirable for the farmer, and we discuss the example of digestate below.

11.4.1. *Socio-technical contexts that leave more or less room for good practices: example of digestate spreading*

The spreading of digestate is a new area of activity, bringing new opportunities and costs to farms. Farmers have a greater volume of material to spread, and depending on their fertilization strategy, spreading requires specific equipment (dribble bars, spreading techniques, etc.) and more or less manpower (see Box 11.1). Their spreading practices are the result of a complex equation between their desire to use digestate as a substitute for mineral nitrogen (economic and environmental motivations), the constraints of their structures (digestate storage volume, soil type), their resources (ability to invest in optimal spreading equipment, available manpower to use the equipment) and their work schedule. Through surveys of 23 biogas farmers in eastern France, we identified three scenarios for the evolution of mineral fertilization on farms:

– Group 1: 0% or slight reduction in mineral nitrogen purchases – farmers spread little or no digestate on cereals in spring, mainly because the soils are not very load bearing, or because spreading would take too long to reach the nitrogen dose required for fertilization. They prefer to spread digestate in summer or autumn, when the work schedule is less hectic and the soils are stronger, and to empty the digestate pit so that it does not overflow during the winter (when spreading is very limited). Crop fertilization in spring is then essentially provided by mineral fertilizer, as before the development of the biogas activity.

– Group 2: between −33% and −60% reduction in mineral nitrogen consumption – farmers have similar practices to those in Group 1 but are slightly more successful in spreading digestate in spring, and thus making the first nitrogen applications to cereals, when soil and climate conditions allow.

– Group 3: 70%–100% reduction in mineral nitrogen consumption – farmers in this group save a very significant amount of nitrogen, either because they have converted (partially or totally) to organic farming, or because they have succeeded in implementing optimized techniques for spreading digestate in the spring, to meet plant nitrogen requirements.

These digestate use logics are more or less in line with the good spreading practices recommended in the scientific literature. On some farms, particularly those converting to organic farming, farm biogas production leads to the substitution of mineral nitrogen by digestate; on others, biogas production leads to unsustainable nitrogen management practices on the farm and in the surrounding area, practices that are sources of water (nitrate) and air (ammonia, nitrous oxide) pollution. Long-term soil fertility issues have yet to be investigated (as in Marty et al. 2022). These issues concern both the farms of biogas farmers and those of farmers who supply materials to the biogas digesters, and on whose farms the biogas farmers spread digestate.

A spreading technique is one that allows digestate (or slurry) to be spread on a field without having to tow a slurry tanker across the field. The slurry tanker (or box) containing the digestate remains at the edge of the field and is connected by a hose to a tractor to which a spreading boom is attached. The farmer can then spread the digestate with minimum weight on the field. The slurry tanker must be refueled throughout the process, so as to be able to supply the spreader. Alternatively, the hose can pump directly into a nearby remote pit. The theoretical output is high (approximately 50–130 m^3/h) compared to spreading with a tanker. However, this type of operation requires a lot of manpower to drive the tractor and the slurry tanker. It also requires the availability of the tractor and the various supply tanks. This operation is less costly in terms of manpower and equipment if the spreader can be directly connected to the digestate storage pit.

There are also other types of spreading techniques, where a self-propelled spreader is used in the field to pump the digestate to be spread into an off-field slurry tank. The weight in the field depends on the capacity of the self-propelled unit.

A dribble bar has flexible hoses that deposit the liquid by flowing over the soil at regular intervals, thus limiting ammonia volatilization. It costs between 35 and 50,000 euros for a 12–15 dribble bar (Carton and Bulcke 2021). A specific kind of dribble bar has, at the end of each pipe, a skid to dig a surface trace in which the digestate is deposited.

Box 11.1. *Technical details on digestate spreading equipment (tankless and with a dribble bar)*

11.4.2. How are agricultural models changing as a result of biogas production?

For some farmers, anaerobic digestion is leading to changes in their agricultural production systems (crop and livestock farming). As discussed in section 11.2.1, in the Bas-Rhin and Vosges regions, biogas production helps to partially redirect crop production towards supplying the biogas unit. The effects on rotations are attenuated, or even nullified, if the farmer has stable external sources of highly feedstocks for anaerobic digestion. Thanks to the fertilizing value of digestate, anaerobic digestion can also be a lever in the transition to organic farming: in this case, the transformation of the cropping system is more closely linked to the need to lengthen and diversify the rotation, with a view to controlling pests and weeds. In this way, farm transformations can follow a variety of trajectories, from an adaptation to the ratio of energy crop acreage to a more in-depth reconfiguration of the production system to supply the biogas unit, or to an in-depth transformation of the production system as in the case of conversion to organic farming.

The impact of anaerobic digestion on the livestock farming activities of farmers seems to depend very much on the farmer's strategy: whether they are in the process of reducing, maintaining or increasing their livestock farming, anaerobic digestion is more likely to be an element that reinforces the current trajectory, by offering a remunerative alternative activity or by justifying an increase in livestock farming through the valorization of livestock manure. On farms that internalize the management of the biogas unit, managing the digester can lead to competition with the livestock farming (2 out of 19 cases observed).

These farm trajectories can bring certain agri-environmental benefits (soil cover, reintroduction of temporary grassland with legumes in certain mixed cropping–livestock systems, mineral nitrogen savings), as well as have various negative effects (reversal of grassland in certain mixed cropping–livestock systems, increased input consumption, soil compaction linked to digestate spreading, nitrogen pollution linked to nitrogen surpluses). Digestate management is a key factor in the agri-environmental sustainability of anaerobic digestion.

The environmental impact trajectories of agricultural biogas can thus be very diverse, depending on the objectives that the biogas farmer sets for their farm and the new constraints that the biogas unit places on the farm. These impacts need to be analyzed in the context of scaling up the industry.

11.4.3. *New challenges at the territory level in a scaling-up dynamic*

The development of anaerobic digestion in the Grand Est region shows, through the phenomenon of competition for byproducts, that indirect effects can appear in the territories, and in turn influence biogas production techniques. These effects may concern farms with an anaerobic digester, as well as other agricultural systems in the territory.

Cadiou's work (2023) highlights an initial issue that requires further investigation: competition between biogas production and animal feed. The ability to negotiate biomass can be unequal between a biogas farmer and a local livestock farmer. A report by FranceAgriMer shows that biogas development scenarios, in line with current dynamics, could lead to increased local competition between livestock farming and biogas for agricultural byproducts, particularly in drought years (FranceAgriMer 2022). As mentioned above in relation to competition for byproducts, the dense development of anaerobic digestion in a given territory can lead to changes in the way byproducts for anaerobic digestion are valorized. A lack of access to byproducts for anaerobic digestion at a profitable price could thus lead farmers to consume more crops to supply their digesters.

In the longer term, the development of anaerobic digestion raises the question of its role in maintaining or transitioning production methods: if anaerobic digestion helps to maintain existing activities, does it encourage their transition to more sustainable practices? How sustainable are agricultural activities that depend on subsidies for energy production? An agricultural territory in which biogas is being developed may be partially reconfigured, as a result of investment in these energy infrastructures. This raises the question of the dependence of agricultural activities on energy production, and their future if energy subsidies come to an end. Some of the agri-environmental issues associated with these dynamics are assumed to be addressed by public policy mechanisms. However, a study of the implementation of this public policy, presented in the following section, shows its low effectiveness in relation to certain issues.

11.5. A biogas development policy faced with the challenge of current regulatory dynamics?

National public policy on biogas, and its regional application in the Grand Est, has sought to address certain sustainability issues identified as potentially problematic, and to encourage the development of certain models with what is considered to be a "better" environmental record. A study of the implementation of biogas policy in the Grand Est region shows, however, that economic or energy objectives are more effectively promoted in the territory than the agri-environmental objectives of this public policy.

In the Grand Est region, the main public and private actors involved in developing the sector (the region, ADEME[5], GRDF[6] and the Chambers of Agriculture) are primarily concerned with economic development and energy independence. In recent years, this convergence of interests has led to strong growth in biogas units in the Grand Est region, driven by an ambitious Regional Biogas Strategy and dynamic technical actors. The region is thus one of the French regions where agricultural biogas has developed most intensively in recent years, with 183 biogas units in operation at the start of 2021 (Chambre d'agriculture du Grand Est 2021), out of a total of 1,090 biogas units in France (Service de la donnée et des études statistiques (SDES) 2021a, 2021b). While certain agri-environmental issues (digestate management, competition for use) have been identified in regional policy, public policy instruments to regulate biomass mobilization and digestate use are proving inadequate to deal with them. We present below the dynamics of the

5 A state environmental agency.
6 Main gas distributor in France

implementation of two public policy instruments that illustrate the need for a better approach to supporting the development of the sector.

In France, public policy on anaerobic digestion has been developed with a view, asserted by the public authorities, to avoiding reproducing the strong competition observed in Germany between anaerobic digestion crops and food crops[7] (Emmann et al. 2013; Appel et al. 2016; Lüker-Jans et al. 2017). Anaerobic digestion of energy crops has thus been limited in France with the introduction, in 2016, of a maximum rate of 15% of energy crops (excluding energy cover crops) in the supply. However, as shown by the effects on territories (Grand Est) as well as other territories such as Île-de-France (Carton and Levavasseur 2022), the use of biomass and respect for the hierarchy of uses are still difficult to supervise and evaluate. For example, in the Grand Est region, when examining biogas projects, the DREAL[8] verifies the risks of competing uses for certain CHP[9] projects, but in reality, it has little information to assess these issues. Similarly, ADEME grants subsidies to applicant projects if they comply with input regulations, compliance being assessed on the basis of essentially declarative elements. In terms of monitoring and evaluating the sector, ADEME and DREAL were still finding it difficult in 2021 to provide technical data on the nature of feedstocks consumed. Downstream, the strong development of the sector is limiting the number of inspections carried out on ICPE[10] facilities, as the departments in charge of inspections have few human resources to carry them out. As a result, supervision and control systems are not always effective. Self-regulation at farm and territorial levels does not guarantee a hierarchy of uses, nor the absence of competition between uses, which are determined by variable socio-technical and socio-economic factors. In 2022, the French government issued a decree[11] redefining the notion of energy cover crops, with the aim of limiting the blurring observed in certain territories between energy crops and energy cover crops. It will be interesting to see in the years to come the effects of this policy and the ability of territorial actors to implement it.

7. In Germany, the development of biogas in the 2000s led to an increase in corn monocultures, with negative impacts on the environment and farming activities (reduced biodiversity, water consumption for irrigation, higher prices for agricultural raw materials and land).
8. The departmental administration responsible for assessing biogas projects with regard to the risk of competition for biomass use.
9. Anaerobic digestion units with an installed capacity of between 300 and 500 kW.
10. Installation industrielle classée pour l'environnement: environmentally classified industrial facility.
11. Decree of August 4, 2022, which, among other things, defines as the main crop the crop present on the plot on June 1, even if it was planted after a main crop.

The second example concerns digestate. In France, the spreading of digestate is governed by regulations on the agricultural spreading of organic waste materials (with territorial variations), which include the requirement to draw up a spreading plan. A series of best practices (timing, burial) are also recommended by professional technical bodies and associations. In practice, it is difficult to monitor spreading plans (declarative data), and we have shown that practices can vary from one farmer to another (see section 11.3.1), as the determinants of spreading are not only technical, but also economic or ergonomic. Here again, it appears that current public policy mechanisms are ill-suited to take account of the real dynamics of implementation.

11.6. Conclusion

With the development of anaerobic digestion, these territorial dynamics introduce new challenges for a circular economy consistent with an agroecological transition. To support sustainable anaerobic digestion, we now need to take greater account of socio-technical and socio-economic dynamics at farm and territorial levels when designing public policy measures. In the field of research, this could involve building systems for assessing the systemic effects of biogas at the farm and territorial levels, which we have seen to be highly diverse and often indirect.

11.7. Reference

ADEME and Région Grand Est (2022). Observation régionale de la méthanisation en région Grand Est [Online]. Available at: https://www.climaxion.fr/sites/climaxion/files/docutheque/synthese_ademe_regionale_finale_2206.pdf.

ADEME, GRDF, GRTgaz (2018). A 100% renewable gas mix in 2050 [Online]. Available at: https://www.ademe.fr/sites/default/files/assets/documents/renewable-gas-mix-2050-010521.pdf.

ADEME, Decid&Risk, Boucher, L., Levasseur, P. (2019). Performances et potentiels de diffusion d'unités de méthanisation agricole.

Appel, F., Ostermeyer-Wiethaup, A., Balmann, A. (2016). Effects of the German Renewable Energy Act on structural change in agriculture – The case of biogas. *Utilities Policy*, 41, 172–182. doi: 10.1016/j.jup.2016.02.013.

Atee Club Biogaz (2018). Fertilisation par les digestats en agriculture biologique.

Cadiou, J. (2023). Le déploiement de la politique de méthanisation agricole en France : implications pour la transition agroécologique. Thesis, Paris Saclay University [Online]. Available at: https://pastel.hal.science/tel-04190138.

Carton, S. and Levavasseur, F. (2022). Performances agronomiques et environnementales de la méthanisation agricole dans un contexte de grandes cultures céréalières (sans élevage) et recommandations de bonnes pratiques. Final Report, AgroParisTech/INRAE.

Chambre d'agriculture du Grand Est (2021). Méthanisation agricole et territoriale en région Grand Est—Etats des lieux [Online]. Available at: https://grandest.chambre-agriculture.fr/fileadmin/user_upload/Grand-Est/049_Inst-Acal/RUBR-agro-environnement/ACSE_2020_2022/44_ACSE_methanisation_agricole__GE_janvier_2021.pdf.

Emmann, C.H., Guenther-Lübbers, W., Theuvsen, L. (2013). Impacts of biogas production on the production factors land and labour – Current effects, possible consequences and further research needs. *International Journal of Food System Dynamics*, 4(1), 38–50. doi: 10.22004/ag.econ.164768.

Esnouf, A., Brockmann, D., Cresson, R. (2021). Analyse du Cycle de Vie du BIOMETHANE issu de ressources agricoles—Rapport d'ACV. Report, INRAE Transfert [Online]. Available at: https://www.inrae.fr/sites/default/files/pdf/Rapport%20ACV_Biomethane%20issu%20de%20ressources%20agricoles_INRAE%20Transfert_GRDF.pdf.

FranceAgriMer (2022). Ressources en biomasse et méthanisation agricole : quelles disponibilités pour quels besoins ? FranceAgriMer [Online]. Available at: https://www.franceagrimer.fr/content/download/69402/document/RESSOURCES%20EN%20BIOMASSE%20ET%20METHANISATION%202022%20-%20Maquette%20VF.pdf.

GRDF, GRTgaz, SPEGNN, Terega (2018). Perspectives gaz [Online]. Available at: www.grtgaz.com/fileadmin/plaquettes/fr/2019/Perspectives-Gaz-2018.pdf.

IDELE (n.d.). Compétition sur les coproduits entre méthanisation et alimentation animale.

Lüker-Jans, N., Simmering, D., Otte, A. (2017). The impact of biogas plants on regional dynamics of permanent grassland and maize area—The example of Hesse, Germany (2005–2010). *Agriculture, Ecosystems & Environment*, 241, 24–38. doi: 10.1016/j.agee.2017.02.023.

Madelrieux, S., Grilllot, M., Dermine-Brullot, S., Marty, P., Godinot, O., Ruault, J.-F., Chatzimpiros, P., Gabriel, A., Buclet, N., Lescoat, P. (2020). Rapport final du projet Boat : biomasses d'origine agricole à l'échelle de territoires. Quelles formes de gestion et valorisation : Entre cloisonnement, concurrence ou intégration ? Research Report, ADEME.

Marty, P., Dermine-Brullot, S., Madelrieux, S., Fleuet, J., Lescoat, P. (2022). Transformation of socioeconomic metabolism due to development of the bioeconomy: The case of northern Aube (France). *European Planning Studies*, 30(7), 1212–1229. doi: 10.1080/09654313.2021.1889475.

Moinard, V. (2021). Conséquences de l'introduction de la méthanisation dans une exploitation de polyculture-élevage sur les cycles du carbone et de l'azote. Combinaison de l'expérimentation et de la modélisation à l'échelle de la ferme. PhD Thesis, Université Paris-Saclay.

Möller, K. (2015). Effects of anaerobic digestion on soil carbon and nitrogen turnover, N emissions, and soil biological activity. A review. *Agronomy for Sustainable Development*, 35(3), 1021–1041. doi: 10.1007/s13593-015-0284-3.

Morin, F. (2015). La méthanisation : technique de valorisation des déchets organiques. ADEME, Franche-Comté [Online]. Available at: https://docactor.fr/15181066-La-methanisation-technique-de-valorisation-des-dechets-organiques-florence-morin-ademe-franche-comte.html.

Service de la donnée et des études statistiques (SDES) (2021a). Statinfo—Biogaz [Online]. Available at: https://www.statistiques.developpement-durable.gouv.fr/publicationweb/340.

Service de la donnée et des études statistiques (SDES) (2021b). Statinfo—Biométhane [Online]. Available at: https://www.statistiques.developpement-durable.gouv.fr/publicationweb/341.

12

Databases for Analyzing Relationships Between Farm Holdings and Territories

Philippe MARTIN[1] and Myriam GRILLOT[2]

[1] *Paris Saclay University, AgroParisTech, INRAE, UMR SADAPT, Palaiseau, France*

[2] *University of Toulouse, INRAE, INPT, El PURPAN, AGIR, Castanet-Tolosan, France*

12.1. Interest in and availability of databases for the analysis of farm–territory relationships

Agricultural research and development actors involved in agroecological transition are proposing tools and methods for redesigning farming systems. In order to do this, they need to understand and describe farm holdings in their context, what we refer to here as *farm–territory relationships*. These elements of characterization concern the spatial organization of the farms themselves (location of infrastructures, parcels of land). They can also refer to their biophysical or socio-economic environment. The biophysical environment may, for example, include information on climate, soil and non-agricultural land uses (water bodies, forests, towns and road infrastructures, etc.). Socio-economic aspects include not only administrative boundaries but also land registry and characterization of economic activities in the value chain, from upstream to downstream of agricultural production activities. Downstream activities include agricultural product collection and storage (cooperative silos and trading), as well as primary processing (packaging of field vegetables, flax scutching, etc.). Upstream includes the production and sale of off-farm inputs such as pesticides, fertilizers, livestock feed and water for irrigation.

Farms and Territory,
coordinated by Philippe MARTIN and Solène PISSONNIER.

Information on farm–territory relationships can traditionally come from field surveys carried out on all or part of the farms in a territory, on targeted themes. This approach has the advantage of being able to focus interviews on specific points of interest, but these operations are time-consuming to set up, difficult to carry out exhaustively if the territory is large, and as such, difficult to update frequently. Databases established for different purposes can also be mobilized to provide information on the various dimensions of farm–territory relationships. As we will see below, these databases, most often administrative, are established according to a logic that it is important to be familiar with, in order to properly grasp their value and their limitations.

Each database has its own purpose and collection method (in situ or remote sensing observations/measurements, surveys, mandatory reporting, simulations, etc.). Data can be contained in a variety of formats: tables listing a range of information for given entities (farm, municipality), or data from Geographic Information Systems (GIS) in vector or image format. Vector layers combine attribute data (land use, utilized agricultural area (UAA), number of animals) with geo-referenced elements (e.g. parcel of land or farm head office, livestock building), while image data (raster) are derived from photographic or satellite campaigns. They correspond to an assembly of pixels of the same size (e.g. 10 m × 10 m for Sentinel-2) informed by a raw signal that must then be processed visually or by algorithms to give it meaning according to the targeted themes (e.g. nature of the cover, harvest date according to the appearance of the parcel over the course of the shooting dates).

Data availability and mobilization can vary widely. Some data are produced and made available free of charge to the general public by national institutes (for example France's National Institute of Geographic and Forest Information (IGN), Météo-France, etc.) or international bodies (the European Copernicus program, which produced the Sentinel-2 images). Other data come from surveys and administrative control systems, and the data distributed depend on the user audience. They are generally transmitted in highly aggregated form for the general public and more detailed form for research purposes; some data may be reserved exclusively for government departments and organizations co-managing the database. For individual data (e.g. by farm), statistical secrecy is the rule, and users wishing to access individual information must first submit a request to the secrecy committee to assess its admissibility in view of their data processing objectives. In the event of a positive decision, depending on the case, the data will either be forwarded directly by the department producing the data, or the applicant will be able to access it via the Secure Data Access Center (Centre d'Accès Sécurisé aux Données (CASD)). In the following pages, we focus on open-access data while providing information on certain databases that are accessible under certain conditions.

In the remainder of this chapter, we present the databases used to characterize farms and their inclusion in the value chains of their territories. We then present the SI-BOAT information system, which provides access to these pre-processed data and produces indicators to help diagnose a territory's agri-food systems. Finally, with the RPG Explorer project, we take a closer look at the use of a specific database, the French land parcel identification system called Registre Parcellaire Graphique (RPG), where farms are identified by their parcels and land use. On this occasion, we discuss the possibilities of complementing information on cropping systems with other sources of information on farming practices.

12.2. Databases

In this section, we describe databases providing information on (1) farm holdings and land use; (2) farms and their production activities; (3) relationships between farms, production chains and the territories in which they are located. Finally, we present (4) regulatory zones that may have an impact on farm activities. These databases cover the whole of France and are therefore not region-specific. Most of them are public, and at least some of their content is freely accessible in 2025, at least at the local level, to provide a relationship with the territory. The panorama presented is not exhaustive, and information on accessibility, valid for 2025, may change.

12.2.1. *Characterizing farm holdings and land use with the RPG*

12.2.1.1. *Origin and limitations of the RPG*

The RPG, which translates as "Graphic Parcel Registry" (the French land parcel identification system), provides information on land parcels and land use for all farms receiving Common Agricultural Policy (CAP) subsidies. This information, combining farm holdings and land use, has been produced every year since 2007. Information on farms not receiving CAP subsidies is not included in the RPG. On a national level, it can be estimated that more than 95% of agricultural areas were present in the RPG before 2015 (Cantelaube and Carles 2015). The areas not taken into account by the RPG are those of specialized farms (market gardening, viticulture, arboriculture) that were initially not entitled to a CAP subsidy and therefore did not declare their parcels. Since 2015, CAP subsidies have been accessible to all farms, which increases its spatial representativeness. Not all farms, however, have decided to make area declarations, due to the administrative burden in relation to a level of potential subsidy that is, all things considered, reduced for this type of high-value-added production. To sum up, in 2024, the RPG is a very good tool for analyzing arable and mixed farming territories. However, it may have certain limitations in specialized crop areas.

12.2.1.2. *What RPG data contains*

The RPG data contain a variety of information, not all of which is accessible to the general public and which has evolved over time. Broadly speaking, RPG information includes (1) agricultural parcels, (2) the nature and surface area of crops grown, (3) information on farms and (4) subsidies received for parcels and/or at the farm level. In 2024, the elements available were described in a technical instruction dated 2022[1].

Agricultural parcels were initially provided only at block resolution (from 2006 to 2014). A block is defined as a set of contiguous parcels farmed by the same farmer. In other words, it is possible to have different crops within a block, if, in fact, there are several contiguous parcels. In this case, we only know the surface area of the parcels but not their intra-block boundaries. From 2015 onwards, the parcel is known as the resolution of the crop parcel. By definition, there is now only one main crop per parcel. In other words, if a farmer splits a parcel in two to grow two different crops, this de facto generates two separate parcels in the RPG.

Nature of crops. From 2006 to 2014, RPG data are provided on the basis of 28 crop groups, covering all the crops grown by farmers. As of 2015, the information provided refers to a nomenclature of over 300 crops, which greatly increases the precision of the information. For example, until 2014, the 28 groups did not distinguish between winter and spring barley or even potatoes and melons. Since 2015, these distinctions have been made, as well as others (e.g. identification of certain species mixtures). IGN, among the RPG metadata, gives the correspondence between crop codes and crop groups, as well as the periods of use of the different crop codes[2]. Crop surface areas are those declared within parcels. They may be less than the geometric area of the parcels when these parcels include non-agricultural elements.

Farm holding information is primarily used to reconstruct parcels of land by farm holding, in an anonymous way. In fact, all the blocks or parcels on the same farm are linked by the same number. Until 2014, this was an anonymous number that changed from one year to the next for the same farm. From 2015 onwards, the number will be the PACAGE number, i.e. the unique identification number for each

1. See: https://info.agriculture.gouv.fr/gedei/site/bo-agri/instruction-2022-106.
2. See: https://geoservices.ign.fr/documentation/donnees/vecteur/rpg.

farm in the CAP subsidy file. This number remains the same from one year to the next[3].

Figure 12.1 shows the data structure for the various years of RPG data.

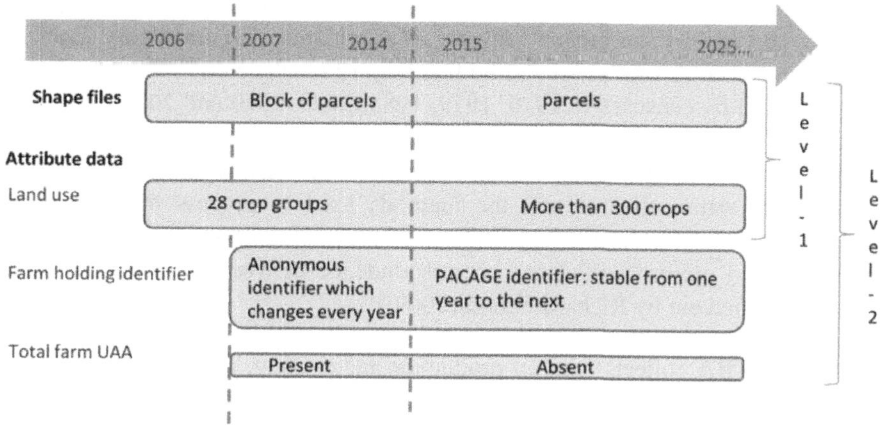

Figure 12.1. *Evolution of Level 1 (first two lines) and Level 2 (all lines) RPG data*

In France, information on spatial division and knowledge of land use is freely accessible on a site administered by IGN[4] and corresponds to the Level 1 RPG data. Data for the current year are only available online one year later (areas declared in May 2022 were put online in November 2023). In addition to Level 1 data, the RPG also includes information on individual farms, known as Level 2 RPG data. Level 2 is not available online. It can only be accessed by structures that have public service missions in their statutes, and only for the area over which they carry out these public service missions[5]. For RPG level 2, only data for 2015 and subsequent years are currently available to entitled parties.

3. However, the number may change if the status of the farm changes. For example, if a farm changes status from a Groupement Agricole d'Exploitation en Commun (GAEC, farming cooperative) to an Exploitation Agricole à Responsabilité Limitée (EARL, limited liability farming), it will be assigned a new PACAGE number without any farmer change.

4. In 2025, the address is https://geoservices.ign.fr/rpg. Please note that on this site (in 2025) and for the period 2006 to 2009, crop information is limited to the main crop on each block. You will find information on all crops on each block at the following address: https://seafile.agroparistech.fr/d/e2908e3b01084b6f93b5/.

5. See: https://info.agriculture.gouv.fr/gedei/site/bo-agri/instruction-2022-106.

12.2.2. *Characterizing farms and their production activities*

12.2.2.1. *Production structure and activities*

The French "agricultural census" (Recensement Agricole, RA) provides an almost exhaustive portrait of all farm holdings in the country. This census, carried out every 10 years by the French Ministry of Agriculture, is compulsory. Each EU country carries out its own census[6]. The first major French census was carried out in 1955, followed by censuses in 1970, 1979, 1988, 2000, 2010 and 2020. The nature and quality of the data collected have evolved from one census to the next, depending on the complexity of the questionnaires, the training of the interviewers, the choice of data collected and the methods used to process the information (Richard-Schott 2009). As a result, analysis of the evolution of agriculture requires precautions and pre-processing (e.g. correspondence between the types of cereals grown, as pointed out by Richard-Schott (2009)).

Today, the RA collects data on production factors (e.g. livestock, surface area, labor, outsourcing of services), quality and environmental initiatives (certifications), as well as diversification of activities (e.g. tourism, energy production) and marketing methods (e.g. short distribution channels). The typology of agricultural holdings based on technical-economic orientations of farm holdings (Orientations Technico-Economiques des eXploitations, OTEX[7]) is determined according to the economic weight of the various activities. The OTEX classification is used consistently in all EU countries. This enables cross-country comparisons and facilitates relationships with production chains.

For open access, RA data are at least aggregated at the municipal level of the farm's head office. Except for general description variables such as number of farms, UAA and dominant OTEX, statistical secrecy applies and data concerning municipalities with fewer than three farms are not disseminated at the municipal level. It is sometimes preferable to work at a higher resolution (canton, EPCI[8]) to benefit from non-secreted information.

SIRENE® (Système Informatique pour le Répertoire des Entreprises et des Etablissements), the "IT system for the directory of companies and establishments" database is a compulsory declaration database for companies and establishments

6. See: https://agreste.agriculture.gouv.fr/agreste-web/methodon/S-RA%202020/methodon/ (accessed on October 18, 2024).

7. See: https://agreste.agriculture.gouv.fr/agreste-web/methodon/N.1/!searchurl/listeTypeMethodon/ (accessed on October 18, 2024).

8. Etablissement Public de Coopération Intercommunal, Public Intercommunal Cooperation Establishments.

operating in France, across all business sectors (32 million establishments, including 13 million active ones). As such, it can be used to identify and characterize businesses linked to the agricultural sector. This database is managed by INSEE. Like all businesses, farms are identified by a company number (Siren) and an establishment number (Siret). A company may have several establishments. The address of each establishment is available, as well as the municipality where the company's head office is located.

A Principal Business Activity (Activité Principale Exercée, APE) code is assigned to the company/establishment, for statistical purposes. This code is based on the French Nomenclature of Activities (Nomenclature d'Activités Française, NAF). Therefore, establishments whose main activity is declared as agricultural have an activity code starting with 01 (approximately 751,000 establishments of this type on January 1, 2024). However, the updating of these data is based on declarations and is therefore not systematic: if an activity changes, it will not necessarily be modified in the database. Similarly, some inactive companies may be counted as active, which may lead to an overestimation.

The SIRENE® database provides information on the legal nature of businesses. It can be used to identify establishments whose legal status is agricultural, which are registered for other diversification activities. For example, farms involved in nutrient recycling through composting or biogas plant can be identified from the SIRENE database (division 35 – Production and distribution of electricity, gas, steam and air conditioning).

However, it cannot be used to track changes in legal status. For example, if an individually owned farm holding evolves to form a farming cooperative (Groupement Agricole d'Exploitation en Commun, GAEC) with partners, the individually owned farm holding will be closed, then the GAEC opened, with no possibility of linking the two entities in the SIRENE® database, as the numbers assigned will have no relationship between them.

Updates are integrated on a daily basis by INSEE and re-disseminated in free access on a monthly basis. These data are not distributed exhaustively. Because of the need to protect personal data, an individual registered as a sole trader, as is frequently the case for farms, can choose whether or not to make the data concerning them public.

Classifying farms according to their OTEX (RA) or APE (SIRENE®) status enables us to situate farms in relation to others. For a given territory, location quotients can be calculated (Gaigné and Letort 2017), indicating whether a type of production is better represented in the territory, compared with the national territory (e.g. share of farms whose OTEX is "cereals, oilseeds" in the territory compared with this share at the national level). This quotient can be calculated in the same way for upstream and downstream companies (e.g. in territory X, milk processing plants are more/less represented than in the national territory).

12.2.2.2. *Specific features of livestock farm holdings*

Livestock farms must be declared for health and safety regulations. The aim is to identify animals and track their movements from birth to death, individually for ruminants and by batch for mono-gastric animals. France's National Identification Database (Base de données nationale d'identification, BDNI) initially listed only cattle but is gradually being extended to sheep, goats and pigs. The BDNI identifies the places where animals are kept. This does not systematically correspond to the head office of the farm, and conversely, the same farm may have animals located in several places in the BDNI (e.g. dairy cows close to the head office and heifers in a more distant stable in a neighboring municipality). Farms are identified by a unique identifier issued by the Departmental Livestock Institution (Etablissement départemental de l'élevage, EDE), which manages sanitary registrations. These data are sensitive, and usually aggregated at the municipal level, so as to provide information on the number of units present. They are neither centralized nor freely available at the national level but may be at the departmental or regional level (e.g. distributed by the state statistical service: Direction régionale de l'alimentation, de l'agriculture et de la forêt, DRAAF).

For environmental reasons, and in particular to monitor the risks of pollution linked to livestock effluents, the French environmental code requires livestock farmers exceeding a certain number of animals to register with the prefecture[9]. Depending on the animal threshold, the farm must be declared, registered (simplified authorization) or apply for an authorization justifying the acceptability of the risks involved (e.g. pollutant discharges into the air and water). For example, in 2024, a dairy farmer must: be declared for between 50 and 150 cows, be registered for between 151 and 400 cows and apply for authorization for more than 401 cows[10]. The associated databases are the Classified Installations for the Environment (Installations Classées pour l'Environnement, ICPE) and the "Pollutant Emissions Registry Database" (Base de Données du Registre des Emissions Polluantes, BDREP). These can be used to find these farms, based on the establishment's SIRENE® identifier (Siret) and/or address. However, matching does not enable

9. See: https://www.ecologie.gouv.fr/installations-classees-agricoles.

10. See: https://aida.ineris.fr/reglementation/arretes-ministeriels-prescriptions-applicables-icpe-soumises-a-declaration.

exhaustive identification, as SIRENE® identifiers and addresses are not systematically entered in the ICPE and IREP (Installations industrielles rejetant des polluants, Industrial Facilities Releasing Pollutants) databases.

12.2.2.3. *Special features of organic materials processing activities: biogas digestion and composting*

The agricultural census of 2020 lists farms producing biogas for sale (excluding self-consumption) on an individual, collective (with several farmers) or centralized (via a local authority or private company) basis. For finer-level monitoring and cross-referencing with SIRENE® (see section 12.2.1.1), the SINOE® waste database lists biogas facilities and composting platforms. It is compiled by ADEME and is not exhaustive. Correspondences exist between SINOE® and IREP. Biogas digestion and composting activities can be classified as "R3 Recycling or recovery of organic substances not used as solvents (including composting and other biological transformations)". An anaerobic digestion plant may also be referenced in the ICPE[11] database, but its heading may vary according to the plant's other activities.

12.2.3. *Identifying relationships with agricultural sectors and collective dynamics*

12.2.3.1. *Quality approaches and marketing strategies*

The agricultural census (RA) provides information on the quality and marketing strategies of farms. As it is only updated every 10 years, for a shorter follow-up period, this information can be supplemented from various sources. These approaches can have an impact on farming practices (via specifications), marketing strategies and the integration of farms into agricultural value chains.

Agence Bio provides data on farms engaged in organic farming (in conversion and certified). In 2025, data are available by municipality (number of farms, livestock, surface area). It also provides data on all certified operators, including processors, distributors and importers. A directory of High Environmental Value (Haute Valeur Environnementale, HVE) certified farms is also available from the French Ministry of Agriculture. For official quality signs supervised by INAO

11. Section 2781 "Biogas facility for non-hazardous waste or raw plant matter, excluding anaerobic digestion facilities for wastewater or urban sewage sludge when methanized on their production site": https://aida.ineris.fr/reglementation/2781-installation-methanisation-dechets-non-dangereux-matiere-vegetale-brute-a.

(a French public body), it is possible to determine, with free access, whether or not a farm is in an area covered by an appellation or not[12]. Its presence in the area does not, however, guarantee that its products are labeled. Under access conditions, the number of producers, the number of processors and tonnages are also available.

With regard to marketing, various directories list farms that are part of *short food supply chains* (e.g. producers' stores, AMAPs[13], farm shop). These directories are supported by a variety of actors, depending on the region or department (e.g. chambers of agriculture, nature parks, actors involved in territorial food plans, etc.). Other directories focus on the diversification of tourism-related activities (e.g. Bienvenue à la ferme).

12.2.3.2. *Industry actors and collective dynamics*

To identify the various actors involved upstream and downstream of agricultural production, we can use SIRENE®, which provides information on all companies in all sectors. NAF codes are used to identify the various sectors. Upstream, division 01 includes ancillary services such as agricultural contractors. SIRENE® can also be used to identify, by their legal nature, agricultural cooperative groups, such as Agricultural Machinery Cooperatives (Coopérative d'Utilisation de Matériel Agricole, CUMAs) and landholding groups. Downstream, in division 10, we find the food industry. For example, in the case of bovine dairy production, we can associate "01.41 Dairy cow rearing", upstream: "01.62 Animal production support services" (e.g. artificial insemination), and downstream: "10.51 Dairy and cheese production" and "10.11 Meat processing and preservation". Provided we identify the possible relationships between the various activity codes and take into account the limitations of the available information, it is possible to propose hypotheses on the structuring of the local sectors in which the farms are integrated.

All operators handling live animals (e.g. veterinarians, transporters, slaughterers) and animal products, in particular by-products (blood, feathers, etc.), are subject to approval and/or authorization. They are registered with the Directorate General for Food (Direction Générale de l'Alimentation, DGAL) with an approval, authorization or registration number issued by the DGAL, their establishment's SIRENE®

12. Controlled Designation of Origin (Appellation d'Origine Contrôlée), Protected Geographical Indication (Indication Géographique Protégée) and Red Label (Label Rouge).
13. Association pour le maintien de l'agriculture paysanne: Community Supported Agriculture.

identifier (Siret) and their address. The Ministry of Food and Agriculture distributes a regularly updated list[14].

Identifying local groups where farmers can exchange views on agricultural practices provides information on local dynamics and issues that can support projects and experimentation. Agroecological collectives[15] include, for example, Economic and Environmental Interest Groups (Groupements d'intérêt économiques et environnementaux, GIEEs) focusing on various themes related to farming practices in the broadest sense (production, processing, marketing), or the 30,000 and DEPHY Fermes groups, which focus more specifically on the development of systems and techniques that reduce the use of phytopharmaceutical products.

12.2.4. *Regulatory zones which may influence farming practices*

A variety of zoning regulations can be introduced by the government or local authorities to encourage and/or restrict certain farming practices. The following are examples of environmental zoning: vulnerable zones and biodiversity protection zones.

Vulnerable zones are zones vulnerable to nitrate pollution from agricultural sources, in line with the European "nitrate" directive adopted in 1991 to preserve water quality. In France, the first delineations in the major river basins took place in the years following adoption of the directive and are revised every 4 years. Each vulnerable zone is the subject of a prefectoral decree and is associated with an action program. Compulsory in application, this program includes nitrogen fertilization management (e.g. spreading schedule, effluent storage, fertilizer doses) and farmland management elements (e.g. soil cover and intercropping). Data for delineating vulnerable zones are referenced by the National Water Data and Reference System Administration Service (Service d'Administration Nationale des Données et Référentiels sur l'Eau, SANDRE).

Protected areas and reserves[16] are listed in the French National Inventory of Natural Heritage (Inventaire national du patrimoine naturel, INPN) managed by the National Museum of Natural History (Muséum national d'Histoire naturelle, MNHN) and the International Union for Conservation of Nature (IUCN). Among these protected areas, the 11 national parks (terrestrial and marine) are attached to

14. See: https://agriculture.gouv.fr/tous-les-etablissements-agrees-certifies-et-autorises-de-la-dgal.
15. See: https://collectifs-agroecologie.fr/.
16. The geoportail site lists various nature protection zones with different statuses: https://www.geoportail.gouv.fr/inspire/annexe-i/sites-proteges.

the French Office for Biodiversity (Office Français de la biodiversité, OFB) and contribute to environmental conservation and to the preservation of biodiversity. The 58 regional nature parks (terrestrial) also aim to preserve environmental heritage and biodiversity, while promoting the economic and social development of the territory (e.g. development of the regional nature park brand)[17]. The IUCN also lists other types of zones. Some, such as the Natural Areas of Ecological, Fauna and Flora Interest (Zones Naturelles d'Intérêt Ecologique, Faunistique et Floristique, ZNIEFFs), are for information purposes only and are not associated with any regulations or management rules. Others, such as Natura 2000, may influence farming practices (e.g., as a condition for defining sensitive grasslands for the CAP[18], which may not be ploughed). In these parks and areas, farming practices may be subject to nature conservation conditions. Financial incentives can also be put in place specifically in these areas (e.g. MAEC[19] for Natura 2000 sites[20]) to enhance certain practices (e.g. planting/protection of rural hedgerows and orchards in the Vexin national regional park in 2020).

In addition, classification as a mountain area[21] makes it possible to identify a priori municipality characterized by constraints linked to altitude, slope and/or climate that may restrict land use and, for example, be favorable to the development of pastoral practices. Compensation and economic aid can also be provided.

12.2.5. *Summary of databases described*

In summary, 15 databases have been described in this section. These make it possible (1) to characterize farms and their production activities, (2) to identify actors in associated agricultural sectors, quality and distribution approaches, and collective dynamics and (3) to identify regulatory zones (see Figure 12.2). Table 12.1 shows the diversity of data collection methods (voluntary or compulsory declarations, surveys, directive-driven reference systems) and the frequency of open-access distribution (from daily to 10-yearly) of these databases.

17. See: https://www.parcs-naturels-regionaux.fr/les-parcs/missions.

18. See: https://agriculture.gouv.fr/pac-2015-cartes-departementales-des-prairies-permanentes-sensibles.

19. Mesures Agro-Environnementales et Climatiques: Agro-Environmental and Climate Measures.

20. See: https://www.natura2000.fr/sites/default/files/references_bibliographiques/analyse_maec_n2000-1.pdf.

21. See: https://www.observatoire-des-territoires.gouv.fr/kiosque/zonage-les-communes-classees-en-zone-de-montagne-zonage-agriculture.

Acronyms used in this chapter	Database name	Main distributor	Data collection	Distribution frequency (free access)
Agence Bio	Agence Bio data on operators, livestock and surface area	Agence bio	Compulsory declaration	Annual
BDNI	National identification database	DGAL	Compulsory declaration	–
Collectifs agroécologiques	Information on agro-ecological collectives	French Chambers of Agriculture	Voluntary declaration	Daily
DGAL	Approved, certified and authorized operators (animal health and protection, animal feed and by-products)	Directorate General for Food (Direction Générale de l'Alimentation, DGAL)	Compulsory declaration	Daily
HVE	Directories of High Environmental Value farms	French Ministry of Agriculture	Voluntary declaration	Annual
ICPE	Classified facilities for environmental protection	INERIS via MTES & BRGM	Compulsory declaration	Bi-annual
INAO	Data for AOC, PDO, PGI and Red Label certified areas	National Institute of Origin and Quality (INAO)	Compulsory declaration	Annual
IREP	Industrial facilities releasing pollutants	INERIS via MTES and BRGM	Compulsory declaration	Bi-annual
RGA	Agricultural Census	CASD, Agreste	Statistical survey	Decennial
RPG	French Land Parcel Identification System	ASP, IGN (since 2015)	Voluntary declaration, remote sensing	Annual
SINOE®	SINOE® – section: directory of facilities treating organic waste	ADEME	Voluntary declaration	Annual
SIRENE®	Systeme Informatique pour le Réperetoir des Entreprises et des Établissements (IT System for the Directory of Companies and Institutions)	INSEE	Compulsory declaration	Daily
Zones de montagne	Mountain areas	Ministry of Agriculture	Reference	Occasional

Zones relevant d'un espace naturel protégé	List of protected areas	French National Inventory of Natural Heritage (Inventaire national du patrimoine naturel, INPN)	Reference	Occasional
ZoneVuln	Vulnerable zones (nitrate directive)	SANDRE – Water France	Reference	Occasional

Table 12.1. *List of databases cited in this chapter with their acronyms, main distributor, collection methods and distribution frequency as of October 1, 2024. Links are available in the webography for further information or direct access*

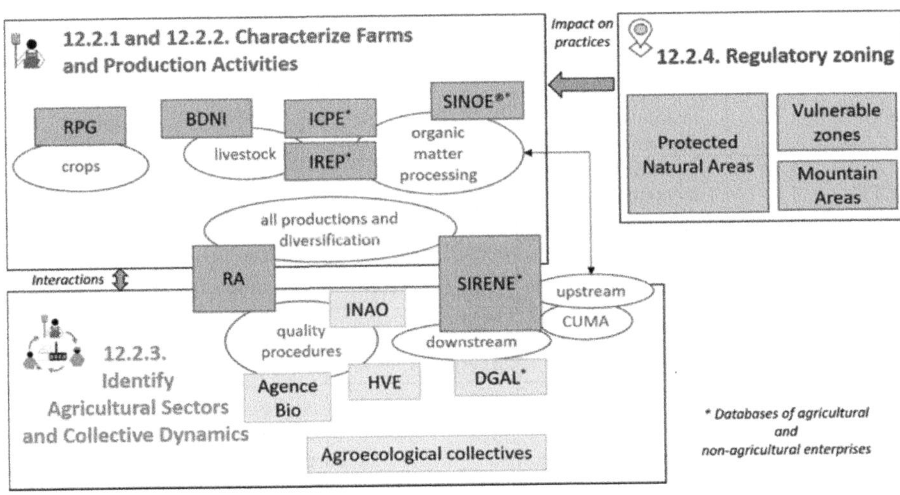

Figure 12.2. *Summary of the various databases described in this chapter, according to their use in characterizing farms (boxes 12.2.1 and 12.2.2), identifying relationships with other actors in sectors and territories (box 12.2.3), or identifying environmental and economic zones (box 12.2.4)*

12.3. SI-BOAT: an information system for agri-food systems

12.3.1. *Principle and function*

The databases described above are, at least in part, freely accessible at the local level for the whole of France (with the exception of the BDNI). By compiling these different databases, it is possible to produce a diagnosis of a territory's agri-food system, and in particular of the diverse production and processing networks involved. To this end, the SI-BOAT[22] Information System has been set up to gather diagnostic data and facilitate analysis, supervised by the user according to its objectives.

SI-BOAT was initially developed in 2019 at Lessem, an IRSTEA (now INRAE) research unit, and is now maintained in collaboration between Lessem and UMR Agir. SI-BOAT is inspired by the SIDDT[23], a "territory information system" created at Lessem in 2005. The principles of both information systems are as follows: the data concern the whole of France and come from open-access databases (such as those described above); anyone interested and asking questions about a given territory can connect to the web interface[24], enter their study area and retrieve analysis elements (see Figure 12.3).

Today, SIDDT is mainly used to study the socio-economic elements of the territory, in addition to the overall description of the agricultural context. SI-BOAT integrates all the actors in the agricultural sector, as well as all the databases mentioned above (except the BDNI). In addition to these databases, SI-BOAT also includes administrative databases, the Official Geographic Code (Code Officiel Géographique, COG) for the delimitation of administrative entities (municipalities, EPCI, region, etc.), and tables specially designed for matching between tables (e.g., from crop codes in annual agricultural statistics to those in the RPG).

At the start of 2024, 20 different databases were listed in SI-BOAT for around a hundred indicators (some indicators are sometimes duplicated with different levels of detail: e.g. lists of companies and numbers of companies by type). These indicators are grouped in a grid around the dimensions of anchoring (spatial, ecological, territorial), autonomy (decision-making, biophysical) and footprints (environmental, energy, socio-economic). An example of an analysis based on this grid is available in Grillot et al. (2021).

---

22. SI-BOAT was initiated as part of the BOAT project (funded by the ADEME GRAINE 2016 call for projects). It focused on Biomasses of Agricultural and Territorial Origin (BOAT). Since then, the information system has been expanded to provide more comprehensive information on agri-food systems, notably as part of the SCALABLE project (financed by the ADEME GRAINE 2019 call for projects).

23. See: https://siddt.inrae.fr.

24. See: https://siddt.inrae.fr/siboat.

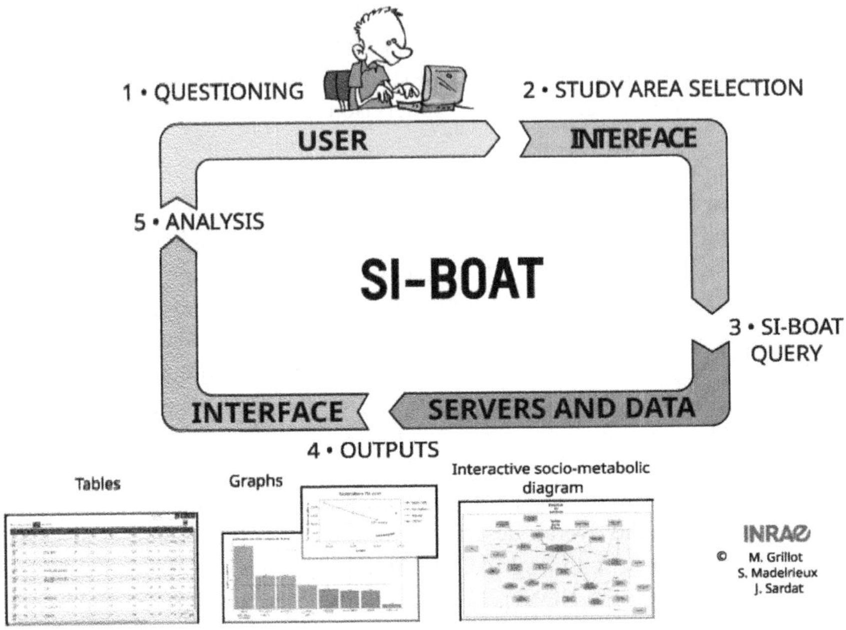

Figure 12.3. *How SI-BOAT works*

12.3.2. *Example demonstration of SI-BOAT*

We present below an example of the use of the databases described above and in SI-BOAT. For example, as part of a project on adapting to climate change, Mathilde, a regional coordinator, is carrying out a diagnostic study in 2023 on the situation of sheep breeders in the EPCI of Haute Ariège (52 municipalities). Mathilde recently arrived in the region and was not yet familiar with the terrain. In addition to contacting experts working in the area (technical advisors, cooperatives, etc.), a preliminary analysis based on existing databases may provide her with information to start to work with.

Visiting SI-BOAT, Mathilde observes the elements described below (see Table 12.2). In the Haute Ariège EPCI, sheep farming seems to be particularly well represented compared to the rest of the department (53% of farms in the zone in 2020, compared to 29% of farms in Ariège). However, the number of farms has been falling since 2010 (-38% in the area). Mathilde takes a quick look at the regional website of the DREAL Occitanie[25] (Direction Régionale de l'Environnement, Aménagement et Logement, Regional Department for the Environment, Planning and Housing), to access regional data from the BDNI. She identifies that sheep production seems to be in the majority (6,057 suckling ewes for 264 milk ewes in 2020). Back on SI-BOAT, she realizes that most livestock is raised with pastoral practices. In fact, 100% of the areas declared in the RPG, in 2021 were classified as forage and pastoral areas (78% predominantly grass with the presence of ligneous forage resources, 19% grazed woodland and ligneous resources and 2% permanent grassland) and 45 pastoral groups are listed (SIRENE® 2021). Numerous animal flows must pass through the area, presenting a number of challenges (health, organization, availability of forage and water resources, etc.).

In terms of farming practices and product development, no farms with sheep were certified or in conversion to organic farming in 2019 or HVE-certified in 2021 in the EPCI. In 2021, the area was covered by 11 designations of origin and geographical indications, but none concerning sheep production. A priori, there are no specifications in place that could have an impact on practices and product development.

All of the EPCI's municipalities are classified as mountain and fire-risk zones. Seven municipalities, representing 28% of the EPCI's surface area, are located in the Pyrénées Ariégeoises regional national park. Around 1/3 of the area is classified as a Natura 2000 zone: 26% as a Special Protection Zone (Zone de Protection Spéciale, ZPS, bird directive) and 13% as a Special Area of Conservation (Zone de Conservation Spéciale, ZSC, habitat directive), mostly located outside the park. Initiatives and measures impacting practices or favoring the presence of sheep may be in place (e.g. to prevent fires). The area is not classified as a vulnerable zone under the nitrate directive.

As far as diversification is concerned, no composting or biogas digesters appear to be present in the area (SINOE® 2021), nor are there any agricultural establishment producing electricity (SIRENE® 2020). The data available on marketing methods from the agricultural census are (as of March 2023) not sufficient to give an idea of the situation (2020 data not yet available). Other sources of information (local directories) are required. In 2010, 26% of farms were involved in direct sales or short food supply chains (RA).

Among the actors in the sector, there are, a priori, no actors identified in the area processing mutton (SIRENE® 2021). The search will have to be extended

25. See: https://www.picto-occitanie.fr/geoclip/#c=home.

geographically. Two small ruminant slaughterhouses have been identified in the department (DGAL 2023). Two veterinary establishments are listed in the area (SIRENE® 2021). In addition, two cooperatives for the use of agricultural machinery cooperatives (CUMA) are listed (SIRENE® 2021), as well as two agricultural land groups and no environmental interest groups (GIEE).

The questions Mathilde is now asking herself concern the following:

– The context:

- What is the reason for the decline in the number of livestock farms? Is there also a decline in livestock numbers? Are there any measures or subsidies in place to encourage sheep farming?

– Organization of the industry:

- Where are the animals slaughtered and cut up? Are there any difficulties in organizing the supply chain (slaughtering, distribution)? Do sheep farms sell their products via short food supply chains or direct sales? Are there any other specifications for increasing the added value of products from the area (e.g. collective brand)?

– Practices:

- How variable is animal density over time, as a result of pastoral activity? What happens to effluents? Are they all spread? How long are the animals kept in buildings? Are other areas (wasteland, etc.) used by the animals in addition to grazing land?

To get to grips with these issues and open up her inquiries to the diversity of actors who can influence sheep farming in the area, Mathilde now knows that she can try to contact various actors: the regional national park and, according to the SIRENE® database, which provides names and municipalities, the two slaughterhouses identified on the Ariège, the 45 pastoral groups, the veterinarians registered in the EPCI. She can also contact the two CUMAs to find out more about the equipment used in common.

This list is not exhaustive, as it is based solely on the above-mentioned databases. A wide range of actors should be considered (technical and advisory institutes such as the French Chamber of Agriculture, the CIVAM network, associations, EPCI, etc.).

Data type and source	SI-BOAT data: EPCI Haute Ariège and Ariège	Observations and open questions
OTEX: sheep, other herbivores (RA)		
- No. of EA (2010)	76 or 56% of FHs (Ariège: 863 FHs or 32% of FHs)	Fewer farms with sheep and other herbivores. These farms are particularly well represented in the study area compared with the department (QL > 1.2). What is the share of sheep?
- No. of farms (2020)	47 or 53% of FHs (Ariège: 655 FHs or 29% of FHs)	
- Evolution (2010–2020)	–38% (–24% in Ariège)	
- Location quotient (LQ 2020)	1.8 (compared with Ariège)	
OTEX: multi-crop or multi-species (RA)		
- No. of farms (2010)	18 or 13% of FHs (Ariège: 323 or 12% of FHs)	Decline in mixed farms. Do mixed farms have sheep?
- No. of farms (2020)	9 or 10% of FHs (Ariège: 256 or 11% of FHs)	
- Change (2010–2020)	–50% (Ariège: –21%)	
No. of establishments with NAF code 0145Z (Sheep and goat breeding) (SIRENE®)	57 (2021) or 30% of FHs	Difference of around ten EA with RA 2020. What is the share of sheep? and mutton sheep?
Establishments declared for environmental risks (IREP, ICPE)	0	No farms exceeding thresholds requiring declaration.
Livestock (via DREAL Occitanie, 2020)	10,739 sheep, including 6,057 suckling ewes and 264 dairy ewes.	Most meat production in the area.

Table 12.2. *Elements of characterization of farm holdings (FHs) raising sheep in Haute Ariège (and comparisons with the department, Ariège) based on SI-BOAT*

12.3.3. *Precautions when using databases: example in SI-BOAT*

SI-BOAT faces a number of challenges that are representative of the challenges that any user of these databases may face. A foremost challenge is the permeability of the study areas over time. The databases are supplied for the COG of a given year. However, the administrative boundaries of the COG may vary from one year to the next, notably as a result of mergers between municipalities. For example, in 2023, a person works in a study area where municipality A merged with municipality B in 2021. If the data she wishes to mobilize is provided to COG 2020, it will be necessary to take data from both municipalities A and B to provide information on the new municipality. This follow-up requires arbitration: (i) in the case of a municipality split (rarer): how to divide the data from municipality X between the two new municipalities? and (ii) in the case of a merger for data that cannot be summed up, such as averages or categorical variables.

A consequent challenge concerns data updates. SI-BOAT is not directly linked to the databases being used. Data are stored on information system servers and are thus downloaded at a given time. However, data suppliers sometimes make corrections to data that have already been distributed. Some data are, in fact, consolidated over time. Data for annual agricultural statistics are known to be first supplied in provisional versions, and then amended and supplied in final versions. Other data may be readjusted according to the COG. This is the case, for example, with RA 2010 census data, which has been readjusted to the same COG as RA 2020 to facilitate comparisons. A manual monitoring process by information system maintenance is in place to track these corrections made by data suppliers and the addition of new tables (Casel and Grillot 2024). Some data providers offer Application Programming Interfaces (APIs) that enable direct connection to their services to access data, making it easier to take updates into account. However, it is still necessary to monitor any modifications. These changes may concern the addition/deletion of variables, modifications to modalities or categorizations, as well as methodological evolutions (e.g. for the agricultural census, inclusion of non-commercial farms before 1988, then only professional farms; addition of new crop codes for the RPG or annual agricultural statistics, etc.).

Another challenge relates to the need to understand the data used. In SI-BOAT, particular attention is paid to the provision of metadata (table date, method, etc.) with, where necessary, elements of vigilance (e.g. relating to the risk of non-exhaustiveness) and indicator calculation methods. SI-BOAT data are supplied in virtually raw form, so that the person carrying out the analysis is free to manipulate the data as required.

12.4. Enhancing the impact of RPG: the RPG Explorer project

In the previous section, we saw with SI-BOAT how it was possible to structure different databases in which the farm was spatially associated with a municipality, without any further details on the spatial organization of its structural elements, i.e. the parcels. In this section, we take a closer look at how the RPG Explorer project can be used to enhance the information contained in the RPG.

12.4.1. *Ambition and scope of the RPG Explorer project*

The RPG data provides a wealth of information on the spatial and temporal organization of crops (crop rotation, crop succession), as well as on the dynamics of farm holdings, provided you are a public service organization with access to the farm identifier. However, these data remain complex to use, requiring both agronomic knowledge to imagine how to structure this information and a good command of GIS tools to implement this data structuring.

The RPG Explorer project was launched with the aim of facilitating the processing of RPG data, with or without farm information. Initially focusing on the development of free software to facilitate data processing (RPG Explorer), the project has gradually expanded to offer freely downloadable pre-processed data (crop sequences), as well as face-to-face and distance learning courses. There are also plans to develop an online version of the RPG Explorer tool. Work is also underway to enhance the use of RPG equivalents from other EU countries (currently Italy and Luxembourg).

12.4.2. *The core modules of the RPG Explorer project*

The RPG Explorer tool can be used to enhance the use of data available to the general public on the IGN website (RPG Level 1), as well as data relating to farm holdings (RPG Level 2).

12.4.2.1. *Land use data: from crop proportions to rotation proportions*

The developments in this section apply to both Level 1 and Level 2 RPGs. In this section, we distinguish between two approaches: one called "year by year", in which the land use information for a given year is used to calculate crop proportions, which can then be compared with those of other years. The second is referred to as "inter-annual", since the information processed at parcel level is no longer land use but crop succession (integration of different years of RPG data).

In the year-by-year approach, it is possible to calculate crop proportions, i.e. the surface areas of different crop groups in a given area, which may be, for example, a municipality, a water catchment area or a cooperative supply basin. This calculation can be carried out over several consecutive years to note changes (development of one or more crops at the expense of one or more others). It should be noted that the transition from old data (2006–2014) to recent data from 2015 poses a methodological problem, as we go from 28 crop groups to over 300. The difficulty is only apparent insofar as for each of the 300 new categories, the French administration gives the corresponding term in the system of 28 crop groups. Over a period of integrating old and new data, it is therefore possible to make all calculations on the basis of the 28 groups. For data from 2015 onwards, RPG Explorer also offers the option of calculating crop rotation by intermediate groups (49 groups). The idea is to reduce the noise generated by information on over 300 crops while at the same time achieving greater precision than with the 28 groups alone. As previously mentioned in footnote 4, the IGN data provided in 2025 for the period 2006-2009 are truncated and only give the main crop for each block. Furthermore, the RPG Explorer tool is currently unable to process IGN data prior to 2015. If we wish to use RPG Explorer to process data for years prior to 2010, we must use the data available at https://seafile.agroparistech.fr/d/e2908e3b01084b6f93b5/, which are complete and compatible with RPG Explorer.

In the inter-annual approach, knowledge of the crops grown each year on each parcel enables us to establish crop sequences at any point covered by a surface declaration. Crop sequences are important indicators of the intensity of farming systems. The length of time it takes for the main crops to return and the frequency with which they appear in the sequences are key determinants of phytosanitary risks. Crop sequences also enable cooperatives collecting a particular crop to know how that crop fits into the different crop sequences, so they can adapt their advice and collection zones to create batches of differentiated quality. Crop sequences can be used to identify and quantify areas that have been plowed or planted with grassland. The sequences also make it possible to distinguish between short and long intercrops, which have a significant impact on nitrate leaching.

The elementary space unit used to define a crop sequence depends on the period during which this space unit (parcel/block) has remained unchanged. If we consider a parcel of land that has remained unchanged since 2015, the crop sequence will be defined at the level of this parcel and will be made up of the succession of crops it has borne since that date. If the parcel, after several years of cultivation in its original form, is then split into two parcels, the result will be two crop sequences assigned to two distinct parcels, known as "intersection parcels", over the study period under consideration. As a result, the intersection parcels (i.e. the set of intersection parcels) are much more fragmented than the RPG parcels of a given year. In addition, the longer the study period, the more fragmented the intersection parcels, as one year's parcel redistributions are added to those of previous years.

The fairly simple principles we have just outlined apply to recent data. The situation becomes more complex with older data, for which only block boundaries are known, and not those of individual parcels. In this case, crop sequences are only known with certainty for blocks containing a single land use (i.e. a single parcel) over the duration of the sequence. In this case, we return to the situation of recent data. In all other cases, we need to mobilize rules for identifying sequences, based on the correspondence between surfaces in the block in year n and those found in year n+1. RPG Explorer mobilizes a series of rules presented in Levavasseur et al. (2016) to attempt to reconstitute these sequences. One of them, for example, is to consider that if we have a block occupied in year n by 5 ha of maize and 6 ha of wheat and this same block is occupied in year n + 1 by 5 ha of wheat and 6 ha of maize, then we will say that on this block there are two sequences – one of 5 ha of maize-wheat and one of 6 ha of wheat-corn. The different rules are mobilized one after the other. As the sequences are established, the crop areas used to create the sequences are removed from each year. The process of establishing sequences is stopped if the mobilization of rules no longer produces new sequences. With older data and for sequences of 6 to 7 years, we rarely exceed 70% of sequences reconstituted with the proposed set of rules. This reconstitution is around 98% for post-2014 data. The 2% of non-reconstitution corresponds mainly to parcels that are not declared in every year of the sequence. Establishing crop sequences with RPG Explorer can require considerable calculation time, depending on the number of departments processed, the number of years and the power of the machine used. To save time, the crop sequences established with RPG Explorer by department are available for free download from https://entrepot.recherche.data.gouv.fr/dataverse/sadapt.

Establishing sequences generates a large amount of information that can be used directly, for example, to determine the surface area and location of permanent grasslands converted to annual crops. In some cases, however, it may be advisable to synthesize the information in order to exploit it to the full. In the following, we propose two ways of synthesizing information: crop rotation simplification (parcel level) and crop rotation proportions (set of parcel level).

The principle underlying crop rotation simplification (inspired by Maxime et al. 1997) is to assimilate each crop sequence obtained into a rotation that can be presented in a simplified form. For example, the crop sequence wheat-barley-rapeseed-wheat-barley-sunflower could be assimilated to the following rotation wheat-barley-(rapeseed/sunflower), i.e. a succession of wheat then barley, followed by either rapeseed or sunflower. This rotation can then be simplified to cereal-cereal-(rapeseed/sunflower), i.e. 2 years of cereal followed by rapeseed or sunflower. A final simplification leads to cereal-(rapeseed/sunflower), which corresponds to an alternation of cereals (without specifying the number of years) with rapeseed and sunflower. The advantage of this representation is that GIS can be used to map simplified rotations, choosing the most appropriate level of simplification to highlight the preferred locations of a given rotation within the area under study. These areas may be the subject of specific advice depending on the theme being addressed locally.

The principle underlying crop rotation proportions (inspired by Schönhart et al., 2011) is to define for a given area, made up of numerous parcels, a single composition of different rotations with their associated proportions that best reflects the crop structure observed in the sequences. Here, rotations are given with explicit crops (not "cereals" but "wheat" or "barley") in a single fixed order (wheat-barley-rapeseed-wheat-barley-sunflower and not wheat-barley-(rapeseed/sunflower)). Whereas the "rotation simplification" approach provides an overview of the diversity of major types of crop succession in a given area, the "crop rotation proportions" approach is primarily used in conjunction with agronomic models to characterize the effect of crop sequences on the environment (water balance, organic matter storage, etc.). There are often several thousand crop sequences in a given territory. As it is not possible to model the effect of each of these crop sequences on each parcel, the "crop rotation proportions" approach aims to reduce too many sequences to a limited number of rotations, the importance of which is quantified, and which we ensure are representative of field reality by means of a constrained linear optimization procedure. These limited rotations can then be used in agronomic models. For rotation simplification, it is possible to map the rotation frameworks for each parcel. For crop rotation proportions, we have information that is valid for a territory as a whole but which cannot be reused for each parcel. While it is not possible to disaggregate crop rotation proportions by parcel, RPG Explorer does have a module for disaggregating crop rotation proportions by soil type. This procedure can be implemented in cases where the territories on which crop rotation proportions are established are Soil Map Units (SMUs), themselves made up of different proportions of Soil Typological Units (STUs). As was the case for crops within blocks prior to 2015, the boundaries of the STUs within the SMUs are not known. The allocation of the various crop rotations to the SMUs is done by constrained linear optimization, mobilizing expert knowledge of the affinities between the different crops in a rotation and the soil types. The procedure for establishing crop rotations is presented in Levavasseur et al. (2016). This procedure, integrating the disaggregation step at the STU level, was implemented as part of the ABC'Terre project as a preliminary step enabling the calculation for each of the rotations their soil carbon storage potential[26].

As mentioned above, the crop sequences for each department are available online. The data are available in two formats at https://entrepot.recherche.data.gouv.fr/dataverse/sadapt. A "classic" format is used to retrieve the parcel shapefile with the sequence table (e.g. Girault and Martin 2023a[27]). An "archive" format is also proposed, which can be loaded into RPG Explorer and used to calculate simplified rotations and crop rotation proportions (e.g. Girault and Martin 2023b[28]).

12.4.2.2. *For a better use of farm data*

In the previous section, we examined how to make the most of freely available land use data. We now present the type of processing that can be envisaged on the Level 2 RPG, which integrates the farm number (anonymous number or PACAGE since 2015). Other information can be provided, but the farm number remains the richest piece of information in the Level 2 RPG, on which we focus in this chapter[29]. As in the previous section, we distinguish between a "year-by-year" approach, where each parcel is qualified only by the farm number of the year under study, and an "inter-annual" approach, where intersecting parcels are characterized by the sequence of farm numbers.

In the year-by-year approach, knowledge of farm identifiers enables calculations to be aggregated by farm for each of the years under consideration. These aggregations may concern land use (crop rotation), but they may also concern parcel structure. In this way, we can compile statistics on parcel area (average, min, max, median), as well as on distances between parcels. This information is particularly useful for assessing the suitability of land parcels for reducing farmers' working hours. It can also be used to explain land-use patterns. In some cases, farmers may oversimplify the management of their crops, which are located on blocks of parcels far from the farm's head office. In the most extreme cases, this simplification can lead to a single crop being grown on all the parcels concerned. It is also possible to calculate the relationship between farms and a given study area: this involves calculating the proportion of the farm's total UAA located in the study area; this is how we calculate the involvement level, i.e. the proportion of total farm UAA

26. See: http://www.agro-transfert-rt.org/abcterre/.
27. See: https://doi.org/10.57745/PMD42J.
28. See: https://doi.org/10.57745/FMXKTF.
29. Note, for example, that the Level 2 RPG also provides the organic or non-organic nature of the cultivated parcel, but in 2023, this information was also freely available from the organic agency (https://www.agencebio.org/cartobio/ or https://www.data.gouv.fr/fr/datasets/parcelles-en-agriculture-biologique-ab-declarees-a-la-pac/).

located in the study area and the contribution of farms, which corresponds to the proportion of the study area occupied by each farm.

Involvement and contribution are important indicators for territorial agricultural coordinators who work on issues leading to very precise spatial delineations. This is particularly the case for the protection of drinking water and the implementation of action programs in catchment protection areas. The contribution makes it possible to identify the farms that have the greatest impact on the study area and on which it may be in our interest to concentrate for reasons of efficiency (e.g. those whose total farm area in the study area represents 80% of the total area). Farms with a high level of involvement (high proportion of their UAA in the study area) are also those that will be most affected by decisions taken to regulate farming practices in the study area. Farms with a high level of involvement and contribution are generally the most important to mobilize in such territorial studies.

In the inter-annual approach, the aim is to analyze the dynamics of farm holdings in order to determine, for example, whether the number of farms in a given territory remains unchanged over time, or whether the number of farms is decreasing, with a more or less equal reallocation of the freed-up areas to the remaining farms. To monitor the evolution of farm holdings from one year to the next, we look at the intersecting parcels to see how the farm numbers of year n are grouped in relation to those of year n + 1. For example, if all the intersection parcels with farm number 125 in year n all have a single number 126 in year n + 1, and this number 126 is not found on any other intersection parcel, then we can say that the farm with number 125 in year n and 126 in year n + 1 is a stable farm. If we had found other intersection parcels with number 126 in year n + 1, then we could have concluded that the farm had expanded. In the situation where some of the parcels bearing number 125 in year n would not bear number 126 in year n + 1 without the latter being found on other intersecting parcels, and then the farm would have reduced in area. This may seem a complex approach, and we might think that it would be sufficient to use the PACAGE number as the basis for determining parcel changes, given that it has remained unchanged from one year to the next since 2015. In fact, the method presented above has the advantage of taking into account situations not covered by a simple tracking of PACAGE numbers. These are (1) changes in land parcels prior to 2015, when the farm number changed from one year to the next and (2) situations where the PACAGE number changes (from one tax status to another) while the land parcel remains unchanged. Tracking parcels of land therefore requires a combination of PACAGE numbers (from 2015) and changes in farm number associations at the level of intersection parcels.

Above, we have only presented the simplest cases of parcel evolution between two years. Other more complex cases of evolution are detailed in Barbottin et al. (2018). Once the farm trajectories have been defined in this way, it is possible to reconstruct, for each farm, the evolution of its crop rotation as well as the crop successions specific to these parcels. This type of information is very useful before a farmer survey, to save time in understanding their cropping systems and take advantage of the survey to go into greater depth on points that may raise questions.

By aggregating information specific to each farm, it is possible to obtain an overall view of farm dynamics on a territorial level: number of stable, reduced or increased UAA, with associated statistics such as average surface area, average reduction or increase in surface area for the farms concerned. Using the same data, it is also possible to identify what might be termed farmland clusters, i.e. farms linked together by the temporary or permanent transfer of parcels or blocks. These exchanges reflect social relations between the farms concerned. A case in point is the year-round exchange of parcels by farms specializing in certain cash crops, such as potatoes or flax, who exchange parcels in order to meet the return periods of these crops (Dongmo, 2023). The existence of such exchanges on a given territory reflects a habit of interaction between farmers over land, which can facilitate negotiations in this field with non-agricultural operators (e.g. in the case of the protection of water catchment areas supported by local authorities, mentioned above). Farmers who do not trade with their neighbors are likely to be even less inclined to do so with a municipality.

12.4.3. *Data processing with RPG Explorer*

In section 12.3.2, we showed how Mathilde could use the SI-BOAT tool on a climate change theme. We now find her grappling with a water quality management issue tackled at the level of a watershed. This watershed is very large (several thousand hectares), making it impossible to acquire information on farming practices through surveying alone. She therefore decided to use RPG data processed with RPG Explorer.

Mathilde is an employee of a local water authority. This syndicate has public service missions in its statutes, which means that she has been able to request the extraction of Level 2 RPG data from the *state statistical services* of her region from 2015 onwards for her study area. She also downloads the archive of crop sequences specific to her department (e.g. 2015 to 2021: https://doi.org/10.57745/FMXKTF).

She also has the watershed shapefile on which she intends to work. This shapefile incorporates zoning to characterize areas that are more or less sensitive to nitrate leaching within the watershed.

She then checks on the dedicated website (https://rpgexplorer.fr/) that she has the latest version of RPG Explorer. She integrates the shape of her watershed, Level 2 RPG data and "archive" sequence data into the tool's local database (postgre/postgis database). Not knowing the area well, and wishing to understand its overall dynamics, she establishes the crop rotation year after year in 28 crop groups and 48 intermediate groups. She could also use the details of over 300 crop codes if she wished to analyze, for example, the presence of low-input crops (which would be masked in the intermediate groupings). You can also check whether crop rotations are relatively stable from one year to the next, as well as from one area of the watershed to another.

She can then use the "simplified rotation" module to map them and visually identify whether they are evenly distributed or whether there appear to be particular spatial concentrations. In particular, she analyzes the plowing/sowing of permanent grasslands, given the important role played by these surfaces in nitrogen flows. In addition, the crop rotation modules can be used to quantify the main crop rotations modeled for each zone.

The Level 2 RPG data then enable her to find out how many farms are affected by the watershed and how many of them need to be mobilized to reach 50 or 80% of the watershed's surface area, distinguishing between sensitive areas. It can also determine whether some farms are already known because they have land in another watershed that it manages. It can also analyze recent land dynamics to see whether some areas have recently changed farmers, and whether this has led to changes in cropping systems. By cross-referencing Level 2 information with cropping sequences, she will also be able to obtain details of cropping successions and associated rotation frameworks for each farm. This information will be important to have on hand to save time when Mathilde needs to talk directly to farmers about possible changes to their systems to improve water protection.

12.4.4. *Limitations of the RPG and the use of other sources*

12.4.4.1. *The RPG is limited*

In terms of descriptive data on agricultural land use, the RPG has three main limitations. The first has already been mentioned but deserves to be reiterated: the RPG does not include the surface areas of farms that do not apply for CAP benefits.

This can be particularly problematic in territories dominated by market gardening, vineyards or arboriculture, production systems for which farmers make fewer subsidy applications than for field crops. Secondly, data are only available with a minimum one-year delay. This can pose a problem for "live" monitoring of a large territory. The third limitation concerns the single land use available for each parcel. This corresponds to the main crop declared on May 15 of each year. This information may be insufficient when several crops are grown on a parcel in the same year. This is the case for double cropping, where field vegetables are harvested early (May/June) and a second crop (corn) is harvested in autumn. This is also the case for intermediate crops, which can have different functions (forage catch crops, nitrate traps, energy cover crops, etc.) and are not or poorly identified in the RPG.

12.4.4.2. *The RPG can be extended*

The missing information is potentially accessible at the level of Sentinel-2-type satellite images (European Copernicus program) available free of charge since 2017 with a resolution of 10 m. The processing of these images remains complex, but, as with the RPG, research teams make the results of their work available to the public in the form of shapefiles or rasters, giving the Land Cover Map (LCM) each year (www.theia-land.fr/product/carte-doccupation-des-sols-de-la-france-metropolitaine/#toc8). While this does go some way to completing the land cover, it does not solve the problem of the time needed to obtain it. In fact, this map is only available a year late and with a nomenclature that is still currently simplified compared with what has been available in the RPG since 2015. For example, in 2023, the LCM does not distinguish between wheat and winter barley. The parcel boundary and land use data delivered in this way are the result of image processing using machine learning models, which means their reliability is less than that of the RPG data derived from declarations that can be checked. However, these data provide a very interesting basis for completing the RPG's land use data, as proposed by Cantelaube and Lardot (2022)[30] with "completed RPG" data available online (e.g. https://doi.org/10.57745/U44B7L). Note that the use of Sentinel data also enters a new phase with the CAP 2023–2027, which makes their use compulsory for the continuous monitoring of agricultural surfaces as part of CAP controls. At the time of writing, we do not know to what extent these pre-processed Sentinel data will be available for research and development in the field. In principle, these data would be extremely useful, as they would enable continuous monitoring of land use from the autumn of year n-1 to the spring of year n, when the land is declared. In the meantime, attempts to identify intermediate crops based on these data are beginning to emerge (Nowak et al., 2021; Fendrich et al., 2023).

---

30. For further details, readers are recommended the following two references: Cantelaube and Lardot (2021) and Lardot et al. (2021).

12.4.4.3. *The RPG can be combined with the French agricultural census, with conditions*

The 2020 (RA 2020) and 2010 (RA 2010) French agricultural censuses include the PACAGE number as farm information. This is important, as it enables a match to be made with RPG data, which is also informed by the same PACAGE number. This makes it possible to better spatialize the RA 2010 and RA 2020 data, which were otherwise associated with the municipality of the farm's head office, which can now be correctly distributed across the reality of farm holdings. This also makes it possible to associate valuable information such as the address of the farm's head office or the economic and technical orientation (OTEX) with an RPG parcel. Transferring the results of these matches to the public must, however, respect the rules of statistical confidentiality. For example, we can distribute tables showing the dispersion or fragmentation of farms (from the RPG) for a given geographical entity (department, canton or municipality) and by OTEX (from the RA), provided there is a sufficient number of farms in each geographical entity to comply with statistical secrecy rules. However, without the consent of the farms concerned, it will not be possible to distribute mapping by farm to the public, as the location of parcels by farm is considered personal data. It should also be noted that, to the best of our knowledge, this processing is currently only authorized within the framework of the secure data access center (CASD).

12.4.4.4. *Providing information on practices*

The RPG deals only with land use and not with cultivation practices. However, since 2015, information on the "organic" nature of the crop has been available in the Level 2 RPG. In 2023, this information was also freely available from the organic agency. This information will be very useful for the diagnosis carried out by our friend Mathilde on her watershed!

Another piece of information, the irrigable nature of a parcel, was temporarily available with Level 2 RPG (from 2007 to 2009). This information disappeared with the end of specific subsidies for irrigated crops. For their part, the water agencies centralize farmers' annual water abstraction declarations for payment of water charges. Since 2023, this information has been grouped together in the BNPE[31]. These abstractions are most often geo-referenced (with an indication of the precision of this location), which makes it possible to identify a certain number of irrigating farms by cross-referencing with the RPG, on the assumption that if a catchment, with a high precision of location, is located on a declared parcel, then the farm is irrigating. It is more difficult to identify farms whose water supply point is located

31. Banque Nationale des Prélèvements quantitatifs en Eau (National Bank of Quantitative Water Withdrawals): https://bnpe.eaufrance.fr/acces-donnees.

outside their parcel. It should be noted that the same approach can be adopted for assigning biogas digesters to farms, insofar as these units are described with georeferencing in the SINOE® database mentioned above (Levavasseur et al. 2023).

Other information on technical itineraries used on crops is not freely available throughout France at parcel resolution. Some information is accessible through the above-mentioned matching with RA 2010 and RA 2020 under CASD conditions (confidentiality rules).

However, some information from the "cultivation practices" surveys is freely available on Agreste, the agricultural statistics website, but aggregated at the regional level. For example, the Treatment Frequency Index (Indice de Fréquence de Traitement, IFT) values for the main crops are available. These are average data, sometimes masking considerable variability due to strong intra-regional differences. The AGROSYST database contains data collected on DEPHY Fermes, which aim to reduce the use of phytosanitary products as part of the ECOPHYTO program. There are also databases held by chambers of agriculture and other extension institutions, which include farmers' electronic field notebooks. These databases are not currently distributed. Other databases provide indirect information on the agricultural use of plant protection products. These include the plant protection product sales database, which has been available by professional end-user zip code since 2013[32] (Martin et al., 2023). The quality of the information available in this field is set to improve considerably over the next few years. Indeed, since 2009, European regulation 1107/2009 has required professional applicators of plant protection products to keep a register of applications, which may be inspected by government departments. By 2026, these registers must be in digital form. The European regulation on the sustainable use of phytosanitary products (Sustainable Use Regulation) provided for a national centralization of digital registers. However, this was abandoned at the end of 2023, and France did not include the idea in its ECOPHYTO 2030 strategy published in May 2024.

Information on cropping practices could also be supplemented by new sources of information. These include the satellite data we have already mentioned for identifying crops or intermediate crops, but which could also provide information on harvest dates or crop residue management (Barnes et al. 2021). This also includes data from information systems developed to help farmers make technical decisions. This information can be coupled with information on the state of the environment, whether climate, soil or biodiversity, linked or not to pathogens (Debaeke et al. 2022).

32. See: https://ventes-produits-phytopharmaceutiques.eaufrance.fr/ or even to spatialize sales https://odr.inrae.fr/intranet/carto_joomla/index.php/reseaux/reseau-sdc-sdp/pdt-phyto.

Another area poorly covered in terms of publicly accessible data is that of livestock breeding practices, partly for reasons of protection of individual and health data (e.g. BDNI). However, the Institut de l'Elevage and the French Chambers of Agriculture have access to the INOSYS network[33]. This network enables them to keep an observatory of the technical and economic practices and results of typical farms, known as "reference" farms and representative of their production basin. This data feeds the DIAPASON information system. National financial support is provided for the "national base" and, depending on the region, funding may be made available to cover a greater diversity of systems. These case study data are very useful for adapting studies to different regions. Whether for crops or livestock, access to practice data, necessarily framed by the General Data Protection Regulation (GDPR), would considerably increase knowledge of how farming works in territories.

12.5. Conclusion

We have shown in this chapter that there is a diversity of databases produced and available at the national level, enabling us to characterize farms as well as the sectors and territories in which they are integrated. Each database provides its own wealth of information, but matching between databases is often even richer. These matches are not self-evident, due to the differences that can be found between databases in the definition of variables, their dates and the conditions under which they are mobilized. When data directly concern farms, it is also important to take into account the GDPR in processing outputs and their dissemination.

We have only dealt with some of the available databases. A more detailed treatment of farm-territory relationships would require, for example, the mobilization of additional data on climate, soil and non-agricultural land uses (water bodies, forests, towns and road infrastructures, etc.).

In addition to the databases themselves, we presented tools developed by research teams to facilitate analysis of the relationships between agriculture and the land, either based on land use and agricultural parcels or on the description of farms, sectors and various zoning areas, notably environmental. In particular, we presented an information system (SI-BOAT) inspired by another (SIDDT), as well as an IT tool to facilitate the processing of RPG data (RPG Explorer). It should be noted that the information provided in this way can be supplemented by that provided by other information systems integrating information on farms, such as that developed by the

33. See: https://chambres-agriculture.fr/informations-economiques/inosys-references-systemes/inosys-reseaux-delevage/.

Observatory for Rural Development (Observatoire du Développement Rural, ODR[34]). An INRAE service unit set up in 2009, the ODR is developing an information system on agricultural policies and systems, based on various sources of administrative data concerning, for example, agricultural and rural policies, employment trends (in partnership with the Mutualité Sociale Agricole), or quality signs (in partnership with INAO). The information drawn from these databases will be all the more relevant if it is complemented by solid fieldwork expertise.

12.6. References

Barbottin, A., Bouty, C., Martin, P. (2018). Using the French LPIS database to highlight farm area dynamics: The case study of the Niort Plain. *Land Use Policy*, 73, 281–289. doi: 10.1016/j.landusepol.2018.02.012.

Barnes, M.L., Yoder, L., Khodaee, M. (2021). Detecting winter cover crops and crop residues in the midwest us using machine learning classification of thermal and optical imagery. *Remote Sensing*, 13(10), 1998. doi: 10.3390/rs13101998.

Cantelaube, P. and Carles, M. (2015). Le Registre Parcellaire Graphique : des données géographiques pour décrire la couverture du sol agricole. *Cahier des Techniques de l'INRA, GéoExpé*, 58–64 [Online]. Available at: https://hal.inrae.fr/hal-02639660.

Cantelaube, P. and Lardot, B. (2021). Construction d'une base de données géographiques à échelle fine exhaustive sur l'occupation agricole du sol : le "RPG complété". Partie 2 : attribution des cultures aux parcelles susceptibles d'accueillir les surfaces agricoles hors RPG. Version 1. Methodological Note, February. US ODR, INRAE [Online]. Available at: https://hal.inrae.fr/hal-03818008.

Cantelaube, P. and Lardot, B. (2022). RPG complété Région Aquitaine. Recherche Data Gouv, V1 [Online]. doi: 10.57745/Y1DA3K.

Casel, P. and Grillot, M. (2024). Gérer et maintenir un outil de diagnostic des systèmes agrialimentaires accessible en ligne : SI-BOAT. *NOV'AE*, 5, 12. doi: 10.17180/NOVAE-2024-NO-ART05.

Debaeke, P. et al. (2022). Availability and integration of agro-environmental data: The French case. In *Landscape Agronomy*, Rizzo, D., Marraccini, E., Lardon, S. (eds). Springer, Cham. doi: 10.1007/978-3-031-05263-7_3.

Dongmo, Y. (2023). Les échanges temporaires de parcelles entre exploitations : motivations agricoles et intérêt environnemental. Thesis, Université Paris-Saclay [Online]. Available at: https://www.theses.fr/2023UPASB008.

34. See: https://odr.inrae.fr/intranet/carto_joomla/.

Fendrich, A., Matthews, F., van Eynde, E., Carozzi, M., Li, Z., d'Andrimont, R., Lugato, E., Martin, P., Ciais, P., Panagos, P. (2023). From regional to parcel scale: A high-resolution map of cover crops across Europe combining satellite data with statistical surveys. Science of the Total Environment, 873, doi: 10.1016/j.scitotenv.2023.162300.

Gaigné, C. and Letort, E. (2017). Co-localisation des différentes productions animales en Europe : l'exception française ? *INRA Prodcutions Animales*, 30, 219–228. doi: 10.20870/productions-animales.2017.30.3.2246.

Girault, B., Martin, P., Piskiewicz, N. (2022). RPG Explorer crop successions France 2007-2014, 2007-2020, 2015-2020. Recherche Data Gouv, V4. doi: 10.57745/NPP6YM.

Grillot, M., Ruault, J.-F., Torre, A., Bray, F., Madelrieux, S. (2021). Le proto-métabolisme : approche du fonctionnement bioéconomique d'un territoire agricole. *Economie Rurale*, 55–75. doi: 10.4000/economierurale.8908.

Lardot, B., Cantelaube, P., Carles, M., Séard, C., Truche, C., Poméon, T. (2021). Construction d'une base de données géographiques exhaustive à échelle fine sur l'occupation agricole du sol : le "RPG complété". Partie 1 : production de la couche géographique des parcelles susceptibles d'accueillir les surfaces agricoles hors RPG. Version 1. Methodological Note, February. US ODR, INRAE [Online]. Available at: https://hal.inrae.fr/hal-03647500.

Levavasseur, F., Martin, P., Bouty, C., Barbottin, A., Bretagnolle, V., Thérond, O., Scheurer, O., Piskiewicz, N. (2016). RPG Explorer: A new tool to ease the analysis of agricultural landscape dynamics with the Land Parcel Identification System. *Computers and Electronics in Agriculture*, 127, 541–552. doi: 10.1016/j.compag.2016.07.015.

Martin, P., Ramalanjaona, L., Truche, C., Ballot, R., Carozzi, M., Pomeon, T. (2023). Modelling the spatialisation of pesticide sales to monitor environmental policies in France. *Journal of Cleaner Production*, 403. doi: 10.1016/j.jclepro.2023.136880.

Maxime, F., Mollet, J.M., Papy, F. (1995). Aide au raisonnement de l'assolement en grande culture. *Cahiers Agricultures*, 4, 351–362 [Online]. Available at: https://revues.cirad.fr/index.php/cahiers-agricultures/article/view/29909.

Nowak, B., Marliac, G., Michaud, A. (2021). Estimation of winter soil cover by vegetation before spring-sown crops for mainland France using multispectral satellite imagery. *Environmental Research Letters*, 16, 064024. doi: 10.1088/1748-9326/ac007c.

Richard-Schott, F. (2009). Le Recensement Général de l'Agriculture de 1955, une référence pour les géographes ? *Géocarrefour*, 2009, 271–279. doi: 10.4000/geocarrefour.7567.

Schönhart, M., Schmid, E., Schneider, U.A. (2011). CropRota: A crop rotation model to support integrated land use assessments. *European Journal of Agronomy*, 34(4), 263–277. doi: 10.1016/J.EJA.2011.02.004.

12.7. Webography for further information or access to the databases cited

Agence Bio. Available at: https://www.agencebio.org/vos-outils/les-chiffres-cles/.

BDNI. Available at: https://agriculture.gouv.fr/evaluation-de-la-base-de-donnees-nationale-didentification-animale; https://agreste.agriculture.gouv.fr/agreste-web/methodon/D-BDNI/methodon/.

Collectifs agroécologiques. Available at: https://collectifs-agroecologie.fr/.

DGAL. Available at: http://agriculture.gouv.fr/agrement-sanitaire-des-etablissements-au-titre-du-reglement-ce-ndeg8532004.

HVE. Available at: https://agriculture.gouv.fr/ou-trouver-des-exploitations-certifiees-haute-valeur-environnementale-en-France.

ICPE. Available at: https://www.georisques.gouv.fr/donnees/bases-de-donnees/installations-industrielles.

INAO. Available at: https://www.inao.gouv.fr/.

IREP. Available at: https://www.georisques.gouv.fr/donnees/bases-de-donnees/installations-industrielles-rejetant-des-polluants.

RGA. Available at: https://agriculture.gouv.fr/les-chiffres-definitifs-et-detailles-du-recensement-agricole-2020.

RPG. Available at: https://geoservices.ign.fr/rpg.

SINOE®. Available at: https://www.sinoe.org/.

SIRENE®. Available at: https://www.sirene.fr/sirene/public/static/acces-donnees.

Zones de montagne. Available at: https://agriculture.gouv.fr/aides-aux-exploitations-classement-en-zone-defavorisee.

Zones relevant d'un espace naturel protégé. Available at: https://inpn.mnhn.fr/telechargement/cartes-et-information-geographique/nat/natura; https://www.geoportail.gouv.fr/inspire/annexe-i/sites-proteges.

ZoneVuln. Available at: https://www.sandre.eaufrance.fr/atlas/srv/fre/catalog.search#/metadata/8ddc0f01-6708-4b23-a79a-e9bac3beeee6.

Conclusion

Philippe MARTIN and Solène PISSONNIER

Paris Saclay University, AgroParisTech, INRAE, UMR SADAPT, Palaiseau, France

Coming to the end of these 12 chapters, it feels only right that we reiterate what was said in the introduction to this book, namely that the journey we embarked upon can in no way describe all the relationships between farms and territories. What follows is a brief description of what we have not been able to develop, while providing the reader with some ideas for further reading.

In this book, and particularly in the last chapter, we discuss access to data on the relationships between farms and territories, and the use of databases, but the qualitative approach based on surveys is also highly instructive, as shown in Chapters 1, 3 and 6. For a more in-depth look at survey methodologies for the territorial anchoring of farms and their practices, refer to Laurent and Thinon (2005) (especially Chapters 3 and 4). On the dynamics and place of farmers in territories and society, refer to *Sociologie des mondes agricoles* ("Agriculture Without Farmers") by Hervieu and Purseigle (2013). With regard to the recomposition of farms in territories, you can also read "Des entreprises aux allures de firmes" ("Companies that Look Like Corporations") by Purseigle et al. (2019) in the book *Les mutations récentes du foncier et des agricultures en Europe* ("Recent Changes in Land Tenure and Farming in Europe") edited by Gérard Chouquer and Marie-Claude Maurel. Characterizing farms and their sustainability in relation to the territory is a point addressed in the IDEA4 method, which distinguishes between restricted sustainability (at the farm level) and extended sustainability (at the territory level) (Zahm et al. 2023). In English, we recommend a recent summary on landscape agriculture, "Landscape Agronomy", which discusses, among other things, the construction of prospective scenarios to debate the dynamics of land use and how it can be impacted by different factors (Rizzo et al. 2022).

References

Chouquer, G. and Maurel, M.-C. (2019). *Les mutations récentes du foncier et des agricultures en Europe. Les cahiers de la MSHE Ledoux*. Presses universitaires de Franche-Comté, Besançon, and Presses Universitaires Sun Yat-Sen, Canton.

Hervieu, B. and Purseigle, F. (2013). *Sociologie des mondes agricoles*. Armand Colin, Paris.

Laurent, C. and Thinon, P. (2005). *Agricultures et territoires. Information géographique et aménagement du territoire*. Hermès Science Publications, Paris.

Purseigle, F., Anzalone, G., Nguyen, G., Hervieu, B. (2019). Des entreprises agricoles "aux allures de firme". Mutations des exploitations agricoles françaises et nouveaux modes d'accès au foncier. hal-02063962f.

Rizzo, D. (2022). *Landscape Agronomy: Advances and Challenges of a Territorial Approach to Agricultural Issues*, 1st edition. Springer, Cham.

Zahm, F. and Girard, S. (2023). *La méthode IDEA 4 : indicateurs de durabilité des exploitations agricoles principes & guide d'utilisation évaluer la durabilité des exploitations agricoles*. Éducagri éditions, Dijon.

List of Authors

Valentin ASSELAIN
Paris Saclay University
AgroParisTech
INRAE
UMR SADAPT
Palaiseau
France

Christine AUBRY
Paris Saclay University
AgroParisTech
INRAE
UMR SADAPT
Palaiseau
France

Aude BARBOTTIN
Paris Saclay University
AgroParisTech
INRAE
UMR SADAPT
Palaiseau
France

Thierry BONAUDO
AgroParisTech
UMR TETIS
Montpellier
France

Pierre-Etienne BOUILLOT
Paris Saclay University
AgroParisTech
Palaiseau
France

Jeanne CADIOU
AgroParisTech Innovation
IDEAS
UMR SADAPT
Palaiseau
France

Daphnée DURANT
Domaine expérimental
de Saint-Laurent-de-la-Prée
INRAE
UE0057
Saint-Laurent-de-la-Prée
France

Nadège GARAMBOIS
AgroParisTech
UMR PRODIG
Palaiseau
France

Myriam GRILLOT
University of Toulouse
INRAE, INPT, El PURPAN, AGIR
Castanet-Tolosan
France

Hélène GROSS
ACTA – Les Instituts Techniques Agricoles
Paris
France

Sabine HOUOT
Paris Saclay University
AgroParisTech
INRAE
UMR ECOSYS
Palaiseau
France

Anne LAUVIE
Institut Agro Montpellier
INRAE
CIRAD
UMR SELMET
Montpellier
France

Agnès LELIÈVRE
Paris Saclay University
AgroParisTech
INRAE
UMR SADAPT
Palaiseau
France

Florent LEVAVASSEUR
Paris Saclay University
AgroParisTech
INRAE
UMR ECOSYS
Palaiseau
France

Philippe MARTIN
Paris Saclay University
AgroParisTech
INRAE
UMR SADAPT
Palaiseau
France

Romain MELOT
Paris Saclay University
AgroParisTech
INRAE
UMR SADAPT
Palaiseau
France

Jean-Marc MEYNARD
Paris Saclay University
AgroParisTech
INRAE
UMR SADAPT
Palaiseau
France

Kevin MOREL
Paris Saclay University
AgroParisTech
INRAE
UMR SADAPT
Palaiseau
France

Nathan MORSEL
Association pour le Pastoralisme de la Montagne Limousine
Millevaches
France

Caroline PETIT
INRAE
UMR SADAPT
Palaiseau
France

Florence PINTON
Paris Saclay University
AgroParisTech
INRAE
UMR SADAPT
Palaiseau
France

Solène PISSONNIER
Paris Saclay University
AgroParisTech
INRAE
UMR SADAPT
Palaiseau
France

Emmanuel RAYNAUD
Paris Saclay University
AgroParisTech
INRAE
UMR SADAPT
Palaiseau
France

Raymond REAU
Paris Saclay University
AgroParisTech
INRAE
UMR Agronomie
Palaiseau
France

Etienne VERRIER
Paris Saclay University
AgroParisTech
INRAE
UMR GABI
Palaiseau
France

Index

G, I

L, M

N, O

P, R

S, T

Printed and bound by CPI Group (UK) Ltd, Croydon, CR0 4YY

13/09/2026

14961010-0008